COMPUTER VISION
AND
SENSOR-BASED ROBOTS

PUBLISHED SYMPOSIA

Held at the
General Motors Research Laboratories
Warren, Michigan

Friction and Wear, 1959
Robert Davies, *Editor*

Internal Stresses and Fatique in Metals, 1959
Gerald M. Rassweiler and William L. Grube, *Editors*

Theory of Traffic Flow, 1961
Robert Herman, *Editor*

Rolling Contact Phenomena, 1962
Joseph B. Bidwell, *Editor*

Adhesion and Cohesion, 1962
Philip Weiss, *Editor*

Cavitation in Real Liquids, 1964
Robert Davies, *Editor*

Liquids: Structure, Properties, Solid Interactions, 1965
Thomas J. Hughel, *Editor*

Approximation of Functions, 1965
Henry L. Garabedian, *Editor*

Fluid Mechanics of Internal Flow, 1967
Gino Sovran, *Editor*

Ferroelectricity, 1967
Edward F. Weller, *Editor*

Interface Conversion for Polymer Coatings, 1968
Philip Weiss and G. Dale Cheever, *Editor*

Associative Information Techniques, 1971
Edwin L. Jacks, *Editor*

Chemical Reactions in the Urban Atmosphere, 1971
Charles S. Tuesday, *Editor*

The Physics of Opto-Electronic Materials, 1971
Walter A. Albers, Jr., *Editor*

Emissions From Continuous Combustion Systems, 1972
Walter Cornelius and William G. Agnew, *Editors*

Human Impact Response, Measurement and Simulation, 1973
William F. King and Harold J. Mertz, *Editors*

The Physics of Tire Traction, Theory and Experiment, 1974
Donald F. Hays and Alan L. Browne, *Editors*

The Catalytic Chemistry of Nitrogen Oxides, 1975
Richard L. Klimisch and John G. Larson, *Editors*

Future Automotive Fuels — Prospects, Performance, Perspective, 1977
Joseph M. Colucci and Nicholas F. Gallopoulos, *Editors*

Aerodynamic Drag Mechanisms of Bluff Bodies and Road Vehicles, 1978
Gino Sovran, Thomas Morel and William T. Mason, Jr., *Editors*

Mechanics of Sheet Metal Forming — Material Behavior and Deformation Analysis 1978
Donald P. Koistinen and Neng-Ming Wang, *Editors*

Computer Vision and Sensor-Based Robots, 1979
George G. Dodd and Lothar Rossol, *Editors*

Combustion Modeling in Reciprocating Engines, 1980
James N. Mattavi and Charles A. Amann, *Editors*

COMPUTER VISION AND SENSOR-BASED ROBOTS

Edited by
GEORGE G. DODD and **LOTHAR ROSSOL**

General Motors Research Laboratories

Library of Congress Cataloging in Publication Data

Main entry under title:

Computer vision and sensor-based robots.

"Proceedings of a Symposium held at General Motors Research Laboratories, September 25–26, 1978."
Includes index.
1. Robots, Industrial—Congresses. 2. Computers, Optical—Congresses. I. Dodd, George G. II. Rossol, Lothar. III. General Motors Research Laboratories.
TS191.C64   629.8'92   79-18698
ISBN-13: 978-1-4613-3029-5   e-ISBN-13: 978-1-4613-3027-1
DOI: 10.1007/ 978-1-4613-3027-1

Proceedings of the Symposium on Computer Vision and
Sensor-Based Robots, held at the General Motors Research
Laboratories, Warren, Michigan, September 25 and 26, 1978

© 1979 Plenum Press, New York
Softcover reprint of the hardcover 1st edition 1979
A Division of Plenum Publishing Corporation
227 West 17th Street, New York, N.Y. 10011

All rights reserved

No part of this book may be reproduced, stored in a retrieval system, or transmitted
in any form or by any means, electronic, mechanical, photocopying, microfilming,
recording, or otherwise, without written permission from the publisher

PREFACE

The goal of the symposium, "Computer Vision and Sensor-Based Robots," held at the General Motors Research Laboratories on September 25 and 26, 1978, was to stimulate a closer interaction between people working in diverse areas and to discuss fundamental issues related to vision and robotics. This book contains the papers and general discussions of that symposium, the 22nd in an annual series covering different technical disciplines that are timely and of interest to General Motors as well as the technical community at large. The subject of this symposium remains timely because the cost of computer vision hardware continues to drop and there is increasing use of robots in manufacturing applications.

Current industrial applications of computer vision range from simple systems that measure or compare to sophisticated systems for part location determination and inspection. Almost all industrial robots today work with known parts in known positions, and we are just now beginning to see the emergence of programmable automation in which the robot can react to its environment when stimulated by visual and force-touch sensor inputs.

As discussed in the symposium, future advances will depend largely on research now underway in several key areas. Development of vision systems that can meet industrial speed and resolution requirements with a sense of depth and color is a necessary step. A vision system that can handle situations where parts are randomly piled in bins or swinging from moving hooks is another logical step. Improvements in robot technology are also needed. Current robots must rely on control mechanisms slower than a human's and far less flexible. They also require supporting aids such as feeder mechanisms for part presentation and have almost no error recovery capability.

The engineers and scientists who participated in the symposium came from nine countries and were chosen because of their expertise in artificial intelligence, pattern recognition, cybernetics, robotics, computer science, control systems engineering, optics, electronics, and manufacturing technology. Attendance was limited to maintain a closer contact among the participants. The symposium agenda was established so that the first session focused on the fundamental — and perhaps limiting — issues in vision and robotics; the second examined the newest thoughts on robot and vision system development under way today; and the third and fourth addressed future directions for vision and robot systems, respectively.

Symposium papers were presented by authorities in the particular technical areas, and persons of recognized international reputation in the field acted as session chairmen to assure competent direction during the technical sessions. A question and answer session followed the presentation of each paper and is included as part of these proceedings.

While this publication records the original papers and subsequent discussions, a major benefit of the symposium cannot be recorded here. The two-day meeting offered many opportunities for personal interaction among the participants, many of whom were meeting together for the first time. New channels of communication were opened to create an awareness of the concerns and objectives of others, and this will greatly contribute to the direction and success of future research in computer vision and robots.

The symposium could not have been held or these proceedings published without the valuable assistance of many people. We thank Mr. David Havelock for skillfully shepherding the conversion of the edited manuscripts into this volume. For assisting in the transcribing and editing of the question and answer sessions, we recognize Drs. Michael Baird and Walter Perkins and Messrs. Arvid Martin and Mitchel Ward. We also thank Ms. Karen Perkioniemi who expertly assisted in the typing and administrative details of the symposium, and Mr. Thomas Beaman for physical arrangements. The after dinner address by Mr. Frederik Pohl on the role of robots in science fiction was very well received and we thank him for his participation. To our session chairmen, Drs. Harry Barrow, Charles Rosen, Peter Will and Prof. Jerome Feldman, we express our indebtedness for participating in the sessions. Finally, our thanks to Prof. Patrick Winston and to the advisory committee consisting of Professors Marvin Minsky, Azriel Rosenfeld, and Dr. Charles Rosen for helping us select topics and speakers for the stimulating and worthwhile meeting.

George G. Dodd
Lothar Rossol

CONTENTS

PREFACE v

SESSION I —**Fundamental Issues in Vision and Robotics**
        Chairman: A. Rosenfeld,
        University of Maryland 1

Machine Vision and Robotics: Industrial Requirements
  C. A. Rosen, SRI International 3
    References 18
    Discussion 20

Human and Robot Task Performance
  R. L. Paul and S. Y. Nof, Purdue University 23
    References 47
    Discussion 48

Mechanisms of Perception
  R. L. Gregory, University of Bristol 51
    References 68

Artificial Intelligence and the
Science of Image Understanding
  B. K. P. Horn, Massachusetts Institute of Technology 69
    References 75

SESSION II —**Vision and Robot Systems**
        Chairman: P. M. Will,
        IBM Thomas J. Watson Research Center 79

CONSIGHT-I: A Vision-Controlled Robot System for
Transferring Parts from Belt Conveyors
  S. W. Holland, L. Rossol and M. R. Ward,
  General Motors Research Laboratories 81
    References 96
    Discussion 97

An Industrial Eye that Recognizes Hole
Positions in a Water Pump Testing Process
   T. Uno, S. Ikeda, H. Ueda and M. Ejiri,
   Hitachi Central Research Laboratory
   T. Tokunaga, Hitachi Taga Works 101
      References 114
      Discussion 114

APAS: Adaptable Programmable Assembly Systems
   R. G. Abraham, Westinghouse Research and Development Center 117
      References 136
      Discussion 136

PUMA: Programmable Universal Machine for Assembly
   R. C. Beecher, General Motors Manufacturing Development 141
      Appendix 145
      Discussion 149

Programmable Assembly Systems
   M. Salmon and A. d'Auria, Ollivetti 153
      References 163
      Discussion 163

SESSION III — Future Vision Systems
            Chairman: H. G. Barrow,
            SRI International 167

Computer Architectures for Vision
   D. R. Reddy and R. W. Hon, Carnegie-Mellon University 169
      References 184
      Discussion 185

Three-Dimensional Computer Vision
   Y. Shirai, Electrotechnical Laboratory...................................... 187
      References 203
      Discussion 205

Optical Computing for Image Processing
   A. D. Gara, General Motors Research Laboratories 207
      References 234
      Discussion 235

Prospects for Industrial Vision
   J. M. Tenenbaum, H. G. Barrow and R. C. Bolles,
   SRI International... 239
      References 256
      Discussion 256

SESSION IV — Future Robot Systems
Chairman: J. A. Feldman,
University of Rochester 261

Stand-Alone vs. Distributed Robotics
J. F. Engelberger, Unimation, Inc. 263
References 270
Discussion 270

Robot Assembly Research and Its Future Applications
J. L. Nevins and D. E. Whitney,
Charles Stark Draper Laboratory 275
References 320
Discussion 321

Future Prospects for Sensor-Based Robots
R. B. McGhee, Ohio State University 323
References 332
Discussion 334

Symposium Speakers and Chairmen 335

Summarizer and Advisor 339

Symposium Participants 341

Subject Index 349

SESSION I
FUNDAMENTAL ISSUES IN VISION AND ROBOTICS

Session Chairman
A. ROSENFELD

University of Maryland
College Park, Maryland

MACHINE VISION AND ROBOTICS: INDUSTRIAL REQUIREMENTS

C. A. ROSEN

SRI International, Menlo Park, California

ABSTRACT

Machine vision is a major discipline derived from artificial intelligence research. With appropriate simplifying constraints, it is providing a powerful sensory tool for robot control and for important applications to automated inspection.

Two groups of desirable applications are described. In the first group, machine vision, supplemented as required by force and torque sensing, can greatly enhance the performance of first generation robots presently limited to operations based on fixed, predetermined actions. The new capabilities include the identification of workpieces, the determination of their position and orientation, and the provision of real-time visual feedback for effecting adaptive corrections of the robot's trajectories. In the second group, machine vision can replace or aid a human worker in performing visual inspection for quality control, for minimizing production of scrap (by increasing yield), and for safety.

In each group, typical applications selected from real problems in industry will be described, including some which can be implemented within the present state-of-the-art and others which are still in the research stage.

INTRODUCTION

It is hardly necessary to emphasize the importance of visual sensing and interpretation in human activities. We can appreciate the severe handicaps and limitations imposed on the mental and physical activities of the blind, often ameliorated to varying degrees by laborious training of tactile, auditory, and olfactory sensory capabilities. The importance of human vision is a major motivation for the intense interest and significant research effort devoted to machine vision. For the past 15 years

References pp. 18-20.

the Artificial Intelligence community has been slowly developing understanding in this field of research [1-11] and has begun to implement simple but increasingly sophisticated machine vision techniques for use in many fields, including manufacturing processes, medical diagnosis, photo-interpretation, and military missile guidance.

Machine vision research, in common with most of the work in artificial intelligence, is still conducted primarily as an experimental science. Certainly, fundamental principles in physical optics, electronics, and computer science are employed to good advantage in acquiring and processing images, but the interpretation of such images for pragmatic use depends to a great degree on a large and growing number of algorithms and methods, heuristically conceived, rationally extended, and experimentally verified. It has been, and is, the author's opinion that we cannot wait to "solve" the machine vision problem in general by establishing a relatively complete scientific theory. It appears to be more sensible to make full use of what is known, applying the rich accumulation of methods to situations in which simplifying physical constraints can be applied to yield economically viable solutions. Fundamental machine vision research should proceed vigorously, generating new knowledge and techniques because such understanding is basic to the solution of difficult problems. At the same time, applied research and development resulting in solutions to simpler, more constrained but general classes of tasks will, at the very least, give credibility to the field. More importantly, this approach may indicate new strategies and directions to explore in concurrent research on the more fundamental issues of machine vision.

In this paper I have attempted to select major classes of problem areas in industry in which successful application of machine vision will have a significant, if not revolutionary, impact on productivity, product quality, and even the mass-production process itself. Although the grouping of tasks in each class is somewhat arbitrary, each class is real, having been identified and described by competent factory personnel during visits to plants and in subsequent discussions between the author and his colleagues over the past six years. Further, there are sufficient instances with comparable requirements in each class, so that general rather than ad-hoc techniques can be applied.

In the next sections two broad groups of applications are described. The first group includes applications in which machine vision can be an essential part of a manipulative task involving industrial manipulators (robots), or of the control functions in a production process. In the second group, machine vision supplants or supports the human in performing inspection of quality control, minimizing production of scrap, and improving safety. Manipulation may be involved in the second group, but primarily for the purpose of presenting workpieces or assemblies to the vision system.

MACHINE VISION FOR SENSOR-CONTROLLED MANIPULATION

For over 15 years programmable manipulators (industrial robots) have been performing important but fairly simple manipulative tasks, such as loading and unloading other machines, stacking parts, spot-welding, paint-spraying, and so forth [12]. In a few applications, mostly in the laboratory, these robots have begun to be employed for assembly, material handling, and other fabrication processes, with the aid of sensory feedback [13-22]. The great majority of applications involve highly constrained conditions in which the positioning and orientation of workpieces, assemblies, packing boxes, and machines to be served must be known with considerable precision, and usually require expensive jigs, fixtures, and elaborate conveyors. Examples of some applications in factories are the force feedback and a compliant wrist used by Hitachi for assembly [23]; force sensing used by Olivetti for assembly [24,25]; visual feedback used by Cheseborough Ponds for process control [26]; and non-contact eddy-current sensory feedback used by Hitachi for path control in arc-welding [27]. To date there does not exist a commercially available, fully programmable industrial robot capable of using, as needed, *all* of the available sensory feedback systems, and in particular the machine vision system.

Fig. 1 is a summary of desired functions for machine vision applicable to sensor-controlled manipulation. In the succeeding sections the application of these functions to various classes of manipulative tasks will be described.

Representative task areas are summarized in Fig. 2. Machine vision can be applied in an effective and economic manner to permit industrial robots to deal with imprecisely positioned or unoriented workpieces and assemblies, to compensate for buildup of errors in tolerances, and in general to enable the robot to "fine-tune" the positioning and orientation of its end-effector to adaptively correct for unforeseen changes in the position and orientation of workpieces. It should be added that in many instances the use of the compliant wrist and of force/torque and tactile sensing may be also indicated, especially where the positive action of contact sensing provides a precision unobtainable with relatively low-resolution cameras.

- RECOGNITION OF WORKPIECES/ASSEMBLIES AND/OR RECOGNITION OF THE STABLE STATE WHERE NECESSARY.

- DETERMINATION OF THE POSITION AND ORIENTATION OF WORKPIECES/ASSEMBLIES RELATIVE TO A PRESCRIBED SET OF COORDINATE AXES.

- EXTRACTION AND LOCATION OF SALIENT FEATURES OF A WORKPIECE/ASSEMBLY TO ESTABLISH A SPATIAL REFERENCE FOR VISUAL SERVOING.

- IN-PROCESS INSPECTION--VERIFICATION THAT A PROCESS HAS BEEN OR IS BEING SATISFACTORILY COMPLETED.

Fig. 1. Desired functions of machine vision for sensor-controlled manipulation.

References pp. 18-20.

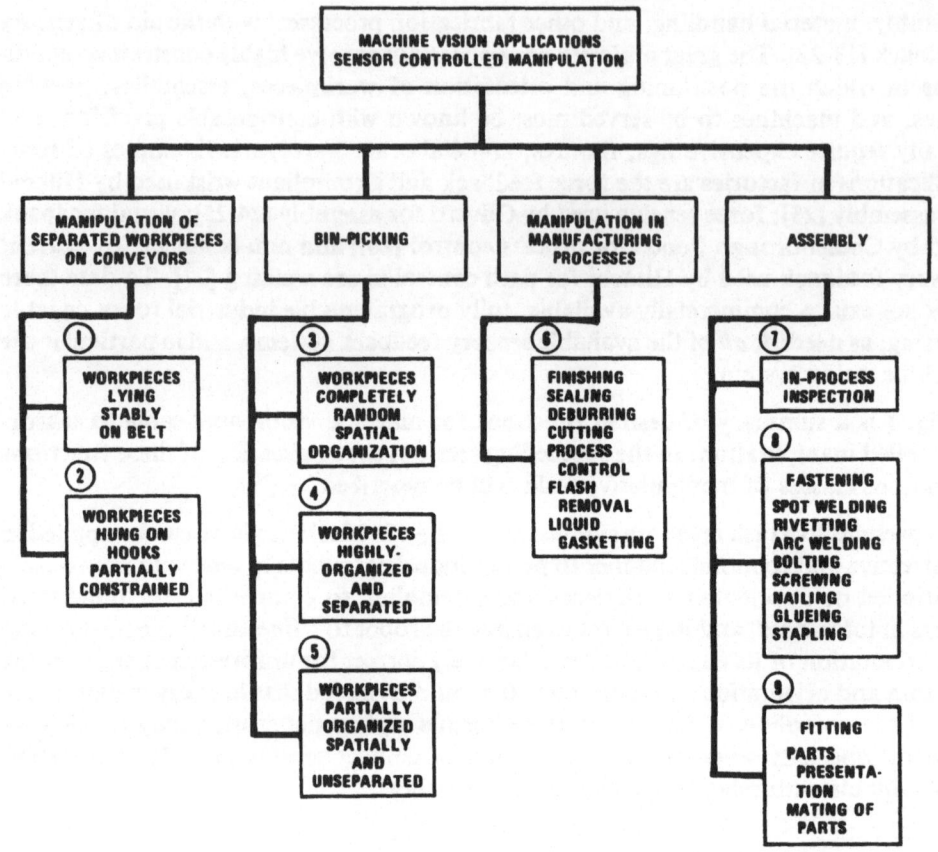

Fig. 2. Machine vision applications — sensor controlled manipulation.

Manipulation of Separated Workpieces on Conveyors — There are many instances in which individually separated parts, sub-assemblies, and assemblies are being transported by overhead or belt conveyors from station to station in the factory. The workpieces are more often than not randomly oriented and positioned because this is the least expensive way to transport them. Occasionally they can be quite close to each other and may touch. In those instances in which they are piled more than one layer thick, simple passive mechanical means can usually be devised to unscramble them and essentially maintain one layer. It may be necessary to acquire each workpiece with a manipulator and to transport it for packing in a prescribed order and state in a container, for feeding into another machine with a prescribed position and orientation, or for sorting in the case of a batch of mixed workpieces. Two general subclasses can be distinguished—separated workpieces lying stably on a belt, and workpieces hung on hooks.

1. Separated Workpieces Lying Stably on a Belt — For many cases in which workpieces are lying stably on a belt and there is an unobstructed view of each separate workpiece (even though they may be touching) it is feasible to apply available simple machine vision techniques to perform one or more of the following:

- Identify the workpiece
- Determine the stable state of the workpiece
- Determine the relative position and orientation of the workpiece.

Successful use of binary (black and white) visual information depends critically on carefully engineering illumination and control of background. This is necessary to provide a high-contrast image for extraction of a two-dimensional outline representing the shape and major internal features of the workpiece for each stable state [28]. These techniques can be successfully applied on a belt moving with variable speed using the equivalent of flash photography. When sufficient contrast cannot be obtained, more sophisticated picture-processing methods are available, but have yet to be applied in industry. These include the potential use of (1) gray-scale imaging, (2) intense patterned or structured light with known geometry projected onto the workpiece to create a contrasty pattern against the background, and (3) ranging techniques to permit extraction of the outline from range rather than intensity data, or if necessary, the use of both intensity and range data.

2. Workpieces Hung on Hooks — If the workpiece is being transported by an overhead chain conveyor or the equivalent, then the machine vision techniques just discussed will not suffice. Usually the workpiece is crudely supported, requiring three-dimensional information to determine its position and orientation *in space*. One solution to this problem is to use more expensive supports, guaranteeing unique positioning and orientation in space, and eliminating sway and rotation as much as possible. A more general solution is to add some form of range sensing to existing two-dimensional intensity sensing. Again, the use of structured light or other optical ranging techniques may be good options.

The transport and handling of workpieces in most factories occupy a great many human workers, and it is rare that the expense of special magazines or jigged fixtures for maintaining precise known workpiece positioning and orientation can be justified. In batch production (and for frequent model change) use of special fixtures may become even more expensive. Thus it is highly desirable to develop techniques using machine vision, readily reprogrammable, and generally applicable to many initial and modified situations.

At present, this application of machine vision is most highly developed for dealing with workpieces of the first subclass—that is, those that are, or can be, separated, and that lie stably but unoriented either on moving belts or on a stationary support structure. A relatively inexpensive vision system, composed of a solid-state television camera (resolution, approximately 128 × 128 elements) and a microcomputer or mini-processor, can perform the required visual sensing for a wide range of tasks [16]. Software developments over the past five years are adequate to meet many performance requirements, such as a total time for acquiring and processing a picture within a half second. By adding specialized preprocessing hardware, on-line processing can be reduced to approximately two hundred milliseconds. There is little doubt that this

References pp. 18-20.

performance can be significantly enhanced by carefully engineered systems designed for one specialized task. One can predict that within two years the cost of such a vision system will not add more than 10% to 15% to the cost of a fully programmable robot system.

For the second subclass of workpieces—those that are hung, partially constrained, on hooks or the equivalent—a considerable research and development effort is needed. Range sensing obtained by passive or active means can help solve this problem in a direct manner, general in application, yielding three-dimensional information on the position and orientation of workpieces *in space* [29,30,31].

Bin Picking — Another common method of transporting and buffer-storing workpieces in factories is in containers or bins. Three major classes can be distinguished.

1. Bin Picking—Workpieces in Completely Random Spatial Organization — The workpieces are jumbled together in a container, sometimes interlocked, randomly positioned and oriented, with no clear unobstructed view of most of the workpieces. This method of transport and buffer storage is the most common in practice, being the least expensive method for rugged workpieces that can withstand some degree of marring or scratching without degradation of quality, performance, or reliability.

2. Bin Picking—Workpieces Highly Organized Spatially and Separated — The workpieces have to be handled carefully to prevent damage to delicate details, or they have exposed, finely machined or finished surfaces that should not be scratched or dented. In this class the workpieces are usually arrayed in separate compartments within a bin lined with protective material to prevent damage during transport (e.g., "egg-crate" compartments). Often there are several layers of workpieces in depth in the same bin with protective material between layers. Each workpiece separated from its neighbors is usually in a known stable state, and although not positioned and oriented precisely, the permitted variations or "slop" in these parameters are small. We will not consider the extreme of this class, which involves the use of expensive jigs or magazines to prevent damage and to maintain precise positioning and orientation during transport and robot acquisition.

3. Bin Picking—Workpieces Partially Organized Spatially and Unseparated — A third class, quite common in practice, includes workpieces disposed with a degree of organization intermediate between the random arrangement of the first class and the very orderly arrangement of the second class. The workpieces are piled in a bin in a crudely ordered array without protective interlayers or separators, often several layers in depth. Usually each workpiece is not in one of the easily recognizable stable states that occur when it is lying on a flat surface. Each workpiece, however, is often aligned with its neighbors, with its position and orientation considerably more predictable than with the completely randomized organization of the first class.

The general problem of automating the handling of workpieces is the same for these three classes—workpieces must be acquired one at a time and then presented to some other location with a predetermined position and orientation in space. Applications include sorting and packing, loading workpieces into another machine or process, and presenting parts for assembly. At present one has to rely on sophisticated human visual and tactile capabilities in those instances in which *bowl* orienters/feeders and other feeders cannot be successfully used due to the size, weight, delicacy, or other properties of the parts. Machine vision, augmented as needed with other sensors and devices, can be successfully applied to a subset of these classes now, with promise of future extension to the rest.

A high level of picture processing and interpretative capability is required for dealing with the completely jumbled and random workpieces. The vision system has to cope with poor contrast, partial views of parts, an infinite number of stable states, variable incident and reflected lighting, shadows, geometric transformations of the image due to variable distance from camera lens to each part, etc. It is a formidable problem in scene analysis, with some initial success reported by General Motors [32]. Progress using this technique alone will be slow, and practical implementation will require considerably faster and less expensive computational facilities than presently available.

An approach that finesses many of these problems is to divide the problem of completely disoriented workpieces into two stages. First, remove a few (one or more) of the workpieces at a time from the bin, deposit them with random position and orientation, separated, lying stably on a flat contrasty surface, and then apply known simple machine vision techniques to determine the identity, stable state, position, and orientation of each separated workpiece. Machine vision will thus provide the necessary information for controlling a second and final acquisition by a robot. The first gross acquisition of a few workpieces can be assigned to an inexpensive manipulator, since there is no need for precise positioning and orientation. Initial work on this approach has been reported by SRI International [16]. A somewhat different approach in which the parts orientation and position in space are determined while the part is held by the robot has been reported by University of Rhode Island [33]. It remains to be shown that these methods can be acceptably implemented for a large class of workpieces. The method of separation first and then recognition will certainly be acceptable for large, heavy parts for which there is ample time between successive operations.

Simple machine vision techniques appear adequate for the second class, described above, in which separated, partially oriented workpieces are arranged in orderly arrays. It will be necessary to devise software to differentiate between the workpieces and the egg-crate enclosing walls (if present)—a complication that appears surmountable. If necessary, inexpensive bins can be used, specially designed for simplifying the visual processing either to enhance contrast or to provide a known geometric background. Control of incident lighting including the use of appropriate filters may also provide a useful means of simplifying the acquisition of suitable images of the workpieces.

The third class of semi-organized workpieces can be handled by the method of first separating the workpieces and then applying simple machine vision techniques, as described above. Since the workpieces are somewhat aligned and partially organized, separating them is simplified to some degree, with the possibility of reducing the time required for a complete operation. In fact, in considering costs of the total system, in

References pp. 18-20.

many instances, it may be feasible to pack parts that do not require surface protection in this semi-organized fashion, since a robot can be used to pack the parts in the bin at the station where the parts originate (the foundry, receiving, or the machine shop). Standardization in methods of handling can reduce the number of different robot handling systems, ensuring economies in maintenance, set-up time, and inventory of replacement parts and machines.

Bin-picking is an ubiquitous process in almost all factories and represents one of the best candidates for the introduction of sensor-controlled manipulation. Assuming the availability of an inexpensive low-resolution vision system and less costly industrial robots, many bin-picking tasks can be accomplished economically within the state of the art today. A promising candidate system would be composed of a vision module, a modified limited sequence robot and a servoed X-Y table, all under micro-computer control [34]. At present unit retail prices, such a system would cost approximately $30,000 to $35,000, with considerable reductions in cost if mass produced.

Vision-Controlled Manipulation in Manufacturing Processes — In batch production involving discrete parts, a large number of important manufacturing processes cannot be economically automated because of cost of specialized jigs and "hard" automation production machines is prohibitively high for the small number of workpieces or assemblies in each batch. Programmable automation based on the use of robots has already proved its worth in application to spot-welding (an assembly process) and in automated paint-spraying (a finishing and/or surface protection process). This type of automation is being introduced into many production lines by major companies in the world. Under pressure of government regulations in this country, and because of the expense of coping with undesirable and unhealthy environmental conditions associated with noxious vapors and fumes, automated paint-spraying equipment will essentially replace human workers, where economically feasible, within the next ten years.

In paint-spray and similar applications, there still is universal reliance on relatively costly control of workpiece positioning during painting. Since the majority of spray-painting lines do not have specialized conveyors to eliminate the uncertainty in position and orientation, there exists a need for sensors, preferably non-contact, to provide these adaptive corrections. Machine vision in its present implementation is applicable if two-dimensional information suffices. However, in most cases there is again a need for three-dimensional processing to locate a workpiece or assembly *in space,* thus requiring range information as well as intensity information. Key requirements for an acceptable machine vision system are again that its cost does not add more than 10% to 25% to the robot cost, and that for moving lines the required sensing information be provided in a fraction of a second (0.2 to 1.0 second). For production lines in which the workpiece or assembly can be stationary during painting, this permissible time for visual processing can be extended to several seconds.

Although spray-painting is a currently popular candidate, similar finishing applications include sand-blasting and spraying of protective chemical coatings on selected areas of the workpieces.

A sub-class of manufacturing processes that have common machine vision requirements includes the application of semi-liquid sealants, deburring and removal of flash from castings and moldings, torch cutting, laser machining, and liquid gasketing. These applications require tool path control along paths that are specified in relation to defined edges, seams, or other features of the workpiece. Due to variations in tolerances and fit, these paths are not precisely predetermined, and therefore each operation is unique within narrow limits, requiring continuous servoing of the tool. Often, for flash removal and sealing, the amount of material to be removed or added, respectively, is variable with position, requiring more elaborate image processing for automatic control of these variables as well as path control. For many of these applications, two-dimensional image processing suffices; with the addition of a proximity sensor, air-pressure sensing, or contact sensing, many contoured workpieces with relatively uncomplicated surfaces can be accommodated. The general case, however, requires three-dimensional servoing, and range information would be most useful. The cost/time requirements are similar to those previously mentioned for paint-spraying.

It should be noted that efficient computational algorithms are available to provide control functions for adaptively correcting the path of a tool carried by a multidegree-of-freedom robot. These algorithms are included in programs that can be executed on-line for:

1. Interpolation—piecewise linear approximations to continuous-path trajectories.
2. Path smoothing—elimination of abrupt changes in the trajectory [16].
3. Transformations from robot to world coordinate systems and the reverse [14], including moving-line coordinate systems.

Vision-Controlled Manipulation in Assembly — The Bureau of Census reported (in Occupation by Industry, PC2-7C) that in 1970, production workers, not including overhead staff, constituted from 40% to 65% of the total workers, dependent on industry, in the manufacture of durable goods. Assembly workers constituted approximately 17% and inspection workers approximately 10% of all of the production workers. These percentages do not include other workers, such as welders, painters, material handlers, packers, etc., who may be affected by the new technology. The development of programmable automation tools and techniques for automating batch assembly has therefore become a worthy major goal with highly significant economic and social consequences. Further, there are interesting and useful by-products of these studies. This is a consequence of the disciplined analysis that is mandatory if a computer is to correctly control every step in assembly. We tend to take for granted the human assembler's accumulation of rich world experience, coordination, dexterity, and adaptability. However, for computer control no detail can be omitted. Thus the analysis needed to implement a total assembly system for computerized robots can lead to an improved and simplified assembly system for humans. Further, when any manufacturing process, including assembly, is brought under precise computer control, it becomes possible to improve and maintain product quality to any practical limit that we can afford.

Machine vision has as important a role in automated assembly as human vision has for assembly by humans. The desired vision functions, described in Fig. 1, are all

References pp. 18-20.

applicable to assembly. Following are brief descriptions of how they can aid and simplify the assembly process:

1. Machine vision enables parts and subassemblies to be identified, acquired, and presented in a predetermined manner to mate with other parts or assemblies. These are under the heading of "fitting" processes.

2. Machine vision can provide feedback to adaptively control the position of robot assembly tools to be used for fastening, eliminating many costly jigs. This function has been labeled active accommodation, inferring that errors in position and orientation are sensed, and the tool (or part) is repositioned as part of an active servo loop. This correction is relatively coarse, and, where necessary, fine tuning of such corrections can be effected by using passive accommodation, in which a compliant wrist [35] corrects for the remaining error in the position and orientation through the action of reaction forces and torques on the tool.

3. Machine vision can continuously control the path of a robot-held tool along a designated line or curve whose trajectory is described in relation to either a workpiece edge or other salient features.

4. Machine vision can provide the essential function of in-process inspection. It can verify that the correct parts are being mated and that each step has been completed satisfactorily without damage. When augmented by contact sensors, such as displacement, force, and torque sensors, the proper seating of mating parts and the desired degree of tightness of bolts and screws can be monitored as well.

For convenience, assembly operations can be classified into two broad groups under the headings of fastening and fitting. Applications of machine vision to these operations will now be discussed.

1. Fastening — The most common fastening methods include spot-welding, riveting, arc-welding, bolting, screwing, nailing, stapling, and gluing. The major requirement for vision is to control the position and orientation of the fastening tool carried by the robot in relation to the workpieces, which may be stationary or on moving conveyors. The same considerations apply that were previously described under the heading, "Vision Controlled Manipulation in Manufacturing Processes." Essentially, machine vision can sense the position and orientation of workpieces or salient features of workpieces, eliminating the need for costly jigging.

For bolting and riveting, the insertion of bolts and rivets in holes can utilize both active and passive accommodation to adaptively correct for alignment errors. Spot-welding, screwing, nailing, stapling, and gluing can utilize the path control function of machine vision when the location and trajectory of the line of fastenings is variable due to buildup of manufacturing tolerances or to poor fit. Corrections to a stored robot program can be generated and new tool trajectories calculated in real time. Again, the cost should not exceed 10% to 15% of the cost of the robot, the total time for sensing and interpretation should be in the range of 0.1 to 1 second.

The automation of electric arc-welding is of particular interest because the working environment is generally deleterious for humans, and this highly cost-effective method of fastening is not likely to be readily replaced. Machine vision is a good candidate for a sensor to control the torch path and provide the information to control weld parameters to correct for poor fit, heat distortion, etc. of the parts to be mated.

In all of these fastening operations, machine vision can verify the satisfactory completion of each fastening step, if such inspection is desired and can be afforded.

2. *Fitting* — For fitting operations in assembly systems in which the use of jigs are to be minimized, the primary functions in which machine vision can be useful are parts presentation and the mating of parts. Identification of each part or of the stable state of each must be effected if the parts, or several different types of parts, are buffer-stored randomly in bins or on endless belts. Further, each part must be sequentially presented and fitted to the rest of the assembly-in-process, requiring the determination of the relative orientation and position of the main assembly and each part of sub-assembly. Finally, in-process visual inspection can monitor the successful completion of each step.

It should be noted that machine vision can be dispensed with entirely, with reliance on contact sensing only for monitoring fitting, step completion, and making minor adjustments in position and orientation. However, this strategy requires elaborate control of the position and orientation of each component part with no means to adapt to errors due to tolerance buildup. Modifications or model change would require reworked jigs or new special jigs to be designed and fabricated, as well as modified software program. This reduced flexibility in changeover must be compared with the additional one-time cost of a machine vision system, especially in terms of lost production time.

MACHINE VISION FOR INSPECTION

Economically priced solid-state television cameras in the form of one-dimensional linear and two-dimensional planar arrays are now commercially available. Their advantages over vidicon type cameras include small size, inherent ruggedness, stability, discrete photoelectric elements that can be individually addressed, and compatibility with computer hardware. Inspection of all kinds occupies approximately 10% of production workers in the manufacture of durable goods, and is an essential function (often implicit) in the service industries as well. Picture-processing software developed over the past 15 years can now be practically applied to many inspection problems in which the information in optical images can be extracted and interpreted in rich detail. We can expect these systems to become as ubiquitous in factory and commerce as the discrete photocell, which is primarily a binary on-off sensory device with no imaging capability.

Machine vision applications to inspection can be classified into two large groups—inspection requiring highly quantitative mensuration, and inspection that is primarily qualitative but frequently includes semi-quantitative mensuration. Many inspection tasks of the second group are implicit—that is, are performed by workers as part of

References pp. 18-20.

some other job (e.g., an assembler makes certain that the right parts, complete and undamaged, are being assembled). This classification and its subdivisions are shown in Fig. 3. Applications in each group will now be discussed.

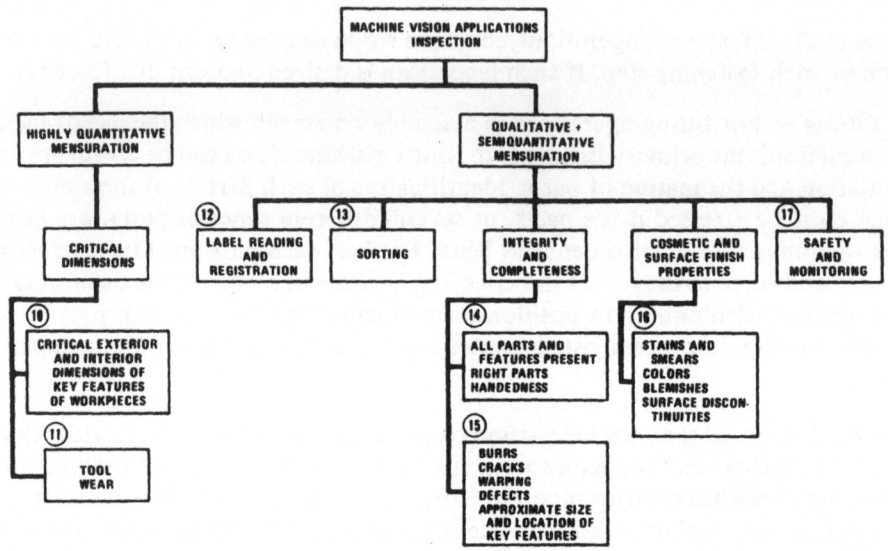

Fig. 3. Machine vision applications — inspection.

Inspection Requiring Highly Quantitative Mensuration — The mass-production process depends, in great part, on maintaining critical dimensions of workpieces within given tolerances. Many methods of measurement using mechanical devices are currently in universal use, including micrometers; dial gauges; vernier calipers; Johanssen blocks; go, no-go gauges; and special measuring jigs. Where hard automation is implemented for large production runs it is sometimes possible to automate these mechanical measurements as part of in-process inspection for quality control, and to limit the production of scrap, as, for example, detecting the breakage of a die in a progressive stamping machine. In many instances, because of the time required for this additional inspection operation, the throughput of an expensive machine is reduced and therefore 100% inspection is not warranted. Quality control then consists of establishing statistical-based techniques, with manual inspection of a small sample of produced workpieces. In batch production, such inspection is generally performed manually, whether 100% or sampled. There is, therefore, a need to develop techniques that are sufficiently precise, easily programmable for modifications or model change, and noncontact such that the sensor is nonintrusive to the mechanical fabrication and/or assembly process. Flying-spot laser scanning techniques are being successfully used for this purpose, but appear to be limited for general application.

1. Machine Vision for Quantitative Inspection — The solid-state linear diode array and the two-dimensional camera interfaced with microprocessors or microcomputers provide powerful general-purpose tools for accurate noncontact, nonintrusive mensuration at potentially low cost. By careful choice of the optical coupling between camera and workpiece, the full resolution of low-resolution cameras can frequently be utilized, ranging from microscopic to macroscopic fields of view. The discrete nature of the optoelectric sensing elements and the insensitivity to line voltage variations, and to variability and nonlinearity of line scan, provide a superior, more precise means of mensuration by substituting digital counting techniques for analog measurements.

The relatively inexpensive linear diode array, available with up to a thousand elements along one dimension, is generally applicable where the workpiece to be inspected is in motion and thus capable of generating a two-dimensional image when scanned orthogonal to the motion. The flow of data can be processed "on the fly", a line at a time or after a complete two-dimensional image has been buffer-stored in computer memory. Extraction of salient dimensions to be measured can be effected with special high-speed hardware or, more generally, by software.

The two-dimensional array camera can obtain a two-dimensional image of a workpiece whether stationary or in motion, with high intensity flash-lamp operation possible for minimizing blurring. Processing of the image can be effected in a manner similar to that of the linear array. Available cameras are somewhat limited in resolution, ranging from under 100 to 500 elements in one dimension. We have found a 128 × 128 element camera quite adequate for a broad range of applications [16].

In applying these cameras for inspection there is always the problem of positioning the workpiece for viewing. Although precise positioning and orientation can often be compensated for by suitable software, the manipulation into coarse position may require specialized manipulation which is not always cost-effective. The combination of an inexpensive limited-sequence robot, a programmable X-Y table, and a camera all controlled by a microcomputer may be a general-purpose solution [34]. In the future, high-resolution cameras, and faster processors, both at acceptable prices, will tend to replace much if not most of the manual measurements now made, and probably will make 100% inspection cost-effective.

2. Machine Vision for Tool Wear Measurement — The measurement of tool wear has been and is of great economic importance. A nonintrusive, noncontact method is desirable, and novel approaches are being developed in the laboratory [36, 37]. In principle, if the tool shape is accessible to optical imaging, wear can be measured using a solid-state camera. In practice, it is often difficult to obtain a clean view of the image. There are cutting fluids and chips in the line of view. It may be possible to use fiber optics to circumvent some of these problems. In any case, by increasing the sampling rate of inspection of critical dimensions on the machined product, monitoring of tool wear can be greatly improved.

Inspection—Qualitative and Semi-Quantitative — Machine vision for this group of applications for inspection is aimed at emulating the human when he visually inspects a workpiece or assembly for qualitative and semi-quantitative properties without the aid of measuring devices. The human, however, may be aided by optical devices for, say,

References pp. 18-20.

magnification or demagnification of images, masking or windowing of part of images, comparison of outlines, etc. Applications are enormously varied and one can only make a crude and incomplete classification.

In Fig. 3, five major subgroups are identified. A sixth subgroup, which could properly be part of this group, "In Process Inspection," has been described previously under "Sensor-Controlled Manipulation."

The five subgroups for applications of machine vision for inspection are under the headings: label reading and registration; sorting; integrity and completeness; cosmetic and surface finish properties; and safety and monitoring. In all of these applications, the sensory and interpretative system of choice consists of solid state cameras, interfaced with microcomputers and specialized hardware where necessary. In many instances it is certainly possible to engineer a special-purpose contact or noncontact solution that may be less expensive; however, it would be much more difficult to modify or adapt to other similar applications. The following discussion, therefore, will assume the availability of the adaptable and readily reprogrammed camera systems.

1. Label Reading and Registration — Printed characters on labels, or directly marked on bottles, cans, cartons, etc. can be "read" by existing optical character recognition (OCR) devices, providing that the position and orientation of these products are constrained to within narrow limits. Similarly, bar-codes can be "read" by available bar-code scanners, under similar constraints. It is often costly to orient and position the products, or the insignia may be randomly positioned on given surfaces. Relatively simple programs exist to locate the insignia and recognize the printed or coded insignia using the camera systems with far greater latitude in accepting variations of position and orientation.

2. Sorting — There are many sorting applications in manufacturing processes in which workpieces of several types that are packed in containers or randomly arranged on conveyors must be recognized and then manipulated in a desired manner. Applications include the packing of workpieces of one type only in orderly fashion in the container, the selecting of a specific workpiece from a random mix of workpieces, or the identification and counting of the number of workpieces of each type for inventory purposes.

The same techniques and requirements described under "Manipulation of Separated Workpieces on Conveyors" and "Bin Picking" apply to sorting operations. Essentially, discriminating one workpiece type from another is similar to the problem of identification of each of the stable states of a single workpiece type. Obviously the system must be able to discriminate more stable states. The level of performance, therefore, depends on the number of stable states for each type. Decent performance has been obtained on up to half dozen different workpiece types, each with several stable states [16].

3. Integrity and Completeness — This subclass of inspection is perhaps the most important and most needed of this group. It includes qualitative and semi-quantitative inspection of individual workpieces for many types of flaws and errors incurred in fabrication and in handling, including burrs, broken parts, pits, cracks, warping, defects in printed-circuit wiring, etc. The approximate size and location of key features, such as drilled holes, bosses, etc., can be measured for gross conformity to specification. Each inspection for the integrity of individual workpieces can be explicit, occurring after fabrication processes, such as casting, machining, punching, bending, drilling, etc., or implicit, as part of assembly.

Inspection for completeness is universally required for assuring that subassemblies and assemblies have all of the right parts present in the right places.

Each type of defect requires the development of a specialized computer program, which makes use of a library of subroutines, each effecting the extraction and measurement of a key feature. In due course this library will be large and be able to accommodate many common defects found in practice. Simple vision routines utilizing two-dimensional binary information, as previously noted, can handle a large class of defects. However, three dimensional information, including color and gray-scale, will ultimately be important for more difficult cases.

Inspection for completeness is usually quite difficult, and to be effective will probably require the use of structured light and three-dimensional information. Such inspection techniques have barely begun to be developed in various laboratories.

4. Cosmetic and Surface Finish Properties — Stains, smears, color, blemishes, dirt, tears, discontinuities, and so forth are all potential defects found on the surfaces of many finished goods. These defects are usually cosmetic rather than functional, are esthetically unacceptable, and adversely affect the marketability of the product. Representative examples are:

- Discernible variations in uniform paint finishes on car bodies and on major consumer appliances.
- Discontinuities in polished ceramic surfaces, and on polymer sheets.
- Oil, stains, dirt, and extraneous marks on toiletries, paper, cloth, cartons, etc.
- Torn and crushed wrappers of cosmetic and food packages.
- Ripeness (by color), rot, scabs, and other blemishes on fruit.

Many of these defects can be detected with the camera systems utilizing two-dimensional information only. In almost all cases, however, specialized incident lighting must be engineered with appropriate choice of spectral band, filters, and direction(s) of incident light relative to the inspected object. Again, each application requires a specialized computer program tailored to the job, making use of a library of common subroutines that extract feature information.

5. Safety and Monitoring — Safety and monitoring functions of machine vision are included for completeness. Modern robot arms moving at high speed with powerful motive forces can be very dangerous to human workers and destructive to workpieces and other machinery. A highly desirable machine vision system, unfortunately very

References pp. 18-20.

difficult to implement, could monitor the spatial volume surrounding the robot, and stop motion when collisions are imminent. Concurrently, new safety machanisms to effect sufficiently rapid deceleration of the moving links would have to be incorporated in the robot design. A very sophisticated vision software system, operating off several remote cameras, would be required to differentiate between the known trajectories of one or more robots and workpieces, and potential intrusions in the workspace by humans. Further, the system would have to detect wild trajectories due to malfunction of the robot [39, 40].

Such a system, even if available today, would not be cost-effective. However, it may be possible to approach a compromise solution with the aid of internal and other sensory modalities.

CONCLUSIONS

Machine vision can be applied to a large proportion of tasks in batch production of durable goods, to automate sensor-controlled manipulation in material handling and assembly, and to automate visual inspection.

Present robots and machine vision techniques are already sufficiently advanced to permit their initial introduction into factories on a pilot basis.

More sophisticated robots and vision systems are now being developed with promise of meeting cost and performance criteria acceptable to industry. It will probably require two to five years and transfer of this new technology to industrial advanced development user groups, as the first step to deployment in factories. There appear to be no fundamental obstacles to the early development of adequate and cost-effective machine vision systems for the many classes of application described in this paper.

REFERENCES

1. Minsky, M.L. An autonomus manipulator system. Project MAC Progress Rep. 111, MIT, and subsequent Project MAC, Artificial Intelligence Series, MIT, Cambridge, MA., July 1966.
2. McCarthy, J., et al. A computer with hands, eyes, and ears. AFIPS Conf. Proc. FJCC (1968), 329-338.
3. Feldman, J.A., et al. The Stanford hand-eye project. Proc. 1st Int. Jt. Conf. on Artificial Intelligence (1969), 521-526.
4. Nilsson, N.J. A mobile automation: an application of artificial intelligence techniques. Proc. 1st. Int. Jt. Conf. on Artificial Intelligence (1969), 509-520.
5. Barrow, H.G., and Salter, S.H. Design of low-cost equipment for cognitive robot research. *In Machine Intelligence 5*. B. Meltzer and D. Michie, (Eds.) American Elsevier Publishing Co., New York, 1970.
6. Ejiri, M., Uno, T, Yoda, H., Goto, T., and Takevasu, K. A prototype intelligent robot that assembles objects from plan drawings. *IEEE Trans. on Computers* C-21, 2 (1972).
7. Rosen, C.A., et al. Application of intelligent automata to reconnaissance. Final Rep., SRI Project 5953, Stanford Research Inst., Menlo Park, CA, 1968.
8. Barrow, H.G., and Tenebaum, J.M. Recovering intrinsic scene characteristics from images. In *Computer Vision Systems*, A. Hanson and E. Riseman (Eds.) Academic Press, New York, 1979.

9. Nilsson, N.J. Artificial intelligence. IFIP Conf. Proc. (August 1974), 778-801.
10. Duda, R.O., and Hart, P.E. *Patter Classification and Scene Analysis.* John Wiley and Sons, New York, 1973.
11. Rosenfeld, A., and Kak, A.C. *Digital Picture Processing.* Academic Press, New York, 1976.
12. Engelberger, J.F. Production probelms solved by robots. Soc. Mech. Eng. Tech. paper MS74-167 (1974).
13. Pringle, K.K., and Wichman, W.M. Computer control of a mechanical arm through visual input. IFIP Congress 68, Application 3, Booklet H (1968), H140-H146.
14. Heginbotham, W.B., et al. Visual feedback applied to programmable assembly machines. Proc. 2nd Int. Symp. on Industrial Robots (May 1972), 63-76.
15. Goto, T. Compact packaging by robot with tactile sensors. Proc. 2nd. Int. Symp. on Industrial Robots (May 1972).
16. Rosen, C.A., et al. Exploratory research in advanced automation. Reps. 1 through 7 prepared for Nat. Science Foundation by Stanford Research Inst., Menlo Park, CA. (December 1973 through August 1977).
17. Nevins, J.L., et al. Exploratory research in industrial modular assembly. Reps. 1 through 4 prepared for Nat. Science Foundation by Charles Stark Draper lab., Cambridge, MA. (June 1973 through August 1976).
18. Olsztyn, J.T., Rossol, L., Dewar, R., Lewis, N. An application of computer vision to a simulated assembly task. Proc. 1st Int. Jt. Conf. on Pattern Recognition (1973), 505-513.
19. Bolles, R.C., and Paul, R. The use of sensory feedback in a pgrogrammable assembly. Stanford Artificial Intelligence Project Memo No. 220, Stanford U., Stanford, CA. 1973.
20. Will, P.M., and Grossman, D.D. An experimental system for computer-controlled mechanical assembly. *IEEE Trans. on Computers* C-24, 9 (1975), 879-888.
21. Rosen, C.A., and Nitzan, D. Use of sensors in programmable automation. *Computer* 10, 12 (1977).
22. Niemi, A., Malmen, P., And Koskinen, K. Digitally implemented sensing and control functions for a standard industrial robot. Proc. 7th Int. Symp. on Industrial Robots (1977), 487-495.
23. Goto, T., Inoyama, T., And Takeyasu, K. Precise insert operation by tactile controlled robot HI-T-HAND Expert-2. Proc. 4th Int. Symp. on Industrial Robots (1974), 209-218.
24. d'Auria, A., and Salmon, M. Sigma: an integrated general purpose system for automatic manipulation. Proc. 5th Int. Symp. on Industrial Robots (1975), 185-202.
25. d'Auria, A., and Salmon, M. Examples of applications of the Sigma assembly robot. Proc. 6th Int. Symp. on Industrial Robots (1976), G5-37 - G5-48.
26. Davis, Ray. private communication.
27. Ando, S., Kusumoto, S., Enomot, K., Tsuchihashi, A., and Kogawa, T. Arc welding robot with sensor "Mr. Aros", Proc. 7th Int. Symp. on Industrial Robots (1977), 623-630.
28. Agin, G.J., and Duda, R.O. SRI vision research for advanced industrial automation. Proc. 2nd USA-Japan Computer Conf. (1975), 113-117.
29. Agin, G.J., and Binford, T.O. Computer description of curved objects. Proc. 3rd Int. Jt. Conf. on Artificial Intelligence (1973), 629-640.
30. Johnston, A.R. Infrared laser rangefinder. NASA New Technology Rep. No. NPO-13460, Jet Propulsion Lab., Pasadena, CA. 1973.
31. Nitzan, D. Brain, A.E., and Duda, R.O. The measurement and use of registered reflectance and range data in scene analysis. *Proc. IEEE* 65 (1977), 206-220.
32. Perkins, W.A. Model based vision system for scenes containing multiple parts. Research Publication GMR 2386, General Motors Research Labs., Warren, MI. 1977.
33. Birk, J., et al. Robot computations for orienting workpieces. 1st and 2nd Reps., U. of Rhode Island, Kingston, RI. (April 1975 to August 1976).

34. Autoplace Inc., Troy MI, has demonstrated such a system in 1977 using their limited sequence robot. General Motors has developed and implemented a vision-manipulation system for automatically bonding and inspecting a semi-conductor chip in 1977. SRI International has developed, in 1978, a vision-controlled X-Y table that is being applied to an assembly program.
35. Drake, S.N., Watson, P.C., and Simunovic, S.N. High speed assembly of precision parts using compliance instead of sensory feedback. Proc. 7th Int. Symp. on Industrial Robots (1977), 87-98.
36. Cook, H.H., et al. Tool wear sensors. Final Rep., Dept. Mech. Eng., MIT, Cambridge, MA., 1976.
37. Mergler, H.W., and Sahajdak, S. In process optical gauging for numerical machine tool and automated processes. 5th NSF Grantees' Conf. on Production Research and Technology, (1977).
38. Paul, R. Manipulator path control. Proc. Int. Conf. Cybernetics and Society (1975), 147-152.
39. Udupa, S.M. Collision detection and avoidance in computer controlled manipulators. Ph.D. Th., Cal. Inst. of Tech., 1976.
40. Sugimoto, N. Safety engineering on industrial robots and their draft safety requirements. Proc. 7th Int. Symp. on Industrial Robots (1977).

DISCUSSION

H. Freeman *(Rennselaer Polytechnic Institute)*

You haven't said anything about the role that tactile sensing would play.

Rosen

Tactile sensing is extremely important. The visual sensor is very good for roughly finding an object but not for finding it with great precision unless one wants to build an expensive system. Tactile sensing is contact sensing. If you want to know where something is precisely and operate on it, such as insertion of a screw in a hole, tactile sensing becomes very valuable and it is not terribly expensive to implement. Researchers at Draper Labs, for instance, have developed good wrists with strain gauges and other devices that permit their system to servo using tactile sensing. I did not mean to give you the impression that visual sensing is the only sensing available. One should use the appropriate sensor for each problem.

G. J. VanderBrug *(National Bureau of Standards)*

I would like to ask you two questions. (1) How often does the situation occur in which more than one part comes down the conveyor belt? I am not talking about the different stable states of a part but distinctly different parts. (2) How does machine tool loading and unloading fit into the framework you described and how big a problem is it?

Rosen

(1) At present, most assembly lines have only one kind of part coming down the conveyor belt. However, if they had systems with the capability of recognizing and manipulating different types of parts on a conveyor, many more places might use the conveyors for buffer storage. That is, you could have a continuous conveyor with overhead lines as exists now in many factories. These conveyors run throughout the factory and have many different parts hanging on them. Workers take off the parts

that they need for some particular process. I think that developing these techniques to solve a specific problem can sometimes lead to the possibility of making a brand new system because you can do something that you couldn't do before. The business of being able to buffer storage on a moving belt is an example.

(2) Almost everywhere I go, people are worried about loading and unloading large machine tools for very good reason. Japan is one place where they are doing this. However, the parts are fixtured before the robot can acquire them and load them into the machine. The next step, to determine where these parts are so they don't have to be fixtured, is possible now. It is no different than recognizing and determining the position and orientation of parts on a belt. To my knowledge it has not been done. If it were done, I think people would have a little more desire to use a robot. On the other hand the problem of handling a three spindled NC machine has been solved recently by a group in Berlin. They eliminated some of the problem because once the machine grabs hold of the part, it can be moved from spindle to spindle. They don't spend as much time acquiring the part and sticking it in and moving it from machine to machine.

T. Vamos *(Hungarian Academy of Sciences)*

Can you tell us about some of your experience in using illumination to help with the recognition process?

Rosen

During this conference you will hear at least one paper by people at General Motors Research on how they solved the illumination problem for parts randomly located on belts. Most of the illumination techniques are ad hoc. Researchers in laboratories have shown the need for an image with high contrast, because they would like to work with a binary image. In our laboratories we always expend a lot of effort on the kind of illumination that would be satisfactory with a simple vision system. In particular we have found that if you want to look at holes, it is desirable to have illumination from all sides becuase the hole appears dark on a white background. I don't know of any paper that deals just with illumination and its intrinsic effect in the application.

L. Rossol *(General Motors Research Laboratories)*

What is the role of some of the more classical areas of artificial intelligence, such as problem solving and understanding, in this problem?

Rosen

The role of AI in this particular topic is in vision and its interpretation. These have been AI subjects from the beginning. The development of vision techniques in research laboratories gives us our tools. I would like to see a good deal more of effort in artificial intelligence on high order robot languages and on natural languages. For instance in planning operations, a program can help select the proper sequence to follow. We are dealing with programmable automation using a robot and vision.

M. Minsky *(Massachusetts Institute of Technology)*

On this question of the variety of parts on assembly lines, it seems to me that assembly lines are silly and when we have good hand-eye robots, they will usually throw the part across the factory to the machine which wants it and that machine will catch it.

Rosen

Your idea contains one of our objectives: separating the parts.

HUMAN AND ROBOT TASK PERFORMANCE

R. L. PAUL and S. Y. NOF

Purdue University, West Lafayette, Indiana

ABSTRACT

The industrial robot, which might be viewed as a combination of a remote manipulator and a numerically controlled machine tool, is now an accepted element in manufacturing with some thousands of robots at work in the United States today. With the addition of task sensors (vision, force, and touch) and the decision making capabilities of digital computers a much greater range of tasks becomes possible for these devices.

In this paper we examine the basic capabilities of robots, their sensors, and the appropriate level of error recovery in order to compare and contrast human and robot task performance. Based on the capabilities of robots we outline a level of task primitives, programming detail, and task performance techniques suitable for robots. Such a set of primitives could form the basis for measurement and selection techniques in order to specify a robot for a class of tasks.

The sensor controlled robot, in spite of its primitive sensors and weak cognitive ability, provides a powerful adaptive system with good error recovery and in-process inspection capabilities.

INTRODUCTION

The industrial robot is already an accepted element in modern manufacturing facilities with several thousands of robots at work in the United States today. With the addition of task sensors (vision, force, and touch) used for feedback from the work scene, and the decision making capabilities of digital computers which utilize this feedback in the control of the robot, a much greater range of tasks becomes possible for these devices.

References pp. 47-48

Work in general and industrial work in particular has always been studied relative to men and machines. Major strides have been taken in this century in designing and analyzing efficient, optimal work methods and systems for human operators.

Attention should now be given to the question of how to incorporate robots into the manufacturing facility with the objective of further increasing its productivity. Can we use the same work methods? Can we use the same design techniques?

Indeed, fundamental differences exist between the capabilities of human workers and robots, and consequently in their work methods. Nevertheless, we should learn to take advantage of our experience with human work performance, and to utilize as much as possible already established techniques and tools, where applicable, in order to plan robot work most effectively. For instance, in previous work [4], it was shown that based on the differences in characteristics between robots and human operators a revised technique for task analysis may be useful in designing effective work methods for a robot and for the selection of an appropriate robot for that method.

The purpose of this work is to investigate the differences between robots and human operators relative to their capabilities and limitations in task performance. Detailed comparison is made by referring to the Methods Time Measurement method [3] which tabulates elementary human motions practiced in performing industrial tasks. Based on this comparison of capabilities a set of elementary robot task motions is developed with the title RTM, short for "Robot Time and Motion". An example of an assembly task is then used to illustrate typical differences in the work method when performed by either a robot or a human operator.

BASIC ROBOT CAPABILITIES IN TASK PERFORMANCE

In the context of task performance as discussed in this paper six basic capabilities of robots can be identified. They include:
1. Workspace
2. Actuation
3. Positioning
4. Orientation
5. Sensing
6. Decision Making

While any discussion of robot capabilities is highly dependent on the current state of rapidly changing technology we believe that the nature of our analysis is of general relevance. For the sake of specificity let us consider a robot model with which significant experience has been gained, the Stanford Arm (SA). Each of the six basic capabilities is described as it is implemented in the Stanford Arm, including control language requirements and examples. The language used is PAL [5] which is being developed at Purdue University with National Science Foundation support.

Stanford Arm (SA) Capabilities in Task Performance

1. Workspace — The workspace of a robot is defined as the space, in which it is capable of operating with its full given motion flexibility. (A robot can reach beyond its workspace but then it may lose some of its flexibility because the arm is extended to its maximum length.) The workspace of SA is a sphere with a radius of 25 cm within which it can grasp objects in any orientation. The arm (see Fig. 1) consists of six links connected together by joints. The first two joints are rotary, followed by a sliding joint, followed by three more rotary joints.

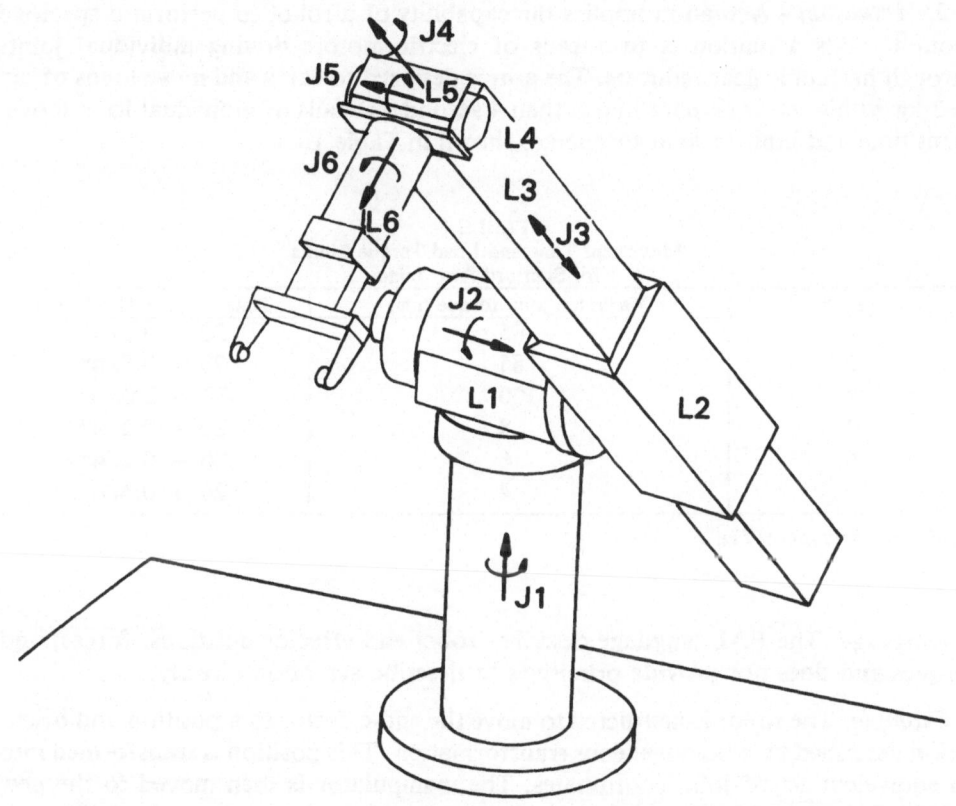

Fig. 1. Links and joints of Stanford Arm (SA).

Language: Positions and orientations are represented in PAL by homogeneous transformations as explained in [5] and are referred to by identifiers. All transformations must be declared in a program as follows*:

   TRN name [, name]

---
* The syntax used in this and following language expressions includes brackets to indicate optional parameters, and braces to indicate a list from which one parameter must be selected.

References pp. 47-48

Once a transform is declared, it may be referred to by name and its value, consisting of a relative position and orientation between two objects, used in expressions.

Example: Objects may be located anywhere within the workspace; three of the joints (degrees of freedom) are required to position the hand over the object. The direction from which the hand approaches an object and the orientation of the hand about this approach direction requires three more degrees of freedom. Objects may be grasped from any direction: from above, from below, from the side.

2. *Actuation* - Actuation implies the capability of a robot to perform a specified motion. SA's actuation is by means of electric motors driving individual joints through harmonic gear reducers. The arm is designed to carry and move loads of up to 3 kg within its workspace in less than 4 seconds. Details of individual joint movement time and limit of load torque are shown in Table 1.

TABLE 1
Movement Time and Load Torque Limits
for Stanford Arm Joints

Joint*	Maximum joint torque (nm)	Time to move (1/60 sec.)
1	85	$20 + 0.5/\theta°$
2	85	$20 + 0.5/\theta°$
3	100	$20 + 2.0/cm$
4	8	$20 + 0.2/\theta°$
5	8	$20 + 0.2/\theta°$
6	4	$20 + 0.4/\theta°$

* *Joint numbers follow Fig. 1*

Language: The PAL language describes robot end-effector positions, forces, and torques and does not provide primitives to describe actuation directly.

Example: The robot is instructed to move the end-effector to a position and orientation described by a homogeneous transformation. This position is transformed into an equivalent set of joint coordinates. The manipulator is then moved to the new position by servoing all joints as a coordinated function of time. Over short distances such motions approximate straight line end-effector motions to a high degree of precision.

3. *Positioning* — Positioning is the robot ability to move to a given point in space. It is measured by two attributes: absolute accuracy and repeatability. Absolute accuracy is defined as the tolerance in each coordinate in reaching any given point in space, and repeatability is defined as the tolerance in reaching a previously visited point in space. For SA the absolute accuracy is ± 0.5 cm and the repeatability is ± 0.05 cm.

The position and orientation of the joints during positioning are measured by encoders or potentiometers. Based on these measurements it is possible to compute (using a digital computer) the Cartesian position and orientation of the hand. It is also possible to compute the joint coordinates necessary to place the hand in a given Cartesian position and orientation.

Language: In PAL, motion is commanded by a MOVE statement with an argument specified as a transform expression. The transform expression is evaluated and then the manipulator is moved to the corresponding set of joint coordinates.

Example: If the Cartesian coordinates position and orientation of an object are determined, for instance, by a camera, the corresponding joint coordinates will position the hand to within ± 0.5 cm of the true position of the object. If the position and orientation of the hand were then corrected to the true position a second object could be placed on the first to within ± 0.05 cm. As a further example, the positions of a 20 cm by 20 cm grid of points spaced at 0.2 cm intervals could be recorded and the manipulator moved directly to any one of the 10,000 points to within 0.05 cm.

4. Orientation — Orientation defines a robot's capabiity to direct its hand in space, usually in order to grasp an object. Similar to positioning, it is characterized by absolute accuracy and repeatability. The absolute accuracy of SA's hand in an approach direction is $\pm 2°$ while the repeatability is $\pm 0.1°$.

Language: Orientation is combined with position in PAL and is expressed by homogeneous transformations. The MOVE statement causes the manipulator to move to a new position and/or orientation.

Example: If the true orientation of an object is determined, the corresponding joint coordinates will orient the hand correctly to within $\pm 2°$. In making small motion with the hand it is possible to maintain the orientation to within $\pm 0.1°$. As a consequence the arm is well suited to making small clearance insertions in which the orientation must be maintained in order to prevent jamming.

5. Sensing — This is the robot's ability to process task related input data from its work environment. Touch compliance and force compliance are the two main senses of SA. Vision can also be provided.

Force Compliance — For many tasks in assembly, such as performing insertions and mating parts, the manipulator motion is defined by external constraints. If the manipulator is position controlled, those constraints appear as external forces. These forces may be used instead of position specifications to control the manipulator. SA can provide force compliance to an external static constraint within 5 N by driving a manipulator actuator at a fixed torque for each degree of compliance specified. It can also provide for any preload force at the hand. In the case of a dynamic constraint, sufficient preload force must be provided to overcome arm inertia in order to maintain contact between the hand and the constraint.

Language: In PAL, compliance is specified by preceding a MOVE statement with a compliance clause, as follows:

$$\text{COMPLY} \quad \begin{Bmatrix} \text{FORCE} \\ \text{TORQUE} \end{Bmatrix} \begin{Bmatrix} X \\ Y \\ Z \end{Bmatrix} \quad [\text{@ level}] \; [\text{WITH K} = \text{constant}]$$

References pp. 47-48

A compliance clause specifies whether force or torque are to be controlled; a direction; the force or torque preload, if not zero; and the compliance K, if not zero. If more than one degree of compliance is required then the MOVE statement is preceded by a series of compliance clauses one for each degree of compliance. Changes in compliance can also terminate a motion by following the move statement with an UNTIL clause. There are two forms of UNTIL clauses. The first terminates a motion on encountering some specified external force. The second terminates a motion when an external constraint disappears.

$$\text{UNTIL} \quad \begin{Bmatrix} \text{FORCE} \\ \text{TORQUE} \end{Bmatrix} \quad \begin{Bmatrix} X \\ Y \\ Z \end{Bmatrix} \quad [@ \text{ level}]$$

$$\text{UNTIL} \quad \begin{Bmatrix} \text{FORCE} \\ \text{TORQUE} \end{Bmatrix} \quad \begin{Bmatrix} X \\ Y \\ Z \end{Bmatrix} \quad \text{ZERO}$$

Example: In order to pull out of a hole a pin which is oriented along the Z axis, until it is free, one would write:

COMPLY FORCE X @ 20
  COMPLY FORCE Y
    COMPLY TORQUE X
      COMPLY TORQUE Y
        MOVE OUT
          UNTIL FORCE X ZERO;

Touch Compliance — Force compliance is too crude for delicate tasks such as locating a small object. For such tasks touch sensors in the fingers are used to guide the hand. The touch sensors are very sensitive to contact forces of the order of 0.05 N.

Language: Fingers are considered separate elements in PAL and are moved in coordination with the arm. A move statement may be preceded by finger statements. A finger statement has a direction of motion followed by a terminating condition.

$$\begin{Bmatrix} \text{OPEN} \\ \text{CLOSE} \end{Bmatrix} \quad \begin{Bmatrix} \text{LF} \\ \text{RF} \end{Bmatrix} \quad \text{UNTIL} \quad \begin{Bmatrix} \text{final coordinate} \\ \text{TOUCH} \\ \text{FORCE @ level} \end{Bmatrix}$$

LF and RF stand for left and right finger respectively. After, or during all finger commands the arm centers itself with respect to the fingers automatically. The commanded position of the arm may change as a result of such a motion.

Example: To center the hand over an object located at OBJ we could write:

MOVE OBJ;

CLOSE LF UNTIL TOUCH

  CLOSE RF UNTIL TOUCH

    MOVE OBJ;

CLOSE LF UNTIL FORCE @ 50

  MOVE OBJ;

6. *Decision Making* — Decision making in robots is the ability to select a proper course of action from a variety of alternatives based on sensory inputs and control information. For instance, a simple form of decision making occurs when a part is grasped, and the distance between the fingers is checked to decide whether a part is present. A more complicated example is a robot using a bar-code reader to identify part types and deciding how to process them.

Decision making is performed by the control computer. In SA computers ranging from a PDP 11/03 to a PDP 10 have been used.

Language: The following ALGOL types of decision making statements are availble in PAL.

1. IF boolean expression THEN statement.

2. WHILE boolean expression DO statement.

3. REPEAT statement UNTIL boolean expression.

In order to simplify programming in PAL any group of statements surrounded with a BEGIN and END are considered a single statement.

Example: The PAL code to perform the decision making after attempting to grasp an object (as described above) is,

CLOSE LF UNTIL FORCE @ 50

  CLOSE RF UNTIL FORCE @ 50

    MOVE HERE; (*GRASP OBJECT AT HERE*)

IF LF-RF < 0.5

  THEN

  BEGIN

    WRITE ("NO PART PRESENT");

    PAUSE;

END;

References pp. 47-48

COMPARISON BETWEEN TASK PERFORMANCE ELEMENTS OF ROBOT AND HUMAN OPERATOR

Robots and human operators differ in numerous characteristics from their manipulation abilities and consistency to their intelligence, sensitivity, and social needs. In [4] a detailed comparison leads to identification of five major differences in task related characteristics: humans possess a set of accumulated and unconscious skills; robots can be built to have no individual differences; robot abilities can be optimized for a task; robots are unaffected by social and psychological effects; for humans, training and retraining are more difficult.

While these differences in characteristics are useful in the general sense of analysis, a different comparison is required in order to investigate the details of task performance elements. For this purpose we will utilize an established methodology in human work analysis, Method Time Measurement (MTM). This method was developed by Maynard, Stegemerten, and Schwab [3, 4]. It includes tables for elementary human task related hand motions such as REACH, MOVE, GRASP, etc.; eye motions, body, leg, and foot motions. Each table specifies the time required to perform elementary motions based on parameters such as distance, accuracy, complexity, etc. Because the time required for individual MTM task motion is very small the time unit used in MTM is TMU (Time Measurement Unit, equivalent to O.036 sec). Using MTM an industrial task can be broken down to its elementary motions performed by each hand and other body members. Accordingly, the time for each motion and total estimated time for the task can be derived. In addition to providing detailed task specification and time estimates MTM tables also indicate whether motions for left and right hands can be performed simultaneously. Further, the MTM user can compare alternative methods per time, complexity, and so on.

Let us now review some important elementary task motions as defined by MTM for a human operator, and compare them to the robot capability.

REACH — The five elementary MTM REACH types which specify an empty hand motion to reach a point under various circumstances are described in Table 2. The robot equivalent elementary motions are compared to these human motions in the third column.

The robot equivalent of REACH RA may be expressed directly in terms of PAL statements. A reach to point A would be expressed as:

    MOVE A;

The MOVE statement causes the manipulator to be moved in a coordinated manner from its present position towards A. An UNTIL clause can be used to cause the manipulator to be brought to rest at A with position and velocity errors less than given bounds. SA execution times for MOVE are given in Table 3 and for UNTIL ERROR in Table 4.

TABLE 2
Elementary REACH Task Motions

Element	MTM Description	Robot Equivalent
RA	Reach to object in fixed location, or to object in other hand or on which other hand rests.	Move end-effector directly to given position and orientation.
RB	Reach to single object in location which may vary slightly from cycle to cycle.	Move to a position close to object; refine position by touch or force feedback.
RC	Reach to object jumbled with other objects in a group so that search and select occur.	Highly inefficient for robot.
	Reach to a very small object or where accurate grasp is required.	Move to a position close to object; reestablish position of object by vision; refine position by touch or force feedback.
RE	Reach to indefinite location to get hand in position for body balance or next motion or out of way.	Indefinite movement to get out of way is equivalent to RA for robot, otherwise serves no purpose.
Spatial reach*	Highly inefficient for humans.	Move to an absolute position and orientation in space such as a storage grid.

Not in MTM Tables

TABLE 3
SA MOVE Times

Distance to move (cm)	Time (TMU)
1	10.5
30	12.5
100	20

TABLE 4
SA MOVE UNTIL ERROR Times

Error bound (cm)	Time (TMU)
1	3.5
0.2	8
0.05	23

References pp. 47-48

The robot equivalent of **REACH** RB to a point A consists of **REACH** RA to a position close to A, say A1, and then a series of force or touch sensing moves to attain the exact position. For example:

MOVE   A1;
MOVE   A
      UNTIL FORCE Z @ 27;
MOVE A
      UNTIL FORCE X @ 10;

SA execution times for UNTIL FORCE or UNTIL TOUCH are given in Table 5.

TABLE 5
SA MOVE UNTIL FORCE or TOUCH Times

Distance moved (cm)	Times (TMU)
0.5	3
3.0	25
15.0	125

As mentioned in Table 2 **REACH** RA which requires handling jumbled objects is an elementary human motion which is highly inefficient for robots, due to their limited sensory perception.

In the robot equivalent of **REACH** RD we assume that the precision required is such that the manipulator cannot be brought into a position so that force and touch sensing can be used. We will follow a **REACH** RA with vision to establish the relative position of the manipulator to the object. One or more force or touch sensing moves will then have to be made to define the exact position. An example:

MOVE A
    UNTIL ERROR <bound> ;
READ (Vision, A);
MOVE A
    UNTIL FORCE Z @ 20;

The time to perform **READ** (Vision, A) will vary greatly with the complexity of the vision task.

One robot capability which is highly inefficient for human performance is reaching to an exact absolute position in space. Such a motion for a robot is actually identical to **REACH** RA.

Table 6 compares execution times of **REACH** RA and **REACH** RB motions for SA and for a human operator. From the table it is clear that for very short reach motions the robot is inefficient compared to a human. This inefficiency is due to the time taken to eliminate errors in RA and RB, plus time for sensing in RB. However, over longer distances these excess times become less significant.

TABLE 6
Comparison of Reach Motion Times by Robot and Human

	Distance (cm)	REACH RA	REACH RB
Human	1	2.0	2.0
	30	9.6	12.9
	100	17.5	25.8
SA	1	18.5*	38.5**
	30	20.5*	40.5**
	100	28.0*	48.0**
Absolute	1	16.5	36.5
Difference	30	10.9	27.6
(TMU)	100	11.0	22.2

*Assume 0.2 cm position tolerance
**Assume one 3.0 cm and one 0.5 cm correction motions

In summary, even though SA is found to reach slower than a human, it can perform four out of the five types of the human reach. In addition, it can perform one reach type unattainable by a human.

MOVE — Three elementary motions are specified by MTM for human task performance of moving objects under various circumstances. They are described in Table 7 and compared to equivalent robot motions.

TABLE 7
Elementary MOVE Task Motions

Element	MTM Description	Robot Equivalent
MA	Move object to other hand or against stop.	Cannot move an object against a stop without deceleration. (Use MC instead.)
MB	Move object to approximate or indefinite location.	Move object to some given position and orientation with given acceleration and deceleration.
MC	Move object to exact location.	Move object close to given position and orientation; move to final position by force or vision feedback.
Spatial Move*	Highly inefficient for humans.	Move object to an exact position and orientation in space.

*Not in MTM Tables

A MOVE MA cannot be performed by a manipulator such as SA. The sudden deceleration at the end of the motion when a stop is encountered would damage the arm and gear reducers. On the other hand, a MOVE MB can be performed by SA as a series of PAL MOVE statements. An initial MOVE accelerates the object away from its initial position. A second MOVE carries the object toward its destination, and a third MOVE decelerates the object to its final position. In many cases the second MOVE will have to be broken down into a series of moves to avoid interference by other objects. The system must be informed of the mass of the object carried to adjust the arm dynamics. An example of a MOVE MB from a point A, accelerating to A1, moving to B1, then decelerating to B:

References pp. 47-48

```
OBJECT MASS = value
BEGIN
    MOVE A1;
    MOVE B1;
    MOVE B
        UNTIL ERROR < bound;
END;
```

SA execution times for moving an object can be computed as the sum of times of the series of individual MOVES as given in Table 3. However, since the arm is loaded, those values should be increased by a weight factor (see Table 8). For instance, if the arm carries a load weighing 2.1 Kg then the time computed for the MOVEs must be multiplied by 2.0, the weight allowance factor, to obtain the actual time.

TABLE 8
Weight Allowance for SA Move Time

Weight (Kg) up to	Factor
0.5	1.0
1	1.5
2	2.0
3	3.0

A robot equivalent of <u>MOVE</u> MC can be viewed as <u>MOVE</u> MB except that it is performed by a move to a position close to destination and then one or more force sensing moves to attain the final position. For example:

```
OBJECT MASS = value
BEGIN
    MOVE A1;
    MOVE B1;
    MOVE B11;
    MOVE B
        UNTIL FORCE Z @ 27;
    MOVE B
        UNTIL FORCE X @ 30;
END;
```

Examining the <u>MOVE</u> types we realize that there is at least one type of task motion that can be performed by a robot but not usually by humans. Since a robot is calibrated absolutely it can perform tasks such as flame cutting of metal plates which require moving a tool in space according to given positions. Such a task cannot be directly performed by a human operator.

For a comparison of move motion time consider the following example: Move an object weighing 2 kgs a total distance of 50 cm including 10 cm for acceleration and for deceleration. A human operator performs this in 19.3 and 23.4 TMU for MB and

MC respectively. SA requires 80.0 and 89.0 for the same operations which indicates that it performs at about 25% of human speed.

GRASP — Grasp is a motion employed to secure control of an object. Five elementary human grasp types are described in Table 9 and compared to equivalent robot motions. Robots cannot perform motions requiring finger dexterity (G1C2, G1C3, G2, G4).

TABLE 9
Elementary GRASP Task Motions

Element	MTM Definition	Robot Equivalent
G1A	Pick Up Grasp — Small, medium or large object by itself, easily grasped.	Close hand
G1B	Very small object or object lying close against a flat surface.	Locate object surface by force, close hand.
G1C1	Interference with grasp on bottom and one side of nearly cylindrical object. Diameter larger than 1.27 cm.	Free object from interference, relocate it and close hand.
G1C2,3	Interference with grasp on bottom and one side of nearly cylindrical object. Diameter less than 1.27 cm.	Impractical for robot.
G2	Regrasp.	Impossible for robot.
G3	Transfer Grasp.	Position object in other hand, release. (Possible only when two hands are available.)
G4	Object jumbled with other objects so search and select occur.	Highly inefficient for robots.
G5	Contact, sliding or hook grasp.	Close hand while in motion.
Measure Grasp	Impossible for human.	Combined grasp and measurement of object.

A robot equivalent to G1A is simple closing of the fingers, specified by PAL as:

CLOSE LF UNTIL FORCE @ 50
CLOSE RF UNTIL FORCE @ 50

In the case of G1B a force sensing move must first be executed to locate the support surface. Then the fingers are closed while complying with the surface, i.e.,

MOVE SUPPORT
  UNTIL FORCE Z @ 27;
COMPLY FORCE Z @ 27
  CLOSE LF UNTIL TOUCH
  CLOSE RF UNTIL TOUCH
    MOVE SUPPORT;

A G1C1 requires some manipulator algorithm and will not be considered further.

References pp. 47-48

A G5 consists of closing the hand while in motion. For example to close the hand while moving to B:

CLOSE LF UNTIL FORCE @ 10
CLOSE RF UNTIL FORCE @ 10
  MOVE B;

One type of grasping motion which can be performed by a robot but is impossible for a human operator is to grasp an object while measuring its width with no special tooling. Such a motion is useful whenever this information is needed for decision making, e.g., in sorting.

Time of grasping motions is given above in Table 5.

POSITION — The MTM *position* types are concerned with the insertion of objects of various degree of symmetry, handling difficulty and closeness of fit. The various types are described in Table 10 and compared to the robot equivalent motion in the last column. The *position* motion is divided into three classes for the robot depending on the difficulty of making the initial insertion.

TABLE 10
Elementary POSITION Task Motions

MTM Definition		Robot Equivalent
Class of Fit	Symmetry	
1-Loose No pressure required	S SS NS	Perform initial insertion of object into opening; then insert completely.
2-Close Light pressure required	S SS NS	Insert edge of object into opening; align object; then insert completely.
3-Exact Heavy pressure required	S SS NS	Insert edge of object into corner of opening; insert edge; align object; then insert completely.

In the case of a loose fit P1 we assume that a PAL MOVE terminated on an UNTIL ERROR will be accurate enough to start the insertion. PAL code for an actual insertion is as follows:

COMPLY FORCE X WITH K = 10
  COMPLY FORCE Y WITH K = 10
    COMPLY TORQUE X WITH K = 100
      COMPLY TORQUE Y WITH K = 100
        MOVE N (*NOMINAL POSITION OF THE PIN FULLY INSERTED*)
        UNTIL FORCE Z @ 30;

SA execution times for *position* motions are actually execution times for SA MOVE UNTIL FORCE or TOUCH which are given in Table 5. In the case of a P2S through P3S the robot must first insert an edge of the object into the hole and then, maintaining contact with the edge of the hole, align the axes before making the insertion as described for P1. The following PAL code must precede the P1 insertion code:

```
MOVE C   (*POSITION OF OBJECT WITH CORNER TILTED
         INTO HOLE OPENING*)
   UNTIL Z @ 20;
COMPLY Z @ 20;
   COMPLY X @ 20;
      MOVE B   (*POSITION OF OBJECT ALIGNED WITH THE AXIS OF
               THE HOLE AT THE OPENING*)
```

In the case of P3SS and P3NS the robot must initially insert a corner, then an edge, align the axes, and then complete the insertion. The PAL code for the insertion is the same as for P1 except that it must be preceded by the following PAL code to achieve the initial insertion:

```
MOVE C   (*POSITION OF OBJECT WITH CORNER TILTED INTO HOLE
         OPENING*)
   UNTIL Z @ 20;
COMPLY Z @ 20;
   COMPLY X @ 20;
      COMPLY Y @ 20;
      BEGIN
         MOVE A;   (*SAME AS ABOVE*)
         MOVE B;   (*SAME AS ABOVE*)
      END;
```

In this section we have compared elementary task motions of a robot vs. human operator in four major classes, <u>REACH</u>, <u>MOVE</u>, <u>GRASP</u>, <u>POSITION</u>. Having discussed the differences and similarities in capabilities we can next analyze how they affect task methods.

HUMAN PERFORMANCE OF A PUMP ASSEMBLY TASK

In order to illustrate some typical differences in task performance by robots and humans let us consider the task of assembling an automobile water pump. This task is described in detail in [2] for a Stanford Arm with sensory feedback.

Each water pump in the example is comprised of a pump base, a pump top, one gasket, and six indentical screws. In order to assemble a pump it is necessary to insert the gasket between the top and base parts, then fasten with six screws.

The complete assembly task can be carred out by a variety of alternative methods. Selection of a detailed work method and workplace for a human operator is usually done by a methods analyst, who investigates several proposed methods. Motion economy principles, time of method and cost of workplace will determine which particular method is preferred. However, all such methods would involve the following main steps:

1. Position a base part (or top part) in fixture

2. Align a gasket on base (or top)

References pp. 47-48

3. Position a top (or base) on gasket
4. Fasten with six screws
5. Test if pump is operational

An example of a workplace for this task is shown in Fig. 2.

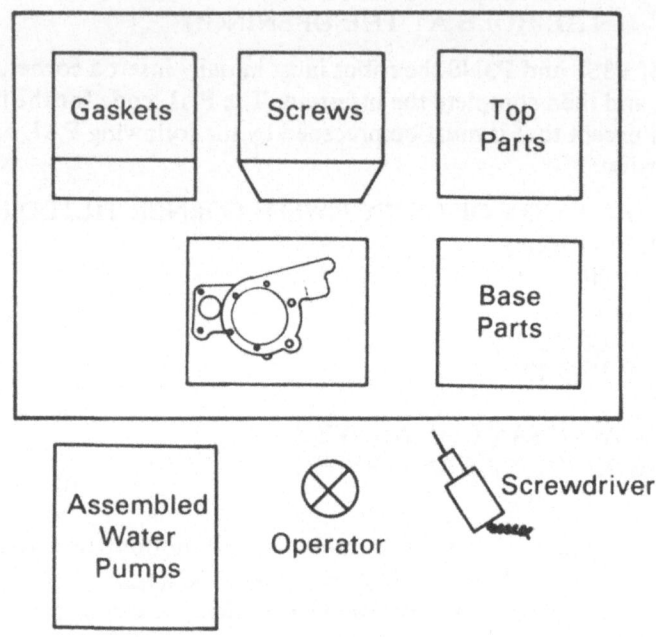

Fig. 2. Workplace for human operator pump assembly task.

For this discussion let us focus on the details of steps 3 and 4. A possible work method is to take a top part in the right hand while the left hand picks up two screws; position and align the top or the base (actually, on the gasket) with aligning support from the left hand; insert the first two screws; pick up and insert another two screws with each hand; use a powered screw driver to fasten all six screws. Motions and times to perform this work (up to the first two screws) are illustrated in Table 11.

In the table several human capabilities and limitations are evident. When the right hand adjusts the top part to align it with the base part (element no. 5) this operation is fast. The human takes advantage of both vision and tactile senses, plus some support from the left hand. The fact that the left hand is useful in this case even though it is holding two screws at the same time indicates the remarkable human use of fingers.

Reaching to a screw in a bin full of screws (element no. 1 by the left hand) exemplifies another human capability, namely, to pick up an object jumbled with other objects. Indeed, using a fixed position screw feeder can change this particular motion to R10A which requires 30% less time (8.7 TMU instead of 12.9). However, the cost of installing a screw feeder is probably not justified.

TABLE 11
MTM Details of Pump Assembly Task by Human Operator

Element No.	Left Hand Element	Symbol*	Time (TMU)	Time (TMU)	Symbol*	Right Hand Element
	Take 2 Screws					*Take the top*
1.	Reach to a screw	R10C	12.9	10.5	R14A	Reach hand to top.
2.	Grasp screw 1.	G4B	9.1	2.0	G1A	Grasp top.
3.	Grasp screw 2.	G4B	9.1	5.6	G2	Regrasp (since it is an odd shape).
	Move to base, support right hand					*Position top to base.*
4.	Move to base.	M10A	11.3	16.9	M14C	Move top to base. (No weight allowance)
5.	Contact grasp top.	G5	0	2.0	M (½) A	Adjustment to align parts.
6.		--		2.0	RL1	Release top.
	Screws					*Screws*
7.		--		5.3	R3A	Reach to take screw 1
8.	Release screw 1.	RL1	2.0	2.0	G1A	Grasp screw 1.
9.	Move screw 2 to hole 2.	M3B	5.7	6.7	M3C	Move screw 1 to hole 1.
10.		M (½) C	2.0		---	
11.	Position screw 2.	P1S(E)	5.6	5.6	P1S(E)	Position screw 1.
12.	Turn screw 2 two turns.	M1A	2.5	2.5	M1A	Turn screw 2 turns.
13.		RL1	2.0	2.0	RL1	
14.		R1A	2.5	2.5	R1A	
15.		G1A	2.0	2.0	G1A	
16.		M1A	2.5	2.5	M1A	
17.		RL1	2.0	2.0	RL1	
			Total Time: 87.1 TMU			

MTM distance is measured in inches

Another noteworthy human capability is the relatively short time a human needs for positioning and inserting screws into their holes (e.g., element no. 9). Again, vision and tactile senses play an important role in turning such an operation into one that is natural and quick. A major human limitation, however, is also shown in this element: vision has to accompany the performance of operations dependent on accurate positioning. Simultaneous movement of screws to their holes by both hands can be achieved, according to MTM tables, only if both holes are within the area of normal vision, and only with practice. This is the reason why in the table element No. 9 specifies M3C for the right hand only, while the left hand performs an M3B and only in element No. 10 positions the screw accurately with an M(1/2)C.

Another human limitation is shown by elements No. 12 through 17 which require over 15% of the total time (for all 17 elements) just to perform initial screw fastening. This limitation stems, of course, from the limited rotation capability of the human wrist.

While some of these human limitations can be overcome when using robots, robot limitations require changes in the method and workplace to perform the task. These will be discussed in the next section.

References pp. 47-48

ROBOT PERFORMANCE OF PUMP ASSEMBLY TASK

The assembly of the water pump as described in [2] differs in method from the human operator assembly for two main reasons. First, the robot has only one arm; second, it has a simple vice grip hand lacking the dexterity of human fingers.

The detailed SA operation in the pump assembly is specified by a control program. However, describing this program in PAL requires lengthy, detailed statements. As a result, the relevant issues concerning the robot work method may be completely obscured. Furthermore, if a program is not yet available and an analyst wishes to document a possible method for evaluation, time can be saved by using a higher level description language.

Based on the comparison of robot and human operators we realize that using MTM for this purpose would also obscure the specification of the robot method. First, its elements aggregate basic robot motions. Second, many MTM functions are impossible for the robot to perform. Therefore, we find it necessary to define new task motion elements for robots as part of a methodology we call RTM, Robot Time and Motion*. These elements can be used in a manner similar to MTM to specify details of task method and estimate their time duration.

In the following paragraphs we describe the structure and contents of RTM. Later in the chapter we return to the robot pump assembly task and utilize RTM for its analysis.

RTM consists of eight elements of task performance by robots, as shown in Table 12. We describe these RTM elements by reviewing MTM and PAL and rationalizing their necessary details.

TABLE 12
RTM Symbols and Elements

Symbol	Element
Rn	n-Segment Reach
SE	Stop on Error
SF/ST	Stop on Force/Touch
Mn	n-Segment Move
GR	Grasp
RE	Release
VI	Vision
TI	Process Time Delay

REACH, Rn — REACH, Rn, is the basic RTM element which describes the motion of an empty hand to a position. Since a robot may not always reach directly it may have to move through a series of (n-1) intermediate points before it reaches the final position.

---

*In order to distinguish between MTM elementary motions, and PAL and RTM elements, we underline the MTM elements, e.g., REACH. PAL and RTM are upper case but not underlined, e.g., MOVE.

HUMAN AND ROBOT PERFORMANCE

Estimated times for SA reach are given in TMU in Table 13. The parameters which determine these times are the number of intermediate points and the total distance to move. For example, reaching a point at distance of 1 cm takes 10.5 TMU compared to 20.0 TMU for a distance of 100 cm, and to 52.5 TMU for R5 (a reach over a segment with 5 points).

TABLE 13
RTM REACH Rn

Distance to move (cm)	Time (TMU)					Description
	Number of path segments, n					
	R1	R2	R3	R4	R5	
1	10.5	21.0	31.5	42.0	52.5	Move the unloaded manipulator the distance indicated. (REACH must be followed by SE or SF elements)
30	12.5	23.0	33.0	44.0	53.0	
100	20.0	26.0	37.5	48.0	55.0	

REACH R1 corresponds directly to the PAL MOVE statement. Rn corresponds to n PAL MOVE statements.

STOP ON ERROR, SE — STOP ON ERROR, SE, is the basic RTM element which describes the manipulator being brought to rest within a given position error tolerance.

Estimated SA times to accomplish SE are given in Table 14, and are relative to the magnitude of the error tolerance.

TABLE 14
RTM STOP ON ERROR SE

Error bound (cm)	Time (TMU)	Description
1.0	3.5	Brings the manipulator to rest to within the position error tolerance specified.
0.2	8.0	
0.05	23.0	

The PAL clause corresponding to SE is UNTIL ERROR < bound.

STOP ON FORCE, SF; STOP ON TOUCH, ST — STOP ON FORCE, SF, and STOP ON TOUCH, ST, are the basic RTM elements describing the manipulator being brought to rest by force or touch sensing.

References pp. 47-48

Estimated SA times for SF and ST are identical, as shown in Table 15. These times depend on the distance over which the manipulator moves while sensing.

TABLE 15
RTM STOP ON FORCE/TOUCH SF/ST

Distance to Move (cm)	Time	Description
0.5	3.0	Stops the manipulator when a force or touch condition is met. Used to define position in one direction per stop.
3.0	25.0	
15.0	125.0	

SF ST elements correspond directly to the PAL clauses:

UNTIL FORCE @ value
UNTIL TOUCH .

MOVE, Mn — MOVE, Mn, is the basic RTM element which describes the motion of a loaded hand to a position. Mn is identical to Rn except that its time is increased depending on the load.

Estimated SA times for MOVE are given in Table 16. These times are the same as in Table 13 for Rn when the load is below 0.5 Kg. Above this load value the time is multiplied by the weight factor. For example, moving 2.2 Kg over 30 cm and 4 points requires 132 TMU (44x3).

TABLE 16
RTM MOVE Mn

Distance to move (cm)	Time (TMU)					Weight Factor		Description
	Number of Path Segments n					up to (kg)	factor	
	M1	M2	M3	M4	M5	0.5	1.0	Move object along path comprised of n path segments. Move must be followed by SE or SF.
1.0	10.5	21.0	31.5	42.0	52.5	1.0	1.5	
30.0	12.5	23.0	33.0	44.0	53.0	2.0	2.0	
100.0	20.0	26.0	37.5	48.0	55.0	3.0	3.0	

GRASP, GR — GRASP, GR, is the basic RTM element which describes closing the fingers. This element is usually used to grasp an object in a given position. It also describes simply closing the hand.

Estimated SA times to perform GRASP are shown in Table 17.

TABLE 17
RTM GRASP GR

Distance to Close (cm)	Time (TMU)	Description
0.5	4.0	Close hand.
3.0	10.0	

The PAL statements corresponding to GRASP are

CLOSE LF

CLOSE RF .

RELEASE, RE — RELEASE, RE, is the basic RTM element which describes opening the fingers. This element is usually used to release an object in a given position. It also describes simply opening the hand.

Estimated SA times for RE are given in Table 18.

TABLE 18
RTM RELEASE RE

Distance to Close (cm)	Time (TMU)	Description
0.5	4.0	Open hand.
3.0	10.0	

The PAL statements corresponding to RELEASE are

OPEN LF

OPEN RF .

VISION, VI — VISION is the basic RTM element which describes the robot obtaining visual input. It is usually used to identify and locate objects and their features. The time to perform VISION is highly dependent on the particular task; however, VISION is usually performed in parallel with manipulation.

PROCESS TIME DELAY, TI — PROCESS TIME DELAY, TI, is the basic RTM element specifying unavoidable process delays during which the robot must wait. For example, the time to drive a screw using a powered screw driver, or to operate any other tool, is described by TI.

RTM for Pump Assembly — As an example of RTM we will describe the robot assembly of the water pump top and the insertion of the first two screws as we did for a human operator in Table 11. The RTM work elements for this task are shown in Table 19. As there is only one robot arm only one column of work element is shown. The detail of picking up the screw driver and replacing it is omitted and only a total time indicated. The method of installing the second screw is identical to the method used to install the first screw and requires the same amount of time.

References pp. 47-48

TABLE 19
RTM FOR TOP ASSEMBLY

Element No.	Work element RTM	Distance (cm)	Time (TMU)	Description
\multicolumn{5}{c}{Pick Up Top}				
1	R2	10.	12.5	Move top over base. (Weight = 2 Kg.)
2	SE	0.05	23	Stop there.
3	ST	3.0	25	Center hand over bearing. Locate position of top.
4	RE	3.0	10	Open hand.
5	R3	3.0	32	Move to correct position for outlet.
6	SE	0.05	23	Stop there.
7	GR	3.0	10	Close hand to align outlet with respect to bearing.
8	RE	3.0	10	Open hand.
9	R3	3.0	32	Return to bearing.
10	SE	0.05	23	Stop there.
11	GR	3.0	10	Grasp top.
\multicolumn{5}{c}{Mount Top on Base}				
12	M3	15	70	Move top over base (Weight = 2 Kg.)
13	SE	0.05	23	Stop over pins.
14	M1	3	22	Place top over pins.
15	SF	0.5	10	Stop when on base.
16	RE	3.0	10	Release top.
17	--	---	139	Pickup screwdriver.*
\multicolumn{5}{c}{Pick up a Screw}				
18	M1	3	11.0	Place screwdriver into feeder guide.
19	SF	1	10	Place screwdriver on screw.
20	M1	.25	10	Lift up.
21	SE	0.05	23	
\multicolumn{5}{c}{Insert Screw}				
22	M5	30	55	Move out and over to pump.
23	SE	0.2	8	Stop with screw over hole.
24	M1	1.0	10.5	Start insertion
25	SF	1.0	6	and stop when seated.
26	T	--	20	Screw in screw.
27	SF	1.0	10	Stop when torque requirement met.
28	SF	1.0	10	Lift driver free of screw.
29	M3	30	37.5	Move back to feeder for next screw.
30	--	--	211	Repeat elements 18-29 for second screw.
31	--	--	110	Replace screwdriver.*
\multicolumn{5}{c}{Total Time 767.5}				

*RTM details are not shown. The time is not included in the total time for the task in order to provide a comparison to the time for a human operator given in Table 11.

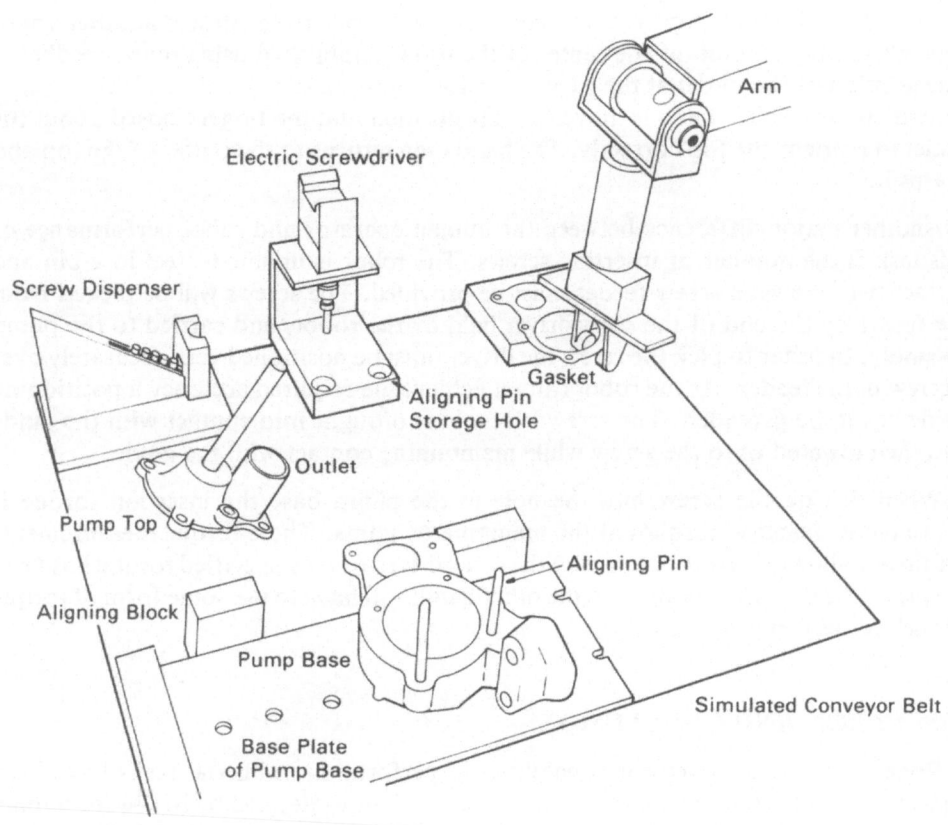

Fig. 3. Workplace for robot operator pump assembly task.

There are fundamental differences between the human operator and robot performance of this task. See Fig. 3 for the organization of the workplace. In positioning the top and gasket over the base the human operator uses vision, tactile senses, and two hands. This method, relying heavily on the subtle interpretation of a mass of sensory data, is far too complicated for a robot. For the robot this step is translated into a simple compliance task by the use of guide pins. Two pins, pointed at both ends, are inserted into screw holes in the base. Both the gasket and top are positioned over the pins. As the pins are pointed the initial positioning tolerance is the radii of the screw holes. When the top is moved down over the pins the robot provides positional compliance. In order to position the top simultaneously over two pins, however, both the position tolerance and the orientation tolerance must be controlled, which the robot can do. When grasping the pump top the relative orientation of the top to the robot hand must be established in order to control the orientation when positioning the top over the pins.

Grasping the pump top with the correct orientation demonstrates another robot capability. The position of the center of the top is established using touch feedback. The position of the outlet at the edge of the top for the actual position of the center is computed. The robot hand is moved to this position and the fingers closed about the outlet to reorient the top correctly. The hand then returns to the center of the top and grasps it.

Another major difference between the human operator and robot performance of this task is the manner of inserting screws. The robot is unable to feel in a bin and extract two screws; a screw feeder must be provided. The screws will be picked from the feeder by the end of the screwdriver held by the robot, and carried to the pump assembly. In order to pick the screw the driver must be positioned very accurately over a screw in the feeder. As the robot cannot achieve the required accuracy a positioning guide has to be provided. The screwdriver tip is brought into contact with the guide and then lowered onto the screw while maintaining contact with the guide.

When driving the screw into the hole in the pump base the insertion torque is measured as reaction torques at the manipulator joints. These torque measurements are linear, allowing the robot to stop the screwdriver when a specified torque has been reached. The human operator, on the other hand, will have to use some form of torque wrench to perform this task.

DISCUSSION AND CONCLUSIONS

Robot and human operator capabilities in performing industrial tasks have been analyzed. Comparing several elementary task motions as defined by MTM for human work with the equivalent PAL statements and robot capabilities, we found that while a methodology similar to MTM may be desirable for robot work method analysis, different motion elements and parameters may be more suitable for robot work. As a result, we derived a new system of standard elementary robot motions; RTM Tables were also developed specifying estimated SA time requirement to perform individual motions based on parameters such as distance, weight, and tolerance.

To compare methods appalied in robot vs. human operator task performance we used the example of water pump assembly. MTM was applied to describe details of the human operator method, while RTM was applied for the original robot task method.

Major differences in method were due to the robot's having almost no finger dexterity, very limited input sensing capability relative to human senses, and having only one arm. Nevertheless, it was shown that by proper methods, placing, and workplace design, these limitations could be overcome. The robot performance of the task example required more than eight times the estimated human performance time.

One important conclusion from this work is that RTM may provide a useful methodology in the analysis and design of robot work, as was illustrated here.

Another important result of this work relates to the transformation from a given human-performed to a robot-performed industrial task. A direct translation from

MTM work elements to RTM is possible and is shown in Table 20. Given a task described in MTM it is possible to translate it directly into RTM using the table. Any MTM elements which translate into an asterisk indicate that the work method must be changed for a robot. A large number of asterisks indicates that the task is probably unsuitable for a robot. Given the list of RTM elements an initial estimate of robot task time can be made. By searching the list for SF, ST, or VI RTM elements the need for various sensor systems can be established.

The RTM elements also provide a method for evaluating different robots. The times given in the paper were for a particular robot (SA) operating under a given operating system. Times for each of the RTM elements could be obtained for other robots. Given any task described in RTM and various robot RTM element times, task performance times could be computed for each robot. The time to perform the task taken together with the cost of the robot would serve to select a robot for a given task. In some cases a task must be performed in a fixed time (on an assembly line for instance); given the task times computed by RTM, robots which are too slow to perform the task can be eliminated from further consideration.

TABLE 20
RTM Equivalent Work Elements for MTM

MTM Work Element	Equivalent RTM Work Element
RA	Rn, SE
RB	Rn, SF, SF
RC	*
RD	R, SE, V, R, SF
RE	R, SE
MA	*
MB	Mn
MC	Mn, SF, SF
G1A	GR
G1B	SF, ST
G1C	*
G2	*
G3	SE, SF, RE
G4	*
G5	GR
P1	SF
P1 and P3S	SF, M1, SF
P3SS and P3NS	SF, M1, M1, SF

Impossible to perform by a robot.

REFERENCES

1. Antis, W., Honeycutt, J.M., and Knock, E.N. *The Basic Motions of MTM.* The Maynard Foundations, 4th Ed., 1973.
2. Bolles, R., and Paul, R. The use of sensory feedback in a programmable assembly system. Memo AIM-220, Stanford Artificial Intelligence Lab., 1973.

3. Maynard, H. B., Stegemerten, G. J., and Schwab, J. L. *Methods-Time Measurement*. McGraw-Hill, 1948.
4. Nof, S. Y., Knight, J. L., and Salvendy, G. *Effective utilization of industrial robots - a job and skill ganalysis approach*. (in print)
5. Paul, R., et al. Advanced industrial robot control system. First Rep., TR-EE 78-25, School of Electrical Eng., Purdue U., 1978.

DISCUSSION

J. M. Tenenbaum *(SRI International)*

I have two questions: What is the potential for exploiting vision to minimize robot stopping times?

Paul

Not too great, because when the robot has stopped, it typically is unable to see any longer. The primary purpose of vision is to help interpret the forces detected by the arm. Vision is needed to locate the objects, so when the hand is over an object and comes down to a stop, the robot knows what is happening.

Tenenbaum

Suppose you had some continuous form of visual servoing, for example, that would allow you to simply close in and gradually decelerate. Would that be useful?

Paul

Yes, it could be, but I think there are alternate methods which are faster.

C. Rosen *(SRI International)*

If I could make a remark, we are now doing something similar to what you showed in the movie — searching for bolts and fastening things together. Our vision time for taking the snapshots so that one can tell the arm to get close to—but not touch—some point is getting to be very small for simple jobs. With a small slow computer, such as an LSI-11, if the part is complex, it may take 1/2 to 1 second to perform the analysis. If the process is put on a little faster computer, we get down to tenths of a second a part. Many of the speeds you have talked about for moves of arms have half a second to one second available. Further, the computer can take a picture on the fly, which will bring the process within the margin of error you mentioned. Then the robot arm can use tactile sensors to do the last part of the process. I think you can shave a great deal of time off those times you showed — which, of course, were very old numbers. Thus, I think what Marty Tenenbaum said is to the point. If we had special hardware to reduce time to a tenth of a second for our visual snapshots, much of what you did and wanted to do could be done today.

M. Minsky (Massachusetts Institute of Technology)

It seems to me that that was an awfully low-level system that you compared to the human system. First of all, a lot of these operations could be programmed today in a goal-directed fashion so that if you described the end result to the assembly system, then a modern program could figure out a variety of ways to do the task and quickly optimize it. The idea that you would take your engineer who has programmed the old robot and look at the new one and translate from that old one to the new one in a step by step way means that the new program will come out a lot worse than you expected. Instead, all those programs should have been generated automatically.

Paul

I am all for someone doing it automatically.

Minsky

Then, why don't you?

Paul

I am waiting for someone to do it. I am struggling at the lower levels. If someone will do the high level development, that would be fine.

J. McCarthy (Stanford University)

I remember that I was always grumbling about those programs that servoed on the trajectory rather than on the path so that if it arrived too soon at a point it would slow down. This means that even with that arm and motor one could get substantially greater speeds by servoing on the path rather than on trajectories.

Paul

Yes. I think those times were very suboptimum. Based on your suggestion, I have a student who is working on such a scheme.

D. E. Whitney (Charles Stark Draper Laboratory)

I want to generalize the previous question just a little bit and ask if you have done an allocation analysis as to the percentage of the time the robot is spending just moving at maximum speed, the percentage of the time it is spending waiting for its processor to tell it what to do next, the percentage of the time it is spending doing the error checking and slowing down. We have done similar things that will show up in our paper tomorrow, and we have found that an inordinate amount of time is spent in slowing down and stopping and in analyzing these conditions. A small fraction of time, smaller than we suspected, was spent moving through space and taking advantage of all the efforts put into building a fast robot. I think what you are doing is beautiful work. I commend you on it, and I would like to know if you are going to pursue some of these allocation questions in order to decide how robot technology and program techniques ought to improve?

Paul

What you say is certainly true. Most of the time is spent in control processing and we are trying to improve the processing so that it all happens much faster. In my talk I meant to show that from this sort of analysis you can see very quickly that control problems are the big issue. Certain things such as using your Remote Compliance Center device, and stopping only on error in position and not stopping on force or touch could be done.

MECHANISMS OF PERCEPTION

R. L. GREGORY

University of Bristol, Bristol, United Kingdom

ABSTRACT

The studies of biological and machine visual perception are becoming happily integrated so that each provides data and concepts for the other; though it remains unclear just how far practical machine perception will be similar to brain processes of biological perception.

In my view, it is useful to distinguish between (1) physical *signals* (action potentials) transmitted in the neural channel from the transducers which are the eyes, ears, touch receptors and so on; (2) filtered and processed signals giving useable *data* (which can only be appreciated through knowing the processing procedures) for identifying objects; (3) employing sensory and stored (memory) data for selecting and correcting *hypotheses* — which are perceptions. We may then ask which of these are most similar in biological and machine perception, and which may be better understood by such comparisons.

Perceptual phenomena, such as the many kinds of illusions in human perception, may be classified and attributed to errors in one or more of these stages (which are conceptually very different) of the perceptual system. Illusory and other such phenomena thus take on deep theoretical significance; and it becomes interesting to ask how far and which phenomena cross the biological-machine boundary so that Artificial Intelligence (A.I.) may be related to, and in its development guided by, the Natural Intelligence of organisms.

This approach has implications for physiological psychology, for it implies that perception cannot be understood purely in terms of recorded neural activity (our level 1, signals). It also has philosophical implications, for it implies that perceptual knowledge is never direct but always hypothetical and subject to errors, which may be due to physiological disturbance of the neural channels (signal distortion, or cross-

References p. 68

talk, etc.) or — very differently — by the nervous system carrying out without error inappropriate data-generating procedures or accepting inappropriate assumptions.

Artificial Intelligence is primarily concerned with level 2: the generation of data from signals—by following processing procedures which may or may not be appropriate, or adequate. This is the central part of human perception, and it is the part we know least about. Some authorities even deny that such processes are required. It is, I think, here that Artificial Intelligence is making important contributions to understanding human perception; and by considering phenomena of human perception within this context we should consummate the marriage of Natural with Artificial Intelligence: to conceive new generations of Seeing Machines while at the same time coming to understand ourselves.

INTRODUCTION

It is clear from physiology that perception depends upon neural signals, trains of action potentials, from the organs of sense. The sense organs are essentially like the detecting instruments of physics and engineering: transducers transforming selected patterns of energy into signals from which external events may be read. It seems perception is like science, in using sensory signals to select and test hypotheses of the external world. If so, perceptions are but indirectly related to the world — related by signals and steps of inference.

The principal founder of the modern experimental study of perception, Hermann von Helmholtz (1821-1894), was concerned to discover physiological mechanisms of the eye, the ear, and the other organs of sense and the nervous system; but he did not believe that a description of these mechanisms would give a complete description of perception. He held that the central nervous system must be carrying out 'unconscious inferences' to make effective use of the sensory signals [12]. Why did Helmholtz, who was a great physiologist, reject the 'straightforward' physiological account that perception is given directly by signals? Why did he hold that elaborate inferences are necessary for perception? It is particularly interesting that he did so before the impact of computers on our appreciation of the power of inference outside the formal proofs of mathematics and the idealized systems of physics. A vital reason is the simple fact (often ignored) that the senses cannot continuously provide adequate and relevant signals — which would be necessary for direct control of behaviour or perception. Behavior follows assumptions, such as that a chair has four legs, though one or more are hidden; or that a table is strong, though the eyes cannot test its strength — and behavior continues through data gaps, as we turn away or objects pass out of sight.

The most generally held account of perception which is still frequently defended by philosophers, is that perception is direct knowledge. It is held today by the distinguished psychologist James J. Gibson, who holds that perception is passive 'pick-up of information from the ambient array of light' and that what we see is determined by stimulus patterns, such as texture gradients and by motion parallax as we move. To cartoon the 'straightforward' physiological account: perceptions are held to be made of neural signals. On Helmholtz's account perceptions are *conclusions of inferences*. He urged that to understand perception it is just as important to appreciate the procedures of inference as it is to understand the physiological processes of signal transmission by the nervous system. Helmholtz's Unconscious Inference has however been strongly resisted: initially because inference was associated with conscious processes, and awareness of the steps was supposedly necessary; and perhaps now — in spite of the obvious power to computers to carry our inferences without consciousness — the dramatic successes of physiology in revealing mechanisms by which neural signals are transmitted and handled which makes 'strategy' or 'cognitive' accounts seem pallid and vague by comparison. This is however beginning to change, with growing success and awareness of A.I. picture-processing procedures. Some distinguished physiologists do however still hold that further elucidation of neural mechanisms will provide the whole story, so that cognitive concepts will fade out as unnecessary.

This is to say that electronics is all we need to know about computers. Programmers take a very different view however — essentially Helmholtz's view — that we must appreciate procedures independantly of understanding of mechanisms carrying out the procedures. They would deny that 'software' accounts can be *reduced* to 'hardware' accounts, and would hold that procedures (or cognitive processes) can be considered quite apart from any mechanisms in which they may be embodied.

So we have three rival Paradigms of perception. The *first* is direct or 'intuitive' knowledge, as held by Bishop Berkeley in the Eighteenth Century and by James J. Gibson today [3, 4]. In the *second* paradigm, the sense organs are regarded as transducers, providing neural signals which are supposed, somehow, to add up to perceptions. So in the first paradigm perceptions are selections of the world, and so direct knowledge, while in the second paradigm perceptions are patterns of neural signals derived from the world via the sense organ transducers. On this second paradigm perception is not direct, and is open to errors through malfunctioning of the transducers and the mechanisms of the signal channels.

The *third* paradigm of perception is essentially Helmholtz's; that perceptions are still more indirectly related to the world: being inferences from stored and currently available signalled data. On this paradigm there is no simple one-one relation between neural activity and perceptual phenomena (which is another reason why it is none too popular with physiologists). On this paradigm perceptions are extrapolations from available data, and are generally much richer than the data.

I like to think of perceptions as 'hypotheses' — essentially like the predictive hypotheses of physics [7]. Both make effective use of limited signals or data, to predict and control events of the environment. Both make use of knowledge derived from the past to predict and control the future. Both depend on inference. It is interesting to compare scientific hypotheses, and how they are derived, with perceptions in rather more detail.

References p. 68

SCIENTIFIC HYPOTHESES COMPARED WITH PERCEPTIONS

Signals from Transducers — Both depend on signals from transducers being processed as data. This involves selecting signals from noise, integrating and sometimes coincidence gating to increase the signal/noise ratio; and comparing sources for discrepancies. There may be, additionally, more sophisticated procedures for accepting signals as data such as statistical significance tests. In any case, we cannot recognize signals as data without knowing the coding.

Calibration — Calibration of the transducers is necessary in both cases, and may involve subtle procedures including: (i) Comparison with signals or data from other transducers (other instruments or other organs of sense) given the same input; (ii) Comparison of current with expected signals or data (which requires stored knowledge of what to expect).

Instruments, or sense organs, may receive calibration corrections. When these are inappropriate, systematic errors (illusions) are generated, though the mechanisms may have no fault. (Thus, after adapting to distorting spectacles, the world appears distorted, in directions opposite to the original distortion).

Scaling According to Assumed Invariances — This is a kind of calibration, but involves specific assumptions about the situation in which the instrument or sensory system is being used. For example, a pilot may assume that the texture over which he is flying is roughly uniform in average density, to estimate his height and his approach path when landing — as well as assumptions that the runway is parallel and the buildings square and so on. Such assumptions, though usually very useful, can generate serious problems. For example, helicopter pilots flying over oases in deserts have to be careful, for the density of the vegetation falls off regularly with distance away from the source of water: so he appears to be climbing when in fact he is approaching the ground. Distances of stars are judged by similar assumptions for setting the scaling constants, as for vision.

Entertaining Alternatives — Data is always, strictly speaking, ambiguous and may be accepted as evidence for a wide variety of alternative hypotheses. (Indeed, variety is limited only by limitations of imagination). Perceptions can switch spontaneously between alternatives with no change of data; and much the same happens in science, especially with changes of accepted paradigm [13].

Guidance by Conditional Probabilities — Data may be accepted, rejected, or interpreted according to the likelihoods of alternative hypotheses. It is thus easier to perceive a likely than an unlikely object in a given context (Fig. 1). But this implies (and it is not difficult to demonstrate) that unlikely objects bearing marked similarities to highly probable objects are readily confused with them, and may be impossible to recognize for what they are. (For example, the hollow mold of a face, appears with its nose sticking out — though it is hollow — against all manner of texture and other sensory data simply because a hollow face is so unlikely). There is a similar limitation in testing unlikely scientific hypotheses: when extremely unlikely, the instrument is more likely to be in error than that the event has occurred. (Thus it would be impossible to

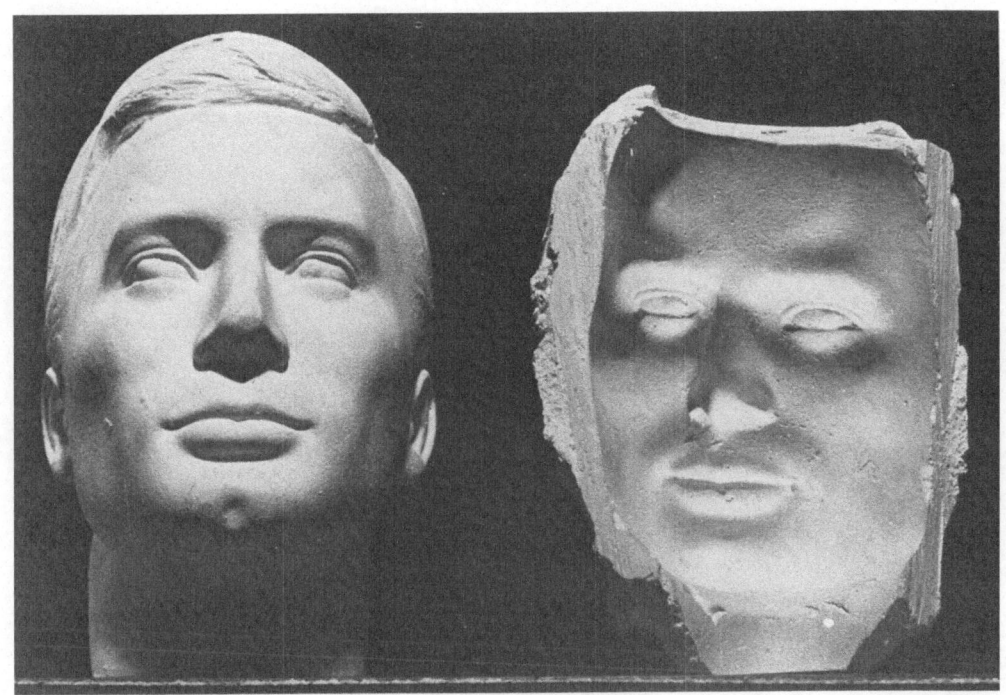

Fig. 1. The left hand face is as it appears; but the right hand "face" is in fact its hollow mold. A hollow face is so unlikely that texture, and even stereoscopic data signalling to the eye that it is hollow, are discounted. (It is virtually impossible to present such an unlikely object vertically with any available electronic display.)

test the prediction that, very occasionally, a pan of water will freeze when put on the gas stove — for observational error or delusion of some kind is more likely than the event, even though this is predicted as a possibility).

Handle New Cases — It is striking that we can perceive objects from unfamiliar positions, and see unusual objects — though there are limitations — and similarly, scientific hypotheses handle new cases but with limitations.

Hypotheses based on wide-ranging generalizations have great power to deal with new cases; and where there are powerful heuristic rules, highly surprising conclusions may be derived.

Generate Fiction — Both scientific hypotheses and perceptions may land into dramatic error — which may however suggest some new and useful solutions.

Errors or illusions may be due to failure, or loss of calibration, of the transducers of the signal channel or breakdown of the computing systems — but they may also be due to inappropriatness of the hypothesis-generating procedures. Although signal distortion through mechanical or electrical trouble with the channel (including overloading and cross-talk) is extremely different, conceptually, from errors due to inappropriate procedures or from false assumptions of invariances; yet it is surprisingly difficult to

References p. 68

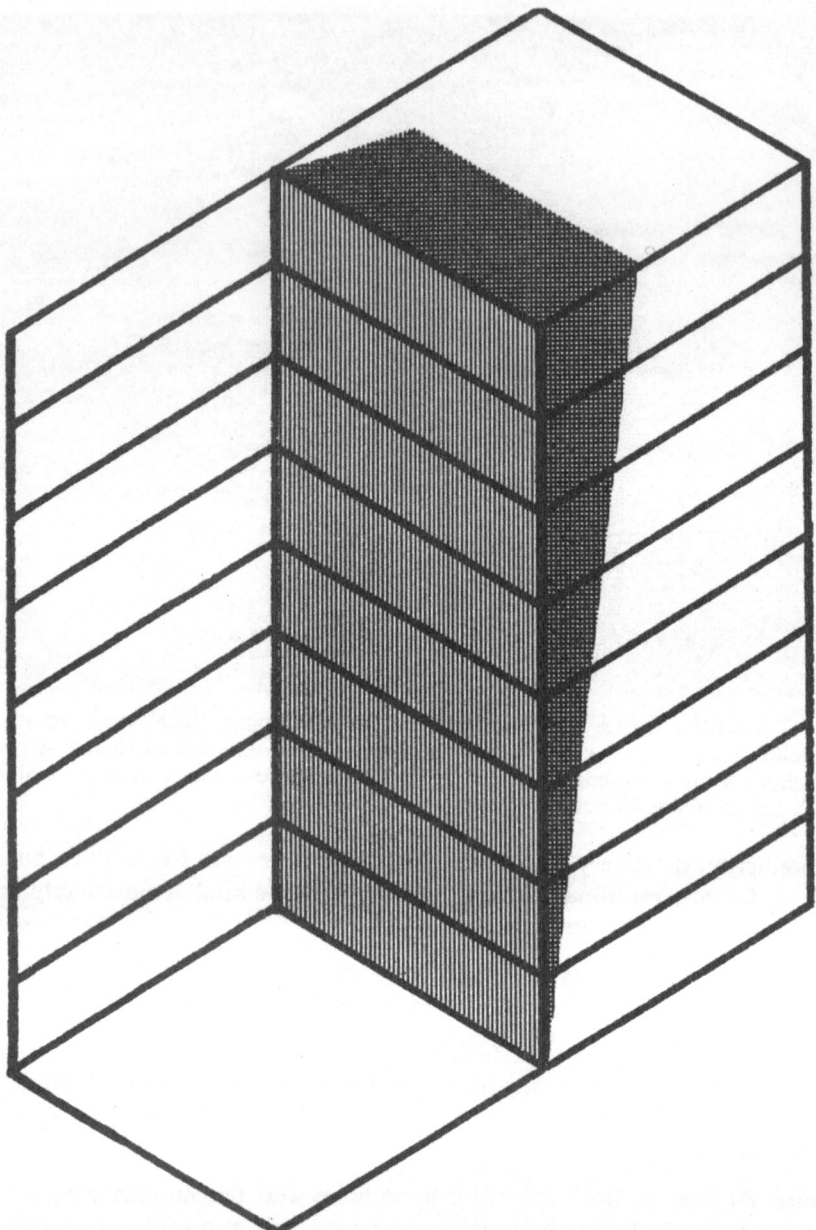

Fig. 2. This is a depth-ambiguous figure with a "shadow". It may be noted that the "shadow" appears dark with one depth-orientation of the figure, and light with the other orientation. It is light when a plausible shadow, as cast by top lighting. We thus see "Brightness Constancy" triggered internally by the prevailing depth-hypothesis. (This is an example of "top-down" scaling; as the stimulus pattern, or input to the eye, remains constant.)

(a)

(b)

Fig. 3a, b. Although this exists as a three-dimensional object, it appears three dimensionally impossible, or paradoxical: from critical viewing positions. (The actual object appears just as paradoxical when viewed directly.)

References p. 68

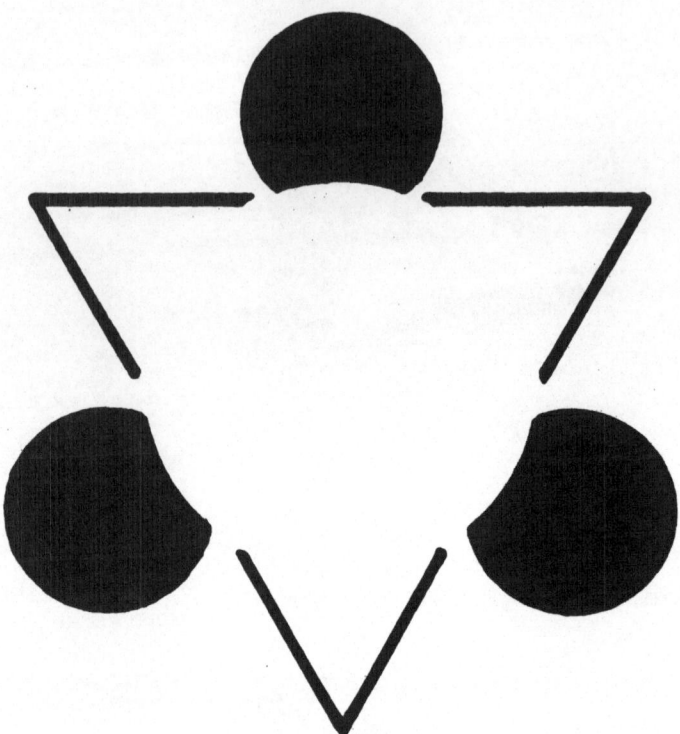

Fig. 4. This illusory white disc seems to be produced by the unlikely gaps suggesting the presence of a nearer masking object. This is evidently *postulated* by the visual system.

distinguish in practice between *physiological* and *cognitive* errors, or illusions. I shall illustrate this by considering with a recent experiment the classical problem of the origin of visual distortion illusions. There are, however, as well as distortion illusions other kinds of illusions notably: *Ambiguity, Paradox,* and *Created* or Hallucinatory features such as "cognitive contours". Examples are shown in Figs. 2 to 4.

Distortions can clearly occur in two ways: first, *"upwards"* by certain stimulus patterns (either because they disturb the signal channel or because they provide data which are inappropriate to the situation) and secondly, *"downwards"* from the prevailing assumption or perceptual hypothesis of relative distances or depth of the perceived object or display. It is a most important fact that with apparent alternations of apparent depth in objects, such as wire cubes, the apparently further features are expanded, and shrink when the depth reverses so that they appear nearer. This shows conclusively that apparent size can be set purely by apparent distance *independently of the stimulus pattern*. It is thus clear that not *all* distortions can be due to channel characteristics, and there is at least this tie-up between apparent distance and apparent size — which is no doubt part of the scaling processes normally giving Size Constancy. If perceptual scaling giving size constancy were to be set inappropriately size distortion must result.

CREATING AND DESTROYING SIZE-DEPTH DISTORTION ILLUSIONS

The classical problem of the origin or cause of visual distortion illusions has at least three rival candidates worth serious consideration: (i) neural interactions, such as lateral inhibition, or other neural channel distortions; (ii) spatial frequency characteristics of the visual system, associated with Fourier-like transforms, supposedly important for pattern recognition: low pass filter characteristics are held to produce the distortions, especially with converging lines; (iii) size scaling, set by perspective features (such as convergence) which when presented on flat picture planes give inappropriate scaling. It has proved surprisingly difficult to distinguish experimentally between *physiological* channel distortion and more *cognitive* accounts, though the logical distinctions between these alternatives seem reasonably clear [8]. The present experiments were designed to provide evidence for or against a cognitive account, such as the inappropriate constancy scaling theory which requires the organism to act on knowledge or assumptions of features of the object world. This theory is, unlike the alternatives mentioned above, a cognitive account because it supposes that the visual system is acting (correctly or inappropriately) from stored knowledge of characteristics of the world. The illusion is attributed to applying knowledge to situations inappropriately. The ways in which the knowledge is stored and used — the physiological mechanisms involved — need not be specified for this to be a legitimate and for many purposes adequate account. The first kind of theory, on the other hand, supposes that a modification of the physiological mechanism or channel is responsible. The second class of theory [11] makes no such demand or claim on detailed physiological understanding; and neither does it call on cognitive processes, involving knowledge of or assumptions about external objects. The three kinds of theory are so conceptually different that the experimental difficulty of deciding between them is surprising. Perhaps this difficulty is due to the reasonable biological assumption that even peripheral mechanisms will be adapted to handle typical object situations. We should therefore expect to find crucial differences only in situations requiring stored knowledge of specific classes of objects or visual features. The inappropriate constancy scaling theory does refer to specific classes of features — especially converging of parallel lines by perspective. Retinal perspective (given by optical projection at the retina), used as information for depth, depends on assumptions of the sizes of objects and typical shapes especially parallel and right-angular features, which hold more generally in some environments than in others. Considerable anthropoligical data are now available supporting the suggestion that distortion illusions are greatest for people living in 'carpentered' environments [15,1].

The notion that distortion illusions such as the Muller-Lyer and Ponzo figures are perspective drawings goes back at least to Thiéry did not realize that typical objects (such as outside corners of buildings, inside corners of rooms, or receding parallel lines) must be accepted as paradigm objects, to infer depth from retinal perspective. (Thiéry's example of the legs of a saw horse is misleading — for the legs could be of almost any angle). Neither did he suggest a modus operandi relating depth features to distortion on picture planes. This was suggested by Tausch [16] as Size Constancy; but was later rejected by him, as depth distortions occur though depth may not be *seen* in

References p. 68

these figures. Gregory [5] suggested that constancy can be given by size and shape scaling set *directly by depth cues* even though depth is not seen. This would be somewhat like 'releasers' in ethology and would be "automatic," and quite low level in the nervous system.

I shall now describe two experiments on size distortions: the first aimed at testing the notion that perspective produces distortions when the convergence of lines is inappropriate to the depth of the picture, or display; the second to explore distortions having a more immediately physiological basis and which do *not* depend on stored object-knowledge. The first experiment was performed by the author with John Harris [8]. In both experiments the aim is to *destroy* illusions; and infer from how they can be destroyed to what their origins may be. It is, also, useful to know how to destroy or remove illusions from displays.

This loss of distortion is also found (though not so readily investigated) when wire corner models are viewed directly; though flat Muller-Lyer illusion figure models, having the same fin angles as the retinal images given by the corner models, are distorted. Since the retinal images are identical — though the corner models appear in depth while the Muller-Lyer models appear flat — the distortion cannot be due to signal channel distortion. We thus have strong evidence that this illusion is not physiological but is cognitive: due to misscaling when depth cues are presented without depth, so the scaling is inappropriate.

EXPERIMENT ONE — DESTROYING PERSPECTIVE ILLUSIONS

This experiment was designed to test a specific prediction from the Inappropriate Constancy Scaling theory. We tried to change the significance of the stimuli without changing what (on other theories) would be characteristics producing distortion physiologically, as by lateral inhibition. The critical prediction is that the illusory distortions, should vanish when the perspective, and the perception of depth, are appropriate. This prediction follows directly from the theory that these distortions are due to inappropriate size scaling. We have said above that scaling is both 'upwards', from certain stimulus patterns typical of depth, and 'downwards' from the prevailing perception (or "hypothesis") of depth — which does not always follow stimulus patterns, and so can be experimentally distinguished. We should expect distortions to vanish when both the 'upward' and the 'downward' scaling are appropriate. 'Downward' scaling led us to use stereoscopic projection, to set apparent depth. The importance of 'upward' scaling led us to precisely controlled perspective.

To attain precise perspective, and geometrically correct stereoscopic displays, we employ point-source shadow projection, onto a translucent screen, by a pair of horizontally separated miniature lamps cross-polarized to the eyes [6, 14]. They are used to project stereo images of wire models of typical three-dimensional objects. These are right angular corners, for the Muller-Lyer illusion (Fig. 5).

When the wire models are projected with a single source, they give two-dimensional perspective figures, which are the usual illusion figures. The usual distortions are

MECHANISMS OF PERCEPTION

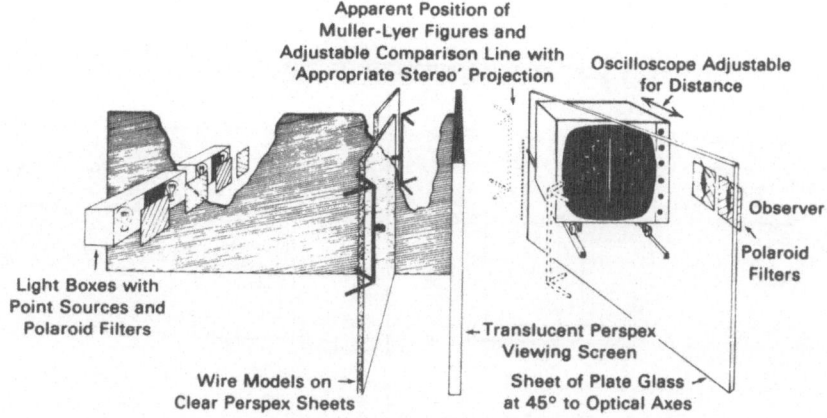

Fig. 5. The shadow projection apparatus set up for illusion-destruction. The distortion illusion — or lack of it — is measured with an adjustable vertical line displayed on the oscilloscope.

observed, and can be measured. When the wire models are projected in stereo depth (with the pair of cross-polarized point sources) the figures are seen as three-dimensional — as though the models were viewed with two eyes directly. So we now have a situation in which both the perspective and the perception of depth are appropriate for these objects. There should therefore be *no distortion*.

The geometrical conditions for providing appropriate perspective and appropriate depth perception are easily defined and attained by this shadow projection technique: (1) the projection distance must equal the viewing distance; (2) the horizontal separation of the projection point sources must equal the observer's eye separation. It is also necessary that the observer has adequate stereoscopic vision, so that he does perceive these projections as three dimensional though they lie on the (textured) picture plane of the screen. Other means of attaining accurate depth perception could be used, such as luminous figures with a textureless background, but such alternatives will not be considered here as stereoscopic projection is readily controlled and most convenient experimentally for giving precise appropriate (or inappropriate) apparent depth.

DETAILS OF SHADOW PROJECTION APPARATUS

The layout of the apparatus is shown in Fig. 6. The point sources are 10 W 6V halide miniature filament lamps. They are individually switched, for monocular or stereoscopic projection, and mounted in a pair of light boxes fitted with Polaroid filters. Monocular projection is given by a lamp placed centrally between the pairs of stereo projection lamps, to preserve symmetry. (We call single-lamp projection 'monocular projection' because it gives the view of a single eye, but in the experiment this is viewed with both eyes, as for the normal viewing of pictures or illusion figures). The Polaroid filters for stereoscopic projection are arranged with their planes of polarization mutually at 90°, and 45° from the horizontal, so that disparities to the eyes can be switched

References p. 68

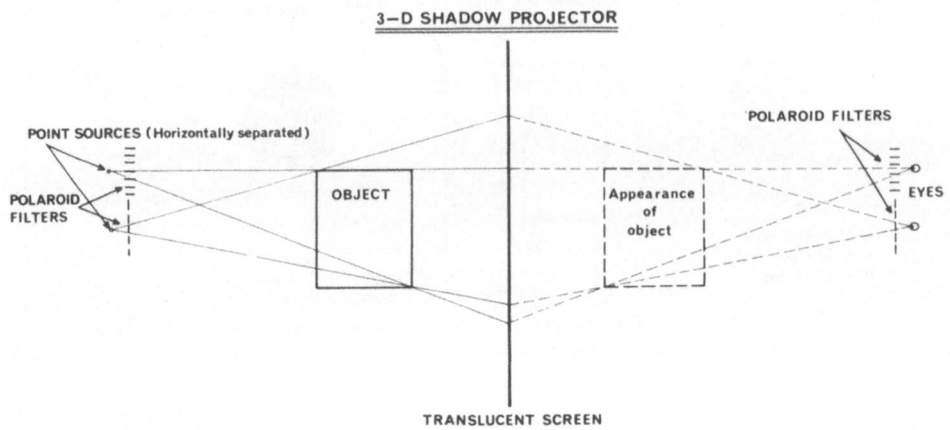

Fig. 6. A pair of point sources (which are polarized orthogonally) can be used to project models in three-dimensional visual space: giving stereoscopic images. Perspective is correct when the projection and viewing distance are equal. Disparity can be controlled by setting the separation of the point sources: so the projection can be precisely correct, or controlled errors can be introduced.

by reversing the Polaroid glasses worn by the observer. The central 'monocular' projection lamps were fitted with vertically polarized filters, to give the same colour temperature as for the stereoscopic projection; the Polaroid glasses being worn in all conditions, including monocular projection. The subject's viewing position — which is critical — was maintained by an adjustable chin rest, with measurements made directly from his eyes to the screen to ensure accurately determined viewing distances.

The wire models were made of 3 mm diameter welding rod. They were mounted on transparent Perspex sheets. One corner model faced the screen (giving the 'in-going' Muller-Lyer figure) while the other model faced away from the screen, toward the projection lights, (giving the 'out-going' figure as a perspective projection.

Two sets of projection lamps were used, as stated above, one for each of the corner models. One pair of lamps would have served; but with the use of two sets (with central 'monocular projection' lamps spaced apart by the distance (14 cm) separating the verticals of the models) symmetry was given for each projection, and the separation of the projections was independent of the projecting distance. The viewing screen was a sheet of frosted Perspex 30 cm high and 35 cm wide, which does not depolarize and so allows stereoscopic projection. In this experiment, the luminance of the screen was about 34 cd m^{-2}. The wire models were 11cm in length, giving projected vertical shaft lengths of 11.5 cm and 'fin' lengths of 3.5 cm.

The Muller-Lyer distortion — the difference in apparent length of the projected verticals — was measured with an adjustable comparison line, given by the adjustable line trace of a large-screen oscilloscope (35 cm × 30 cm screen, Lan-Scope). This was placed at right angles to the projection screen and viewed by reflection from a sheet of plate glass placed at 45° to the line of sight, as shown in Fig. 5. (The glass sheet was sufficiently thick to produce a pair of horizontally separated lines from the single oscilloscope line, the separation roughly equalling the line thickness of the shadow projection lines of the illusion figures). The comparison line was placed centrally between the Muller-Lyer figures and was adjusted in length by a smooth potentiometer controlled by the subject. The length was indicated by a digital voltmeter visible only to the experimenter, calibrated for length by measurements from the screen. The oscilloscope could be moved, in and out, to measure apparent depths.

By varying the projection distance (the distance of the lamps from the wire models) perspective can be controlled. For these experiments a single fixed distance of 40 cm was adopted throughout. This gives an angle of about 8° from the horizontal for each 'fin'. It is most important to note that perspective is correct for the retinal image only when the viewing distance equals the projection distance. In this case it is correct for a viewing distance of 40 cm. We therefore predicted that the illusory distortion should only vanish (or be minimal) at around this critical viewing distance equal to the projection distance. Alternative viewing distances (35, 37.5, 42.5, and 45 cm) were adopted to discover effects of slightly inappropriate perspective of the retinal images. (It is also possible to vary independently disparity — by changing the separation of the projection lamps — and convergence angle — by placing deviating prisms before the eyes — but these were held constant in this experiment).

We have three projection modes: (1) monocular projection. (2) appropriate stereoscopic projection. (3) reversed (pseudoscopic) stereoscopic projection. Monocular projection is the usual illusion figure, viewed with both eyes as for normal pictures.

Each subject was given a different order of the five viewing distances. The three projection modes were presented in random order for each viewing distance. This means that the subjects were not required to shift their position with changes of projection mode. This was important as the distance of the observer's eye was measured by ruler with each change of viewing distance, a somewhat distracting procedure.

The subjects were instructed to match the comparison line to the left hand (outgoing) or the right hand (ingoing) figure, following a random sequence. He was told to look wherever he wished within the relevant display. It was emphasized that three-dimensional perception was needed for the two stereo modes. For each projection mode, five readings were taken for each of the pair of figures. Each subject therefore produced 150 readings, in single sessions of somewhat over one hour's duration. Before each judgement the subject rotated his control potentionmeter to produce a far too small or far too large comparison line. Upon making his comparison setting he was allowed to 'overshoot' and correct as much as he wished, and was allowed unlimited time for each comparison.

References p. 68

RESULTS

The graph (Fig. 7) shows an illusion of about 6% for the 'monocular projection' and for 'reversed stereo' modes. The illusions (means) at the critical viewing distance of 40cm are, respectively +5.5% (standard error 0.67%) and 6.1% (standard error 0.7%).

The 'appropriate stereo' curve is clearly separate from the other curves, and shows *zero illusion* at the critical viewing distance of 40cm. This is the prediction confirmed.

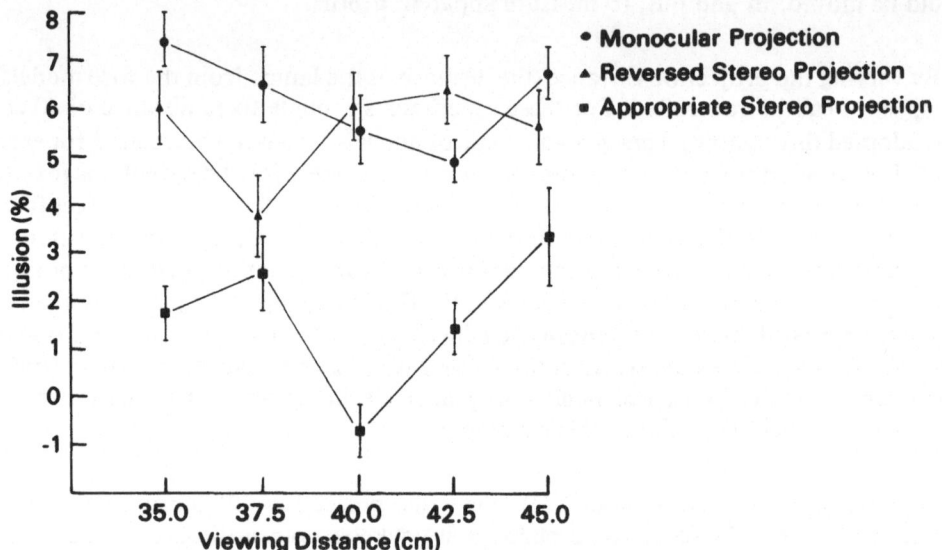

Fig. 7. The horizontal axis gives the viewing distances. Correct perspective is at 40cm. At this distance the illusion distortion (shown on the vertical axis) is entirely *absent*; and is insignificant when perspective is nearly correct at the other viewing distances. The illusion is however present when the *seen depth* is incorrect — given by monocular or reversed stereo projection — though the angles of the figure are identical in all three projection modes for each viewing distance. So the distortion can hardly be due to neural *signal distortion*.

EXPERIMENT TWO — FEATURE-LOCKING DISTORTIONS

Since we can demonstrate the main phenomena of this experiment in figures which can be printed by normal means, whereas the phenomena of Experiment One require displays allowing three-dimensional perception, we can be briefer here.

The components of the nervous system are, by comparison with electronic components, subject to drift and are generally labile, in ways intolerable to electronic engineers. Also, the retinal receptors (the rods, and the three types of cones serving colour vision) have different time-constants, so retinal signals are variously delayed according to intensity and colour. In spite of this, regions of different colour and brightness are very seldom seen with mis-registration of borders. This is surprising, for electronic displays such as colour TV show marked discrepancies between regions of different colour or intensity, especially with motion. The integrity of TV displays depends upon the stability of the components. The nervous system has no such stability, and yet we seldom see discrepancies at borders — though even TV is prone to registration errors in spite of the high stability of its components.

This suggests that there is some special principle active in the nervous system to maintain registration at borders.

It has recently been found [18] that brightness, colour, and movement are represented in the cerebral cortex in different cortical "maps". This makes the fact that registration is normally maintained within visual acuity even more surprising.

We may suppose [9] that registration is maintained by locking signals at contrast borders. I think we can see our own locking system at work in the illusion shown in Figs. 8a, b, c. This consists of alternating brightness-contrasting regions (squares) separated by narrow parallel lines. In (a) the intensity of the separating lines is less than the darker squares; in (b) it lies between the darker and lighter squares. In (c) it is brighter than the light squares. In the conditions (b), when the separating lines lie within the brightness range of the squares, a marked distortion is seen: the parallel separating lines do not appear parallel but seem to converge alternatively to form wedges. The distortion illusion is destroyed by taking the intensity of the separating lines outside the range of intensities of the alternating squares.

We find, also, the neighboring regions of contrasting colour but the same brightness (different colours set to isoluminant match) are not positionally stable, but shift relatively with motion; and the colour-contrast border is unstable and "jazzy". This distortion illusion does *not* occur with coloured squares (red and green) when the colours have equal brightness [9].

I suggest that colour borders, which is absent in the special case of isoluminant colours; and that the positions of intensity borders are set, within the lability of the system, by the relative intensities either side of them around isoluminance.

The illusion of Fig. 8a, b, c, (which we call the "Cafe Wall Illusion", we first observed in the Victorian tiles of a cafe near our laboratory in Bristol) may be explained by supposing that the borders of the squares "see" locking signals across the separating lines (the "mortar"), when this has an intensity between the dark and light squares. When the "mortar" is varied continuously, in the model from which these photographs were made, it is observed that the *dark* squares move towards the lighter squares while the "mortar" is dark, and the *light* squares move towards the dark squares while the "mortar" is light — within the intensity range of the dark and light squares. This would be expected with locking by intensity signals "seen" across the separating "mortar" lines.

This distortion does seem to be essentially different from the classical Distortion Illusions which we investigated in Experiment One. They, if we are correct, arise from

References p. 68

(8a)

(8b)

MECHANISMS OF PERCEPTION

Fig. 8a, b, c. The "Cafe Wall" Illusion. The parallel "mortar" lines appear wedged — when the "mortar" intensity lies between the intensities of the dark and light squares (b). The suggestion is that there is border-locking in the visual system, which normally maintains registration of borders, intensity, brightness and colour regions. When the separating "mortar" lines are of neutral intensity, locking signals are "seen" across the "mortar", and squares of contrasting intensity are then pulled together by the registration locking. (The wedges are evidently produced by integrating the successive local distortions. This mechanism is known, from other distortion illusions such a some of the Fraser figures [2].

In (a) the "mortar" is brighter than the light squares and in (c) it is darker than the dark squares. In both cases the distortion illusion is destroyed.

inappropriately applied knowledge of the world: especially perspective convergence normally associated with distance applied to a picture plane. Since this is flat, the scaling set by the picture perspective features is inappropriate and produces distortion. The distortion of the "Cafe Wall Illusion", if we are correct, arises from inadequacies of the neural components of the visual channel. This has nothing to do with knowledge of the world — and so is not a cognitive illusion — while the classical perspective distortion illusions are cognitive illusions.

The first experiment illustrates *data mismatching* angle features of this illusion figure (and the other classical distortion illusions of this type) are evidently accepted as data for depth, which in this case is misleading, to set Size Constancy Scaling inappropriately. This is very different from a *signal* mismatch, which depends purely on physical characteristics (such as filter characteristics) of the channel, and not at all on the significance of the message. The second experiment shows a rather special but quite dramatic example of signal mismatch when the channel distorts the retinal signals,

References p. 68

due, we may suppose in this case, to inadequacies of neural components and mechanisms which counter these inadequacies in normal conditions.

It is to be hoped that sorting out these issues for human vision may be of some help for designing robots with perceptual capabilities. It is an intriguing thought that robots will share some of our illusions.

REFERENCES

1. Deregowski, J. Illusion and culture. In: *Illusion in Nature and Art* Gregory and Gombrich (Ed.) Chas. Schribner & Sons, New York, 1974.
2. Fraser, J. A new visual illusion of direction. *Brit. J. Psychol.*, 2 (1908), 307-320.
3. Gibson, J. J. *The Perception of the Visual World.* Greenwood Press, Westport, Conn., 1950.
4. Gibson, J. J. *The Senses Considered as Perceptual Systems.* Houghton Miffen Co., Boston, 1966.
5. Gregory, R. L. Distortion of visual space as inappropriate constancy scaling. *Nature,* 199 (1963), 678.
6. Gregory, R. L. Stereoscopic shadow images. *Nature,* 203 (1964) 1407.
7. Gregory, R. L. *The Intelligent Eye.* McGraw-Hill, New York, 1970.
8. Gregory, R. L. Do we need cognitive concepts? In *Handbook of Psychobiology.* Blakemore and Gazzinaga, (Eds.) Academic Press, 1975.
9. Gregory, R. L. Vision with isoluminant colour contrast: 1. A projection technique and observations. *Perception* 6 (1976), 113-119.
10. Gregory, R. L. and Harris, J. P. Illusion-destruction by appropriate scaling. *Perception* 4 (1975), 203-20.
11. Ginsburg, A. P. Psychological correlates of a model of the human visual system. *Proc. IEEE NAEOON* Dayton, Ohio (1971), 283-390.
12. Helmholtz, H. L. F. von *Handbook of Physiological Optics.* Trans. 1924 by J.P.C. Southall, Dover, 1962.
13. Kuhn, T. *The Structure of Scientific Revolutions.* Univ. of Chicago Press, 1970.
14. Lee, D. N. Theory of the stereoscopic shadow-caster: an instrument for the study of binocular space perception. *Vision Research* 9, 1 (1969).
15. Segall, M. H., Campbell, T. D. and Herskovitz, M. J. *The Influence of Culture on Visual Perception.* Bobbs Merril, New York, 1966.
16. Tausch, R. Optische Tauschungen als artifizelle Effekte der Gestaltungspozesse von Grossen and Formenkonstanz in der naturlichen Raumwahrnehmung. *Psychol. Forsch.* 24, 299 (1954).
17. Thiery, A. Ueber Geometrisch-optische Tauschungen *Phil. Stud.* 12 (1896), 67.
18. Zecki, S. M. Colour coding in the superior temporal sulcus of rhesus monkey visual cortex. *Proc. R. Soc.* Lond. B. 197 (1977), 195-223.

ARTIFICIAL INTELLIGENCE AND THE SCIENCE OF IMAGE UNDERSTANDING

B. K. P. HORN

Massachusetts Institute of Technology, Cambridge, Massachusetts

ABSTRACT

Advanced automation promises to increase productivity, improve working conditions and assure product quality. Some computer-based systems perform tasks blindly, without elaborate sensor-based feedback. In many applications however visual input can speed up an automated system by eliminating search or the need for costly fixtures that maintain exact alignment of parts. In still other situations, many inspection jobs for example, there may be no alternative to machine vision.

Unfortunately, image analysis turned out not to involve just a simple extension of some well-known subfield of computer science or optics. A long history of frustrations with techniques borrowed from other domains mixed with clever special case solutions based on *ad hoc* techniques has brought the field to a point where it is finally considered worthy of serious attention. The foundations of a science of image understanding are beginning to be built on the ancestral paradigms of image processing, pattern recognition and scene analysis.

One component of this new thrust is an improved understanding of the physics of image formation. Understanding how the measurements obtained from the vision input device are determined by the lighting, shape and surface material of the objects being imaged helps one develop methods for "inverting" the imaging process, that is, exploit physical contraints to allow one to built internal symbolic descriptions of the scene being viewed. Another ingredient of the renewed optimism is a better understanding of the computations underlying early stages of the processing of visual information in biological systems. Aside from providing existence proofs that certain aspects of a scene can be understood from an image, this also suggests computational architectures for performing such tasks in real time.

A focal point of recent work is the choice of a representation for the objects being

References pp. 75-77

viewed and their internal prototypes. The internal descriptions must be tailored to expedite the computations involved in spatial reasoning. This has turned out to be a challenging new area, with problems and methods quite different from those found in the more serial, linguistic kind of reasoning we can introspect about and that artificial intelligence research has concerned itself with in the past.

Industrial problems, such as the visual determination of the position and orientation of a part in a bin provide new challenges for machine vision researchers. At the same time reductions in computational cost make some of the more complex techniques developed now feasible for industrial exploitation.

WHY USE MACHINE VISION TECHNIQUES?

Computer-based automation appears about to induce the next quantum jump in productivity, while also contributing to improved working conditions and higher product uniformity. The flexibility of the computer leads to greater complexity in systems and increased degrees of freedom in design. One important and difficult choice the designer has to face is whether or not to employ machine vision.

Many computer-based systems perform their tasks blindly, relying on accurate positioning and other methods of traditional "fixed" automation. The added complexity of machine vision will often pay off in improvements in speed by eliminating search or costly pallets used to hold parts in precise relationships to each other and the tools. In some existing systems people load parts into these fixtures, employing *their* vision system to determine the position and orientation of the incoming parts. A boring job indeed!

Besides, there are many cases where there is no alternative to some form of non-contact sensing. Many inspection tasks seem to require visual analysis of one kind or another for example. Sometimes a small number of sensors in strategically located areas will do; but more often than not an imaging system is called for together with the necessary electronic hardware and software for interpreting the image. The applications open to less sophisticated approaches appear to be limited.

WHAT IS MACHINE VISION?

An optical system forms an image of some three-dimensional arrangement of parts. The two-dimensional image is sensed and converted into machine readable format. It is the purpose of the machine vision system to derive information from this image useful in the execution of the given task. In the simplest case the information sought will concern only the location and orientation of an isolated object—more commonly, objects have to be recognized and their spatial relationships determined. This can be viewed as a process in which a description of the scene being viewed is developed from the raw image. The description has to be appropriate to the particular application. That is, irrelevant visual features should be discarded, while needed relationships between parts of objects must be deduced from their optical projection.

One of several reasons why this is non-trivial is that one dimension is lost in the imaging system. Depth has to be inferred from a variety of cues or determined by special techniques employing optical triangulation or time-of-flight measurements [1 - 3]. Fortunately our visual world is very special and it is possible to infer a great deal of information about the scene being imaged from one, or a small number of images. This is because in most cases one is observing opaque, cohesive objects immersed in a transparent medium. In this situation the entities of interest are the *surfaces* of the objects and surfaces are essentially two dimensional. (If *volumes* were imaged a form of tomography based on large numbers of images would be needed.)

Much still has to be learned about what one can reasonably expect a vision system to derive about a three-dimensional reality from an image, and what permanent properties of the surfaces can be calculated from the raw image intensity readings. It is clear that any attempt at recognition, matching or classification ought to be based on these estimates of the permanent properties rather than directly on the image intensities, which reflect these only indirectly.

ROOTS OF MACHINE VISION

Ideas from many related fields have contributed to early progress in machine vision. We are all familiar for example with the work on recognition of printed text. Unfortunately many methods developed in such specialized areas do not generalize smoothly to more complicated situations. Characters for example appear as fixed, two dimensional patterns in essentially only two values of gray. Images of machined parts on the other hand contain many levels of gray, call for high resolution and produce image intensity "patterns" that not only vary with the attitude of the part in space, but depend on the distribution of light-sources. Specular reflections, gloss and mutual illumination further complicate the picture. Nevertheless, machine vision has built, in part, on ideas from three related fields: image processing, pattern recognition, and scene analysis.

Image processing [4 - 10] concerns itself with the production of new images from existing images. Usually these new images are obtained by application of a technique from linear systems theory and are enhanced in some fashion so as to improve their appearance to a human viewer. In machine vision we are more interested in the generation of symbolic descriptions and the use of these descriptions to permit a computer controlled system take appropriate action. Nevertheless, some of the techniques developed for processing raw image intensity values are of importance.

Pattern classification [11 - 17] deals with the mapping of feature vectors, containing measurements of objects, into class numbers. Any connection with our discussion here seems remote until it is remembered that frequently the components of the feature vector represent measurements obtained from parts of an image. Useful segmentation methods and simple measures of shape have been developed for this purpose, although many apply only to patterns of an essentially two-dimensional nature.

Scene analysis [18 - 20], finally, is largely pre-occupied with the transformation of

References pp. 75-77

descriptions into more abstract descriptions. When faced with the problem of vision in a world of toy blocks for example, the transformation of a line-drawing into a description in terms of three dimensional solids and how they relate spatially would be a typical task for a scene analysis system [21 - 28]. Developing the line drawing from the raw image in the first place would be a job left to some other subsystem [29 - 31]. Similar comments can be made regarding the complementary approach of region growing [32 - 34]. The methods that were developed for manipulating descriptions and relating them to known information about the world being imaged have proved invaluable to workers in the fields of machine vision.

DIFFICULTIES AND SUCCESSES

Something important has been learned: vision is difficult. This is surprising because we find it hard to believe that something that is so immediate to us could be hard. If computers can interpret mass spectrograms, perform complicated manipulations on symbolic mathematical expressions and even perform medical diagnosis in limited disease domains [20], why can't they see? Many challenging problems are succumbing to the methods being developed in artificial intelligence, yet some have turned out to be surprisingly resistant to concerted attacks. While computers become expert problem solvers in areas such as electronic circuit analysis, less impressive progress has been made on the problem of common sense reasoning, for example.

All this means is that we don't have a good idea of how difficult tasks are that we cannot introspect about. Vision is one such task. As we learn more about biological vision systems we also become more aware of the prodigous amount of computing power necessary. At this stage then we have to be very clever in our use of existing computer hardware or depend on *ad hoc* solutions applicable only in limited domains. Underestimation of the difficulties of vision led to many attempts, foolish in retrospect, to apply simple methods from other fields such as linear systems analysis, information theory or statistics.

We now know that the field demands its own methodology. In view of what has been said, it is impressive to note that machine vision has already had a significant impact. Several systems incorporating machine vision techniques have progressed beyond the laboratory demonstration stage. Systems for positioning and orienting integrated circuit chips are an example. The major cost component of a *simple* integrated circuit (I.C.) is the packaging, and the major component of that used to be the manual alignment of each I.C. to prepare it for lead bonding. The visual alignment of these chips is a fatiguing, uninteresting task possibly even harmful to the operator's vision.

As a result of pioneering work at the M.I.T. Artificial Intelligence Laboratory [35], General Motors Research Laboratory [36], Hitachi Central Research Laboratory [37] and Texas Instruments this task is now in many cases performed by a computer vision system. Curiously, these systems, developed independently, use quite different techniques for processing the raw image intensities, reflecting to some extent different target components, but also the richness of the storehouse of methods that is beginning to develop in this field. One of these systems for example depends on a careful analysis of one-dimensional profiles of intensity along strategically chosen lines in the image plane. The requisite speed and robustness is obtained in this case without recourse to special-purpose, parallel hardware. Other systems inspect printed circuit cards [38], visually guide a machine that tightens bolts on telephone-pole molds [39] and direct manipulators to pick isolated objects off a conveyor belt [40]. Recently, manufacturers of "industrial robots" have introduced computer-control for their products. This permits them to be programmed in more flexible ways. The use of global, rectangular coordinate systems made possible by this advance prepares these devices for input from machine vision systems such as the ones described here.

CURRENT TRENDS AND BASIC ISSUES

Many aspects of machine vision are currently being explored. I can only focus on a few issues that seem important to me. One new approach stems from a view of the machine vision system as a device for "inverting" the imaging process—that is, developing a description of the scene being viewed from an image. It seems reasonable to suppose that understanding the physics of the imaging process would be helpful in this endeavor, since this determines the "forward" transformation [41 - 46]. What has to be understood is how the measurements obtained from the vision input device are determined by the lighting, shape and surface material of the objects being imaged. Such understanding will permit the exploitation of physical constraints to allow one to build internal symbolic descriptions of the scene based on information about the physical properties of the surfaces rather than raw image intensities. It is possible for example to calculate the shape of a smooth object from a single image if enough is known about the surface reflectance properties and the distribution of light-sources [41, 42].

An extension, of interest in the case of industrial application of machine vision, depends on multiple images obtained from the same viewpoint under different lighting conditions [47-49]. In certain cases this can lead to a simple look-up table computation of local surface orientation and surface "albedo", thus producing just the information needed to determine the attitude of a body in space. Thus these "photometric stereo" methods may play a role in solving the "bin-of-parts" problem [50]. Fortunately the task of understanding the physics of light reflection and imaging has been made simpler by the recent introduction of a systematic formalism for the description of the reflectance properties of surfaces by the National Bureau of Standards [51].

Progress is also being made in developing better models for biological vision systems [51-54]. The central thrust of the new approach is the use of computers in evaluating competing theories with less reliance on raw neurophysiological and experimental psychology data. Instead the emphasis is on what information can be extracted from the image and what representations are appropriate for various intermediate forms of the symbolic descriptions. Suggestions for computer architectures to support the huge amounts of computation needed also may result from this work.

References pp. 75-77

An important issue concerns the appropriate representation for objects and space. This is particularly important if a computer controlled manipulator is to interact with the objects being viewed. Representations of objects in terms of mathematical equations defining surface patches, a common method in computer graphics, has not found much of a following here, where it may be necessary to reason about objects in terms of the space they occupy. Considerable progress has been made in developing representations in terms of volumes [55 - 62]. In fact, the whole question of spatial reasoning is only now drawing attention. We are once more surprised at the computational complexities involved, since we cannot easily introspect into our own ability to think about objects and space.

A representation for surfaces of objects, that arises naturally from the form of the output of the photometric methods described above, is one in terms of local surface normals. This method of capturing the information regarding the shape of the objects leads to a representation that transforms more easily with rotation than more obvious approaches. One can think of the object as if covered with "spines". A mapping of these spines onto the sphere of possible directions often uniquely specifies the particular object and may lead to methods for determining its attitude in space. A similar representation has been recently proposed as an intermediate form between the very crude symbolic description computed directly from the image and the more elaborate ones in terms of volumes described above [54 - 59].

The use of prior knowledge about the objects being viewed is a thorny issue. Where should this knowledge be used and how should it be represented in the computer? On the one side one finds advocates of a "non-purposive" approach, where one computes as detailed a description as possible based on the image, independent of the overall task at hand. Others would prefer to use the image only to verify hypotheses generated by an analysis of prior knowledge about the likely arrangement of objects in the scene. This leads to a paradigm of "controlled hallucination". Probably, future systems will embody a compromise, where simple operators are applied over the complete image using massive parallelism. These operations would not be influenced much by the current purpose of the overall system. The crude symbolic description computed this way would then be used by processes more goal-directed and more appropriate to the task at hand.

Other recent advances have been made in the processing of multiple images of the same scene taken from different viewpoints [63 - 67], and the use of parallel "relaxation" computations that process images by iteration much as numerical methods for solving elliptic partial differential equations [68, 69].

A SCIENCE OF IMAGE UNDERSTANDING

A better understanding of image formation is leading to a better exploitation of physical constraints that allow interpretation of the two-dimensional image in terms of a three-dimensional reality. Simultaneously the success of special purpose systems is supporting a more serious interest in vision problems. The simple-minded application of assorted methods borrowed from other fields has been discredited. We no longer underestimate the problems and are resigned to work hard to get solutions implemented that will be efficient enough to operate successfully on present-day hardware. A science of image understanding is beginning to emerge.

REFERENCES

1. Nitzan, D., Brain, A. E., and Duda, R. O. The measurement and use of registered reflectance and range data in scene analysis. *Proc. IEEE* 65, 2 (1977), 206-220.
2. Nevatia, R. Depth measurement by motion stereo. *Computer Graphics and Image Processing* 5, 2 (1976), 203-214.
3. Shirai, Y. Recognition of polyhedrons with a range finder. *Bul. Electro-technical Lab.* 35, 3 (1971), 209-296.
4. Andrews, H. C. *Computer Techniques in Image Processing.* Academic Press, New York, 1970.
5. Gonzalez, R. C., and Wintz, P. *Digital Image Processing.* Addison-Wesley, Reading, MA, 1977.
6. Huang, T. S. (Ed.) *Picture Processing and Digital Filtering.* Springer, NY, 1975.
7. Lipkin, B. S., and Rosenfeld, A. (Eds.) *Picture Processing and Psychopictorics.* Academic Press, New York, 1970.
8. Rosenfeld, A. *Picture Processing by Computer.* Academic Press, New York, 1969.
9. Rosenfeld, A. (Ed.) *Digital Picture Analysis.* Springer, NY, 1976.
10. Rosenfeld, A., and Kak, A. C. *Digital Picture Processing.* Academic Press, New York, 1976.
11. Cheng, G. C., et al. (Eds.) *Pictorial Pattern Recognition.* Thompson Book Co., Washington, DC, 1968.
12. Fu, K. S. *Syntactic Methods in Pattern Recognition.* Academic Press, New York, 1974.
13. Fu, K. S. (Ed.) *Digital Pattern Recognition.* Springer, NY, 1976.
14. Grasselli, A. (Ed.) *Automatic Interpretation and Classification of Images.* Academic Press, New York, 1969.
15. Tou, J. T., and Gonzalez, R. C. *Pattern Recognition Principles.* Addison-Wesley, Reading, PA, 1974.
16. Watanabe, S. *Methodologies of Pattern Recognition.* Academic Press, New York, 1969.
17. Duda, R. O., and Hart, P. E. *Pattern Classification and Scene Analysis.* John Wiley and Sons, New York, 1973.
18. Hanson, A., and Riseman, E. (Eds.) *Computer Vision Systems.* Academic Press, New York, 1979.
19. Winston, P. H. (Ed.) *The Psychology of Computer Vision.* McGraw-Hill, New York, 1975.
20. Winston, P. H. *Artificial Intelligence.* Addison-Wesley, Reading, PA, 1977.
21. Clowes, M. B. On seeing things. *Artificial Intelligence* 2, 1 (1971), 79-112.
22. Falk, G. Interpretation of imperfect line data as a three-dimensional scene. *Artificial Intelligence* 3, 2 (1972), 101-144.
23. Grape, G. R. Model based (intermediate level) computer vision. AI Memo 201, Stanford U., 1973.
24. Huffman, D. A. Impossible objects as nonsense sentences. in *Machine Intelligence* 6, B. Meltzer and D. Mitchie (Eds.) Edinburgh University Press, Edinburgh, 1971.
25. Mackworth, A. K. Interpreting pictures of polyhedral scenes. *Artificial Intelligence* 4, 2 (1973), 121-138.
26. Roberts, L. G. Machine perception of three-dimensional solids. in *Optical and Electro-optical Information Processing,* J. T. Tippet, et al (Eds.) MIT Press, Cambridge, MA, 1965.
27. Waltz, D. L. Understanding line drawings of scenes with shadows. In *The Psychology of Computer Vision,* P. H. Winston (Ed.), McGraw Hill, New York, 1975.

28. Winston, P. H. The M.I.T. robot. In *Machine Intelligence 7*. Edinburgh University Press, Edinburgh, 1972.
29. Griffith, A. K. Computer recognition of prismatic solids. MIT Project MAC, TR-73, MIT, Cambridge, MA, 1970.
30. Horn, B. K. P. The Binford-Horn line-folder. MIT AI Memo 285, MIT, Cambridge, MA, 1971.
31. Shirai, Y. Analyzing intensity arrays using knowledge about scenes. In *The Psychology of Computer Vision*, P. H. Winston, (Ed.), McGraw Hill, New York, 1975.
32. Brice, C., and Fenema, C. Scene analysis using regions. *Artificial Intelligence* 1, 3 (1970), 205-226.
33. Ohlander, R. B. Analysis of natural scenes. Ph.D. Th., Dept. of Computer Science, Carnegie-Mellon U., 1975.
34. Tenebaum, J.M., and Barrow, H.G. Experiments in interpretation-guided segmentation. *Artificial Intelligence* 8, 3 (1977), 241-274.
35. Horn, B. K. P. A problem in computer vision: orienting silicon integrated circuit chips for lead bonding. *Computer Graphics and Image Processing* 4, 3 (1975), 294-303.
36. Baird, M. L. An application of computer vision to automated IC chip manufacture. GMR-2124, General Motors Res. Labs., Warren, MI, 1976.
37. Kashioka, S., Ejiri, M., and Sakamoto, Y. A transistor wire-bonding system utilizing multiple local pattern matching techniques. *IEEE Trans. on Systems, Man, and Cybernetics* SMC-6, 8 (1976), 562-570.
38. Ejiri, M., Uno, T., Mese, M., and Ikeda, S. A process for detecting defects in complicated patterns. *Computer Graphics and Image Processing* 2 (1973), 326-339.
39. Uno, T., Ejiri, M., and Tokunaga, T. A method of real-time recognition of moving objects and its application. *Pattern Recognition* 8 (1976), 201-208.
40. Holland, S. W., Rossol, L., and Ward, M. R. CONSIGHT-1: a vision-controlled robot system for transferring parts from belt conveyors. GMR-2790, General Motors Res. Labs., Warren, MI, 1978.
41. Horn, B. K. P. Obtaining shape from shading information. In *The Psychology of Computer Vision*, P. H. Winston (Ed.), McGraw-Hill, New York, 1975.
42. Horn, B. K. P. Understanding image intensities. *Artificial Intelligence* 21, 11 (1977), 201-231.
43. Barrow, H. G., and Tenenbaum, J. M. Recovering intrinsic scene characteristics from images. In *Computer Vision Systems*, A. Hanson and E. Riseman (Eds.), Academic Press, New York, 1979.
44. Woodham, R. J. A cooperative algorithm for determining surface orientation from a single view. Proc. 5th Int. Jt. Conf. on Artificial Intelligence (1977), 635-641.
45. Woodham, R. J. Reflectance map techniques for analyzing surface defects in metal castings. TR-457, AI Lab., MIT, Cambridge, MA, 1978.
46. Horn, B. K. P., and Bachman, B. L. Using synthetic images to register real images with surface models. *Comm. ACM* 21, 11 (1978), 914-924.
47. Woodham, R. J. Photometric stereo: a reflectance map technique for determining surface orientation from image intensity. Proc. SPIE 22nd Ann. Tech. Symp. 155 (1978).
48. Woodham, R. J. Photometric stereo. AI Memo 479, MIT, Cambridge, MA, 1978.
49. Horn, B. K. P., and Woodham, R. J. Determining shape and reflectance using multiple images. AI Memo 485, MIT, Cambridge, MA, 1978.
50. Birk, J., Kelley, R., Wilson, L., Badami, V., Brownell, T., Chen, N., Duncan, D., Hall, J., Martins, H., Silva, R., and Tella, R. General methods to enable robots with vision to acquire, orient and transport workpieces. Fourth Rep., GRANT APR74-13935, U. of Rhode Island, Kingston, RI, 1978.
51. Nicodemus, F. E., Richmond, J. C., and Hsia, J. J. Geometrical considerations and nomenclature for reflectance. NBS Monograph 160, Nat. Bur. Standards, Washington, DC, 1977.
52. Marr, D. Early processing of visual information. *Phil. Trans. Roy. Soc.* B 275 (1976), 483-524.

53. Marr, D., and Poggio, T. A theory of human stereo vision. AI Memo 451, MIT, Cambridge, MA, 1977.
54. Marr, D. Representing visual information. In *Computer Vision Systems,* A. Hanson and E. Riseman (Eds.), Academic Press, New York, 1979.
55. Agin, C. J., and Binford, T. O. Computer description of curved objects. Proc. 3rd Int. Jt. Conf. on Artificial Intelligence (1973), 629-640.
56. Binford, T. O. Visual perception by computer. Proc. IFIP Conf. (1971).
57. Binford, T. O. Visual perception by computer. IEEE Conf. on Systems and Control (1971).
58. Hollerbach, J. Hierarchical shape description of objects by selection and modification prototypes. TR-346, AI Lab., MIT, Cambridge, MA, 1976.
59. Marr, D., and Nishihara, H. K. Representation and recognition of the spatial organization of three-dimensional shapes. *Proc. Roy. Soc. London* B 200 (1977), 269-294.
60. Nevatia, R. Structured descriptions of complex curved surfaces and visual memory. AI Memo 250, Stanford U., 1974.
61. Nevatia, R., and Binford, T. O. Description and recognition of curved objects. *Artificial Intelligence* 8, 1 (1977), 77-98.
62. Soroka, B. I., and Bajcsy, R. K. Generalized cylinders from serial sections. Proc. Int. Jt. Conf. on Pattern Recognition (1976), 734-735.
63. Arnold, R. D. Local context in matching edges for stereo vision. Proc. DARPA Image Understanding Workshop, L. Baumann (Ed.), Science Applications, Inc., 1978.
64. Gennery, D. B. A stereo vision system for an autonomous vehicle. Proc. 5th Int. Jt. Conf. on Artificial Intelligence (1977), 576-582.
65. Marr, D. A note on the computation of binocular disparity in a symbolic low-level visual processor. AI Memo 327, MIT, Cambridge, MA, 1974.
66. Marr, D., and Poggio, T. Cooperative computation of stereo disparity. *Science* 194 (1976), 283-287.
67. Quam, L. H. Computer comparison of pictures. AI Memo 144, Stanford U., 1971.
68. Rosenfeld, A. Iterative methods in image analysis. Proc. IEEE Conf. on Pattern Recognition and Image Processing (1977), 14-18.
69. Rosenfeld, A., Hummel, R. A., and Zucker, S. W. Scene labelling by relaxation operations. *IEEE Trans. on Systems, Man and Cybernetics* SMC-6 (1976), 420-433.

SESSION II
VISION AND ROBOT SYSTEMS

Session Chairman
P. M. WILL

IBM Thomas J. Watson Research Center
Yorktown Heights, New York

CONSIGHT-I: A VISION-CONTROLLED ROBOT SYSTEM FOR TRANSFERRING PARTS FROM BELT CONVEYORS

S. W. HOLLAND, L. ROSSOL and M. R. WARD

General Motors Research Laboratories, Warren, Michigan

ABSTRACT

CONSIGHT-I is a vision-based robot system that picks up parts randomly placed on a moving conveyor belt. The vision subsystem, operating in a visually noisy environment typical of manufacturing plants, determines the position and orientation of parts on the belt. The robot tracks the parts and transfers them to a predetermined location. CONSIGHT-I systems are easily retrainable for a wide class of complex curved parts and are being developed for production plant use.

INTRODUCTION

In many manufacturing activities parts arrive at work stations by means of systems that do not control part position. Since present robots require parts to be in precisely fixed positions, their use is precluded at these work stations. To automate these part handling operations, intricate feeding devices that precisely position the parts are required. Such devices, however, are often uneconomical and unreliable.

Robot systems equipped with vision represent an alternative solution. This paper describes CONSIGHT-I, a vision-based robot system for transferring unoriented parts from a belt conveyor to a predetermined location. CONSIGHT-I:

- determines the position and orientation of a wide class of manufactured parts including complex curved objects,
- provides easy reprogrammability by insertion of new part data, and
- works on visually noisy picture data typical of many plant environments.

As a result of these characteristics—and because the vision subsystem does not require light tables, fluorescent conveyor belts, colored parts or other impractical

References pp. 96-97

means for enhancing contrast — CONSIGHT-I systems are eminently suitable for production plant use.

This paper gives an overview of CONSIGHT-I, followed by a description of the hardware and software organization and a more detailed look at the vision, robot, and control software.

FUNCTIONAL OVERVIEW

CONSIGHT-I functions in two modes: a setup mode and an operational mode. During setup, various hardware components are calibrated and the system is programmed to handle new parts. Once calibrated and programmed for a specific part, the system can be switched to operational mode to perform part transfer functions.

Part transfer operates as follows: operators place parts, such as a foundry castings, in random positions on the moving conveyor belt (shown in Fig. 1). The conveyor carries the parts past a vision station which is continually scanning the belt. Position and orientation information is sent to a robot system. As the parts continue to move, the robot tracks the moving parts on the belt, picks them up, and transfers them to a predetermined location. The above sequence operates continuously with no manual intervention except for placing parts on the conveyor.

Fig. 1. CONSIGHT-1 conveyor, camera, and robot arrangement.

CONSIGHT-I is capable of handling a continuous stream of parts on the belt, so long as these parts are not touching. The maximum speed limitation is imposed by neither the vision system nor the computer control, but by the cycle time of the robot arm.

CONSIGHT-I is easily reprogrammed for a new part. The new part to be picked up is passed by the vision station. The vision subsystem determines part location relative to the belt. The belt is stopped and the robot hand is manually positioned at the desired pickup point for that part. The setup programs then automatically determine the transformation between the location determined by the vision station and the pick up point used for grasping this part, and the system is ready to handle the new part.

CONSIGHT-I must also be calibrated when it is first set up, or whenever the camera, the robot or the conveyor is moved. The calibration procedure is simple and requires about 15 minutes. All necessary mathematical transformations are derived automatically and are totally transparent to the operator of the system, that is, he need not understand nor even be aware of the mathematical processes involved.

Both calibration and part programming are described later in the MONITOR section of this report.

SYSTEM OVERVIEW

Organization — CONSIGHT-I is logically partitioned into independent vision, robot, and monitor modules, permitting these system functions to be distributed among several smaller computers. Although the experimental system described here was implemented on a single computer, communication between subsystems was designed to allow easy substitution of new vision modules or new robots. The vision subsystem reports only a unique point (i.e., x and y coordinate) for each part, and an orientation. Neither the location of the point on the part, nor the reference from which to measure orientation, is specified. Our particular vision subsystem reports the part's center of projected area and a direction along the axis of minimum (or maximum) moment of inertia. Other vision modules employing different vision techniques are available which report an altogether different point and orientation for the same part [1,2]. These other vision modules can easily be substituted for the first without changing the control program, the robot system, the part-programming methods, or the calibration methods. More importantly, it is equally easy to substitute another robot or a different monitor subsystem to perform a new and entirely different operation on the part.

Hardware — The hardware for the experimental version of CONSIGHT-I as described in this paper is shown in Fig. 2. The computer is a Digital Equipment Corporation PDP 11/45 operating under the RSX-11D real time executive. The camera is a Reticon RL256C 256x1 line camera. The robot is the Stanford Arm made by Vicarm [3].

References pp. 96-97

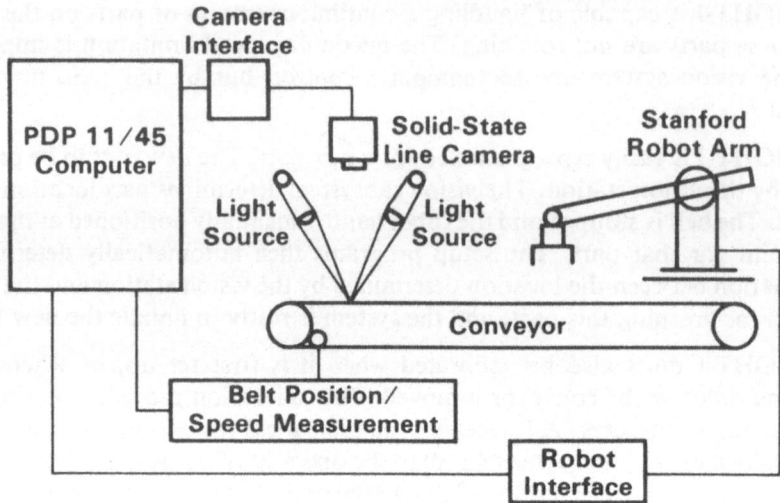

Fig. 2. CONSIGHT-I hardware schematic.

Since its speed is neither constant nor predictable, the belt is instrumented with a position and speed detection device. Position and speed information are necessary for three reasons:

1. The camera scans the belt at a constant rate, independent of belt speed. For each equal increment of belt travel, the vision subsystem records one of these scans. Belt travel increments must therefore be measured precisely.

2. Programs must compensate for movement of the belt between picture digitization and part pickup time. This also requires belt position detection.

3. The robot needs both position and velocity information in order to smoothly track the moving part on the belt.

The specific implementation details for production versions of CONSIGHT-I will differ from those described here. The functions handled by a single computer will be distributed among three smaller computers communicating with each other and the robot will be replaced with one suitable to a production environment.

Software — The software organization for CONSIGHT-I reflects the three major modules of the system.

The monitor coordinates and controls the operation of CONSIGHT-I and also assists in calibration and reprogramming for new parts. The monitor queues part data and the system is thus capable of handling a continuous stream of parts on the belt.

The vision subsystem uses a modified rangefinder approach in which two projected light lines, focused as one line on the belt, are displaced by objects on the belt. The line camera, focused on the line, detects the silhouette of passing objects. When it has seen the entire object, the vision subsystem sends to the monitor the object's position and a belt position reference value. Since the vision subsystem detects only silhouettes, parts must have silhouettes that allow unique orientation determination, or they must be rotationally invariant, (i.e., part orientation cannot be defined or is unimportant). The idea of using structured lighting to simplify visual processing may also be found in other research [4, 5, 6].

The robot subsystem executes a previously "taught" robot program to transfer the part from the conveyor to a fixed position. It accepts information concerning the part's location on the moving belt and uses this data to update the "taught" program. It then monitors belt position and speed to track the part along the moving belt, pick up the part and transport it to a predetermined location.

The vision, robot, and monitor subsystems are described in the following sections.

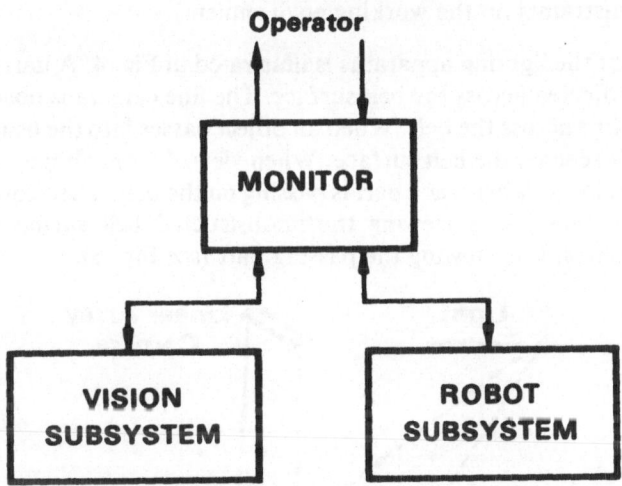

Fig. 3. CONSIGHT-I software organization.

VISION SUBSYSTEM

The vision subsystem detects parts passing through its field of view and reports their position and orientation to the monitor program. Parts may follow in an unending stream. It is also permissible for several parts to be within the field of view simultaneously. Parts which are overlapping or touching each other are ignored and allowed to pass by the robot for subsequent recycling.

The vision subsystem employs a linear array camera. The linear array images a narrow strip across the belt perpendicular to the belt's direction of motion. Since the belt is moving, it is possible to build a conventional two-dimensional image of passing parts by collecting a sequence of these image strips. The linear array consists of 256 discrete diodes, of which 128 are used in the system described here. Uniform spacing is achieved between sample points (both across and down the belt) by use of the belt position detector which signals the computer at the appropriate times to record the camera scans of the belt.

References pp. 96-97

The two main functions of the vision subsystem are object detection and position determination.

Object Detection — A fundamental problem which must be addressed by computer vision systems is the isolation of objects from their background. If the image exhibits high contrast, such as would be the case for black objects on a white background, the problem is handled by simple thresholding. Unfortunately, natural and industrial environments seldom exhibit these characteristics. For example, foundry castings blend extremely well with their background when placed on a conveyor belt. Previous approaches for introducing the needed contrast, such as the use of fluorescent painted belts or light tables [7], are impractical. They would severely restrict the number of potentially useful applications of vision-based robot systems. We developed a unique lighting arrangement which accomplishes the same result without imposing unreasonable constraints on the working environment.

The principle of the lighting apparatus is illustrated in Fig. 4. A narrow and intense line of light is projected across the belt surface. The line camera is positioned so as to image the target line across the belt. When an object passes into the beam, it intercepts the light before it reaches the belt surface. When viewed from above, the line appears deflected from its target wherever a part is passing on the belt. Therefore, wherever the camera sees brightness, it is viewing the unobstructed belt surface; wherever the camera sees darkness, it is viewing the passing part (see Fig. 5).

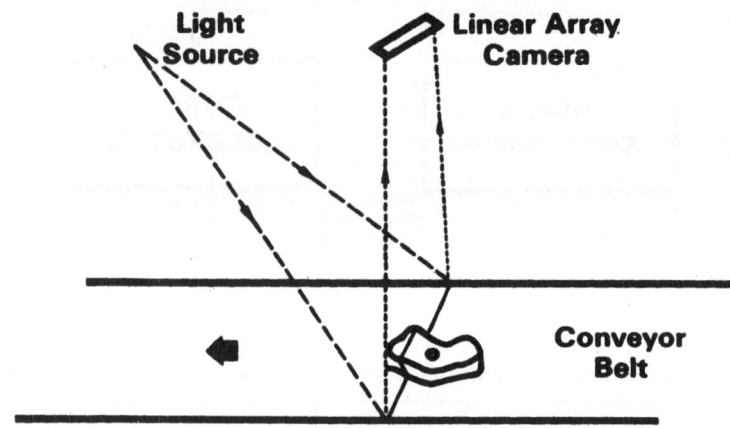

Fig. 4. Basic lighting principle.

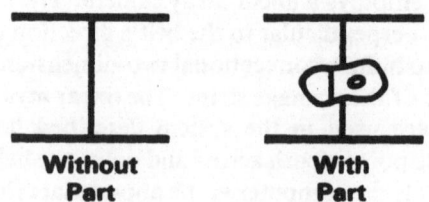

Fig. 5. Computer's view of parts.

Unfortunately, a shadowing effect causes the object to block the light before it actually reaches the imaged line. The solution is to use two (or more) light sources all directed at the same strip across the belt. Fig. 6 illustrates the idea. When the first light source is prematurely interrupted, the second normally will not be. By using multiple light sources and by adjusting the angle of incidence appropriately, the problem is essentially eliminated.

The light source is a slender tungsten filament bulb focused to a line with a cylindrical lens. Fig. 7 illustrates the line of light generation hardware.

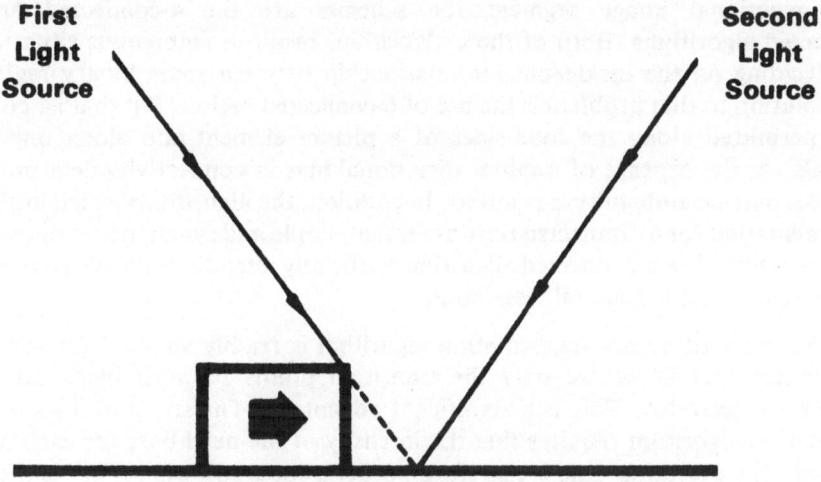

Fig. 6. Improved lighting arrangement.

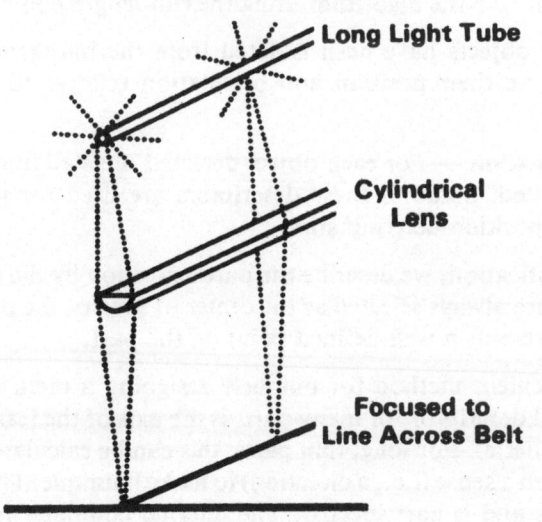

Fig. 7. Line of light generation.

The described lighting arrangement produces a height detection system. Any part with significant thickness will be "seen" as a dark object on a bright background. The computer's view is a silhouette of the object. Note that while the external boundary will appear sharp, some internal features, such as holes, are still subject to distortion or occlusion due to the shadowing effect. These internal features may optionally be ignored or recorded under software control.

It is the computer's responsibility to keep track of objects passing thru the vision system. Since several pieces of an object or even different objects may be beneath the camera at any one time, the continuity from line to line of input must be monitored. The conventional image segmentation schemes are the 4-connected and the 8-connected algorithms. Both of these algorithms result in ambiguous situations [8] when deciding on the inside/outside relationship between some binary regions. A clever solution to that problem is the use of 6-connected regions [9]; that is, connectivity is permitted along the four sides of a picture element and along one of the diagonals. At the expense of a minor directional bias in connectivity determination, the inside/outside ambiguity is resolved. In addition, the algorithms which implement the segmentation for 6-connected regions remain simple and symmetric with respect to black and white. The 6-connected algorithm artificially introduces the pleasing properties gained through hexagonal tessalation.

The 6-connected binary segmentation algorithm is readily adapted for run-length coded input; that is, where only the transition points between black and white segments are recorded. This is a significant advantage. The straightforward binary segmentation algorithm requires that the intensity of the neighbors for each pixel be examined. The execution time is therefore "order n squared" where "n" is the linear camera resolution. Since the number of black/white transitions across a line is relatively independent of the resolution for these types of images, the execution time is reduced to "order n" for the algorithm using the run-length coding scheme.

Once the passing objects have been isolated from the background, they may be analyzed to determine their position and orientation relative to a reference coordinate system.

Position Determination — For each object detected, a small number of numerical descriptors is extracted. Some of these descriptors are used for part classification; others are used for position determination.

For position specification, we describe the part's position by the triple (x, y, theta). The x and y values are always selected as the center of area of the part silhouette. For most parts, this represents a well-defined point on the part.

There is no convenient method for uniquely assigning a theta value to all parts. However, one useful descriptor for many parts is the axis of the least moment of inertia (of the part silhouette). For long, thin parts, this can be calculated accurately. The axis must still be given a sense (i.e., a direction) to make it unique. This is accomplished in a variety of ways and is part specific. The internal computer model for the part specifies the manner in which the theta value should be computed. For example, one method available for giving a sense to the axis value is to select the moment axis direction which points nearest to the maximum radius point measured from the centroid to the boundary. Another technique uses the center of the largest internal feature (e.g., a hole) to give direction to the axis. Several other techniques are also available.

Parts which have multiple stable positions require multiple models. Parts whose silhouettes do not uniquely determine their position cannot be handled.

Reprogramming the vision system for a new part requires entering a description of a new model. Each model description includes information to determine if a detected object belongs to the class defined by the model and also prescribes how the orientation is to be determined.

The vision system sees the world through a narrow slit. As objects pass through the slit, statistics concerning that object are continuously updated. Once these statistics have been updated, the image line is no longer required. Consequently, storage need only be allocated for a single line of the two-dimensional image, offering a major reduction in memory requirements.

The block of statistics describing an object in the field of view is referred to as a component descriptor. The component descriptor records information for every picture element which belongs to that component. This includes the following.

1. external position reference
2. color (black or white)
3. count of pixels
4. sum of x-coordinates
5. sum of y-coordinates
6. sum of product of x- and y-coordinates
7. sum of x-coordinates squared
8. sum of y-coordinates squared
9. min x-coordinate and associated y-coordinate
10. max x-coordinate and associated y-coordinate
11. min y-coordinate and associated x-coordinate
12. max y-coordinate and associated x-coordinate
13. area of largest hole
14. x-coordinate for centroid of largest hole
15. y-coordinate for centroid of largest hole
16. an error flag

Considerable bookkeeping is required to gather the appropriate statistics for passing objects and to keep multiple objects segregated. The primary data structure used for this purpose is the "active line." The active line records the location of each black and white segment beneath the camera and also the objects to which they are connected. For every new segment which extends a previously detected object, the statistics are simply updated. If the segment is the start of a new object which is not in the active line, it will be added to the active line and have a new component descriptor initialized for it. When an object in the active line is not continued by at least one segment in the new input line, it must have passed completely through the field of view. The block of statistics is then complete and may be used to identify the object and compute its position and orientation.

Fig. 8 illustrates the coordinate reference frames used by the programs. It is convenient to consider the x-origin of the vision system to be permanently attached to the belt surface moving to the left. Normally, this would cause the current x-position beneath the camera to climb toward infinity. To avoid this, all x distances are measured relative to the first point of object detection. This, in turn, induces a complication.

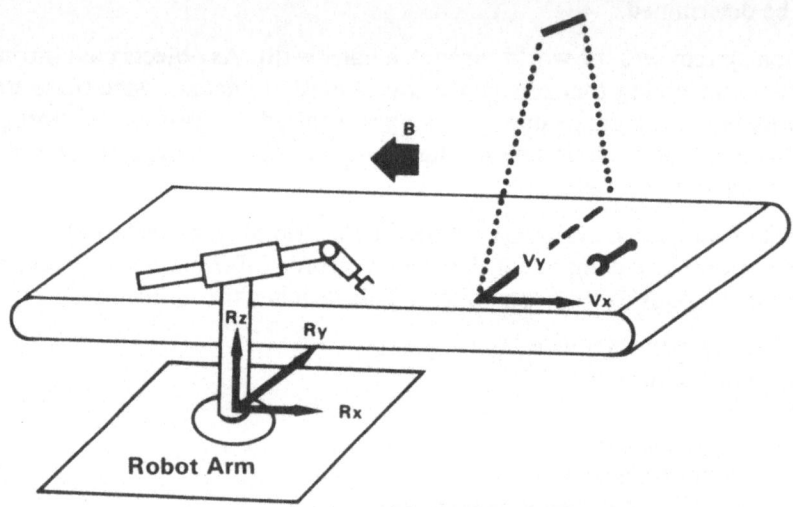

Fig. 8. CONSIGHT-I coordinate systems.

The complication occurs when two appendages of one object begin as two separate objects in the developing image. It then becomes necessary to combine the component descriptors for the two appendages into a single component descriptor which reflects the combination of the two. Moments, however, have been referenced to two different coordinate systems (i.e., the x-origin for each was taken as the point of first detection.) The required shifting of moments is accomplished by applying the Parallel Axis Theorem.

Once parts have passed completely through the field of view, final position determination is made. These computations proceed asynchronously with respect to the processing of the new lines of picture data. The results must be provided to the monitor subsystem before the part travels past the robot's pickup window. To coordinate the scheduling of these final computations, a queue of completed component descriptors is maintained. Component descriptors are removed from the head of the queue and processed as time permits.

Since the belt on which the parts rest is moving, the vision system records the current belt position whenever a new object appears in the developing image. This belt position reference value is obtained from the belt position/speed decoder. Since the leading edge of each part defines the origin of the coordinate system to be used for that part, the position of that part at some future time can then be readily determined by checking the current belt position and adjusting for any belt travel since the initial reference was recorded.

ROBOT SUBSYSTEM

The robot programming subsystem is implemented as two independent tasks. One task is required for robot program development and is necessary only during the programming and teaching phase. The second task, the run-time control system, is required both during the teaching phase and during robot program execution. It interprets and executes a robot program and controls the robot hardware. The execution of this task is controlled by special requests sent from other tasks.

A robot program consists of statements specifying: a position to which the robot should move (setpoint), an operation the robot should perform, or the environment for subsequent execution. Positions to which the robot moves are either taught by moving the robot manually and recording the position or are programmed by entering the specific cartesian coordinates of a point in space from the keyboard.

In addition to this basic programming support, tracking and real-time program modification were developed for CONSIGHT-I. Tracking provides the ability to execute a robot program relative to some moving frame of reference. Program modification provides the ability to modify, in real-time, the robot program under external program control and thereby dynamically modify the robot's path.

Tracking is implemented by defining new reference coordinate systems called FRAMEs [10]. Normally the robot operates in a cartesian coordinate system [R] with its origin at the base of the robot (Frame 0). The robot's cartesian position is described by a matrix [P] which defines the position and orientation of the hand in [R]. The arm solution program then determines a joint vector [J] from [P].

$$[P] \longrightarrow [J]$$

If, however, we want to define [P] relative to a different coordinate system (frame) whose position in [R] is defined by a transformation [F], then the solution program must perform the following:

$$[F]\ [P] \longrightarrow [J]$$

Frames provide a means of redefining the frame of reference in which the robot operates. The robot may be programmed relative to one frame of reference and executed relative to a different frame of reference. For example, a robot may be programmed to load and unload a testing machine. If the testing machine is moved, the entire program can be updated by simply redefining the frame specifying the position of the testing machine without re-programming each individual position point

References pp. 96-97

in the program. The overall effect is to translate/rotate every position to which the robot moves.

In addition to having a position, a frame is defined with a velocity and a time reference. This position, velocity, and time reference are used to compute or predict the frame's position. Each time the run-time system performs an arm solution, (i.e., transforming the position matrix into the corresponding joint angles), it first computes a predicted position for the current frame.

Program modifications are special asynchronous requests sent to the run-time system from other tasks. Via these requests, an external program may modify a robot's programmed path, read the robot's position, start/stop robot program execution, and interrogate status—all while the robot is operating. These requests provide a means for greatly expanding the capability of the robot system without a major effort in developing a powerful robot programming language. Much of the logical, computational, and input/output capability of an algorithmic language (Fortran) is available for programming and controlling the robot external to the normal robot programming and control system.

In CONSIGHT-I, the part position determined by the vision subsystem defines the position and orientation of a frame. The approach, pickup, and departure points are all programmed relative to this frame. The frame is also assigned the velocity of the belt. The robot subsystem does not directly interface to the belt encoder for belt position and velocity data, but receives the data via a request in the same way that the vision data is furnished. Thus, the rate at which the belt position and velocity data are updated is controlled by the monitor program and is a function of the variability of the belt speed. The approach, pickup, and departure points are dependent upon the type of part being picked up as well as its position. Thus, these three programmed points are modified for each cycle of the robot.

SYSTEM MONITOR

The monitor coordinates the operation of the vision and robot subsystems during calibration and part programming as well as during the operation phase. As stated earlier, calibration is required during the initial setup or whenever the camera, the robot or the conveyor have been moved. Part programming is required only when modifying the system to handle a new part.

Calibration — Calibration is the process whereby the relationship between the vision coordinate system [V] and the robot coordinate system [R] is determined (see Fig. 8). In particular we want to compute the position of a part [r] in [R] given the part position [v] in [V]. Taking into account belt travel, this computation is represented by the following equation:

$$[r] = [T] [v] + s \, b \, [B]$$

In the equation above, [T] is the transformation between [v] and [R], s is a scale factor relating belt distance to robot distance, b is the distance of belt travel, and [B] is the belt direction vector relative to [R]. Thus, [v] and b are the independent variables, and [T], s, and [B] are the unknowns to be determined by the calibration procedure.

To determine [B] and s, a calibration object is placed on the belt within reach of the robot hand. The hand is manually centered over the calibration part, the hand position is read, and a belt encoder reading is taken. The conveyor belt is started and the part is allowed to move down the belt. The robot hand is again centered over the calibration part. A second hand position and belt encoder reading are taken. The monitor system can now compute the belt direction vector [B] and a scale factor s, converting belt encoder units to centimeters.

Determining the coordinate transformation between [V] and [R] completes the calibration. The procedure assumes that the plane [Vx, Vy] is parallel to the plane [Rx, Ry] and that scaling is the same in both the x and the y directions.

A calibration object is again placed on the belt and allowed to pass by the vision station. The object position [v1] is determined by vision and a belt reading is taken. The part moves within reach of the robot and the conveyor is stopped. The robot hand is centered over the calibration object, the robot position [r1] is read and the distance of belt travel b1 is computed. This procedure is repeated with the calibration object placed at a different position on the belt resulting in a second set of data, [v2], b2, and [r2]. Combining these two sets of data points into the form above yields

$$[r1\ r2] = [T][v1\ v2] + s[B][b1, b2]$$

This gives us 4 equations for determining the 4 unknowns in the transformation matrix [T].

Part Programming — Part programming is the procedure for defining the gripper position for part pickup. To do this, the vision subsystem must have previously been programmed to recognize the new part as described earlier in the vision section. Generally the pickup position [p] is offset from the part position as determined by vision. Thus, once the vision subsystem locates a part and its position [r] is computed as described in the preceding section, the actual robot pickup position [p] must still be computed.

Fig. 9 illustrates this problem in a general way. The part position [v] and its rotation θ have been determined by the vision subsystem in the coordinate system [V]. The robot operates in [R] and needs to know the pickup position [p]. Since the part may have any orientation on the belt, [p] is a function of both [v] and θ. [V] is rotated and translated from [R] as determined by [T] above.

The procedure is to pass a part by the vision station where its position and orientation are determined and a belt reading is taken. The part moves down the belt to within reach of the robot and the belt is stopped. The monitor subsystem computes the current position of the part in the robot's coordinate system [R] and sends this position and the part direction (as determined by θ) to the robot subsystem as the definition of a

References pp. 96-97

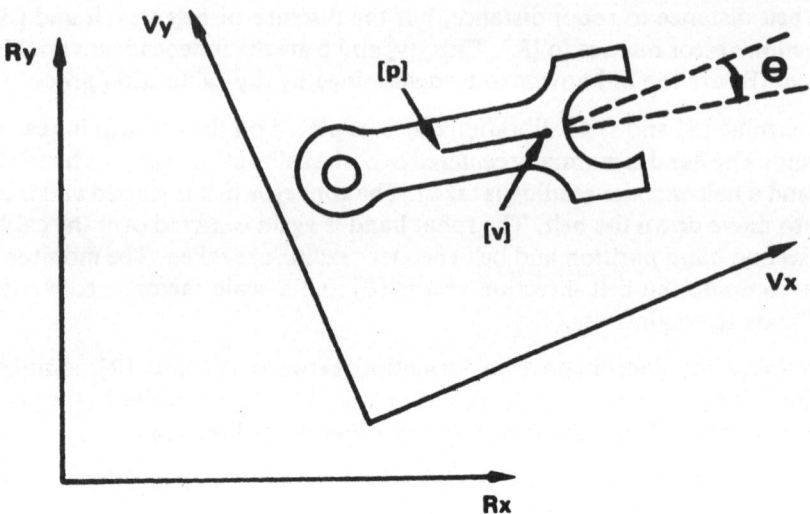

Fig. 9. The pickup problem.

frame. The robot hand is then placed on the part in the desired position for the grasp and pickup. This hand position is read by the monitor, but the robot subsystem returns the hand position relative to the newly defined frame, not the base coordinate system of the robot. Thus, the pickup point is determined relative to the part position and orientation as determined by the vision subsystem.

Later, during the operation phase, the part's position determined by the vision subsystem is again computed relative to [R]. This position and the part orientation defines a frame in which the robot operates during the pickup phase of the robot program. The pickup position is then simply the position determined by the procedure above. The robot subsystem automatically determines the actual robot position from the frame position, the pickup point, and the belt position and velocity.

The Operation Phase — The final function of the monitor subsystem is to control the operation of the CONSIGHT-I system. Fig. 10 illustrates the overall logic of this portion of the monitor. Although the logic is straightforward, the monitor system deals with two rather subtle problems.

The first involves modification of the part pickup point. The robot has a limited range of motion in all of its joints. In particular, the outermost joint, which controls hand orientation when the hand is pointing down, has only 330 degrees of rotation. Clearly, this limited rotation does not permit every possible hand orientation. Since the gripper is symmetric about 180 degrees, we can effectively achieve all rotations by rotating the outer joint by ±180 degrees for some orientations. The monitor system determines when the orientation needs to be changed for a particular pickup position. If the pickup is modified, the putdown orientation also has to be similarly modified so that all parts are put down with the same orientation.

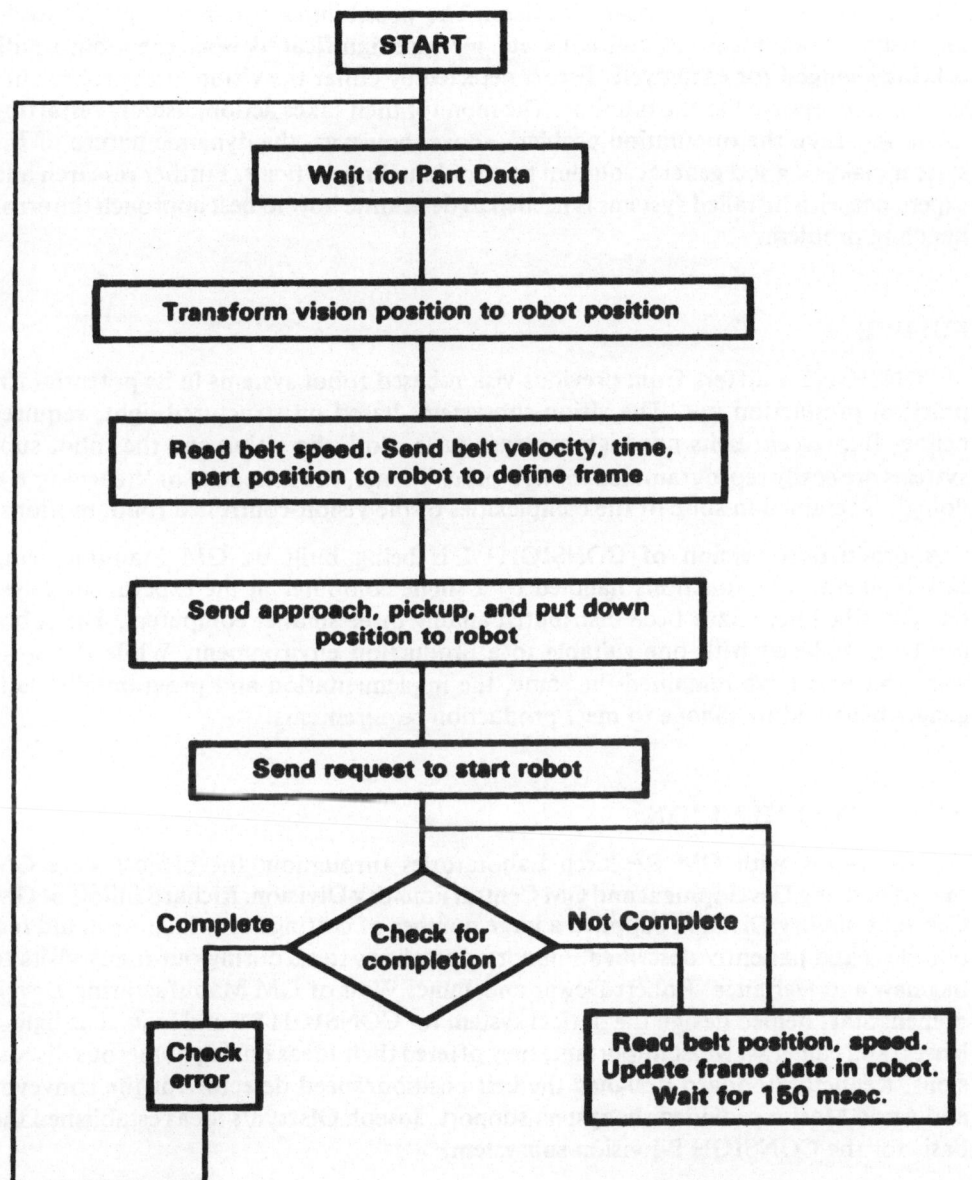

Fig. 10. Overview of the operational phase of CONSIGHT-I.

This orientation problem arises whenever the robot path is dynamically defined. The general problem, for grippers or for parts that are not symmetric or when the gripper is not pointing straight down, is not solvable with commercially available robots. Robots that provide wrists with greater flexibility (greater joint range) are required.

References pp. 96-97

Error recovery is the second problem. The possibilities for parts out of reach, impossible robot positions, collisions, etc. increase significantly when the robot's path is being changed for each cycle. Errors detected by either the vision or the robot subsystems are reported to the monitor. The monitor then takes action, usually restarting the robot. Like the orientation problem above, however, the dynamic nature of the system makes a good general solution to error handling difficult. Further research and experience with installed systems is needed to determine how to best approach the error handling problem.

FUTURE

CONSIGHT-I differs from previous vision-based robot systems in its potential for practical production use. The vision subsystem, based on structured light, requires neither fluorescent belts nor high-contrast parts. Both the vision and the robot subsystems are easily reprogrammed for new parts. In fact, the simplicity of "teaching-by-doing" is retained in spite of the complexities of the vision-controlled robot motions.

A production version of CONSIGHT-I is being built by GM Manufacturing Development. The functions handled by a single computer in the experimental system described here have been distributed among three smaller computers. The robot has been replaced with one suitable to a production environment. While the software concepts have remained the same, the implementation and programming languages have had to change to meet production requirements.

ACKNOWLEDGEMENTS

Cooperating with GM Research Laboratories throughout the project were GM Manufacturing Development and GM Central Foundry Division. Richard Elliott of GM Central Foundry Division supplied a large number of castings on which we could test our ideas and patiently described Foundry operations to us during our many visits to Saginaw and Defiance. Robert Dewar and James West of GM Manufacturing Development Staff helped design the optical system for CONSIGHT-I and loaned us lights, lenses, and cameras. Most important, they offered their ideas during numerous discussions. Kenneth Stoddard designed the belt position/speed detector for the conveyor and Arvid Martin provided the system support. Joseph Olsztyn's ideas established the basis for the CONSIGHT-I vision subsystem.

REFERENCES

1. Perkins, W.A. Model-based vision system for scenes containing multiple parts. 5th Int. Jt. Conf. on Artificial Intelligence (1977), 678-684.
2. Baird, M.L. Sequential image enhancement technique for locating automotive parts on conveyor belts. 5th Int. Jt. Conf. on Artificial Intelligence (1977), 694-695.
3. Ward, M.R. Specifications for a computer controlled manipulator. GMR 2066, GM Research Pub., Warren, Mi. 1976.

4. Agin, G. J. Representation and discrimination of curved objects. Stanford U. Artificial Intelligence Project Memo, AIM-173, 1972.
5. Oshima, M., and Shirai, Y. A scene description method using three-dimensional information. Progress Rep. of 3-D Object Recognition, Electrotechnical Lab. 1977. Japan.
6. Popplestone, R.J., et al. Formation of body models and their use in robotics. U. of Edinburgh, Scotland.
7. Agin, G.J. An experimental vision system for industrial application. Proc. 5th Int. Symp. on Industrial Robots (1975), 135.
8. Duda, R., Hart, P. *Pattern classification and scene analysis,* Wiley-Interscience Publication, p. 284, 1973.
9. Agin, G. J. Image processing algorithms for industrial vision, Draft Report, SRI International, (1976).
10. Paul, R. Manipulator path control, *1975 International Conference on Cybernetics and Society,* (1975), 147-152.
11. Beecher, R. PUMA: Programmable universal machine for assembly. GM Research Laboratories Symposium "Computer Vision and Sensor-based Robots," September, 1978.

DISCUSSION

J. K. Dixon *(Naval Research Laboratory)*

What percentage of connecting rods are missed by the vision system and go off the end of the conveyor belt?

Holland

Very few. The system as you saw it uses a single computer. Sometimes we have conflicts between processing the camera data and controlling the robot. When that conflict arises we give the computer to the robot.

J. L. Mundy *(General Electric R & D Center)*

I noticed that your objects in general were matte and rather flat. What about the case where objects might be more curved such as cylinders and more shiny so that the specular reflection conditions would be met between incident rays and the field of view of the camera?

Holland

We found that not to be a problem. We've looked at chrome plated circular dome type objects and we would experience specular reflection someplace inside the object which would cause it to appear to our system as small holes in the object at those points. But because of the way the software operated, it ignored the reflections. The parts can be virtually any color or texture.

R. J. Popplestone *(University of Edinburgh)*

You mentioned the difficulty or the care you have to take with the computational accuracy. Are you actually using floating point words or fixed point words for this work?

Holland

We used a PDP 11/34 with the floating point processor, and the primary reason for the floating point was because of the extended precision that was available.

Popplestone

So this would have considerable effect in trying to put this in a cheap microprocessor then?

Holland

Not really. Floating point facility is available on computers like the 11/34 which are not extremely expensive, and it's also available on the LSI-11 if you like.

Popplestone

At a rather reduced speed. I mean why do you impose the speed of floating point, if the number is not floating?

Holland

We're using it because we need the precision for the quantities that we are storing. I think you would not want to use an extremely slow, small computer. The 11/34 we've found to be approximately the right size.

C. A. Rosen *(SRI International)*

I'm pretty sure you chose the linear array camera from the point of view of cost, rather than a two dimensional camera and perhaps you got a few more elements across the line. But I think you should consider the possible need and utility of a two dimensional camera even at slightly additional cost. At some point there might not be that much difference. It would be nice to be able to handle parts that would be touching in which you do connectivity analysis. This is a little more difficult than just taking one line at a time. I think that a point to be made is that at present we are awfully cost conscious, every one of us, but maybe we should look just a little ahead for more general solutions.

Holland

I think that's true. In this particular system, because of the way the lighting is arranged, we only require the data along the light line so the line camera was a natural here.

B. H. McCormick *(University of Illinois - Chicago Circle)*

Has anyone tried to do the converse thing of illuminating the conveyor belt, for example making it fluorescent?

Holland

SRI has done things similar to that, such as using fluorescent conveyor belt surfaces and controlling the background. We made an attempt to not impose any restriction on the belt or the hardware and provide the light externally to produce the contrast.

D. A. Seres *(DuPont)*

How do you handle the cases where the parts are touching one another?

Holland

If the parts are touching, the system will detect that there are two parts touching and simply ignore them and let them fall off the end of the conveyor belt. It requires that they not be touching. In practice, in realistic conditions, the parts normally are spaced out and that's usually one of the reasons they are on a conveyor belt in the first place... to pace them out somewhat.

J. F. Engelberger *(Unimation)*

Can you describe one specific application where the actual vision hardware can be now used in a typical automotive application? Or do you have to go further with the hardware before you can apply it to some applications?

Holland

Well, we have seen specific applications, such as castings on a conveyor belt that are loaded into a testing machine or an inspection device or a milling machine.

Engelberger

Without previous handling? How do they get to this position? Where does the saving come in? Is someone putting them on the conveyor?

Holland

A typical application might be where the parts are being placed on the conveyor belt by people and are going through a series of inspection steps during the length of the belt. Towards the end of the belt another person or several people are loading these parts into a testing machine — just picking them off the belt and setting them onto guide rails. The savings is at this point.

D. E. Whitney *(Charles Stark Draper Laboratory)*

Have you considered organizing the processing so that as each scan line goes by you complete the processing and do a summarization of the results and keep in your memory only a running recursive accumulation of what the scan lines are telling you? There are devices on the market for image analysis that operate on this basis. This means you can process any number of images simultaneously as long as you can keep track in separate registers of information on separate parts. Maybe you can get by with less computer power.

Holland

We do keep only a running total of the results on each part currently being seen by the camera. I think certainly that if you were to go to hardware implementation for some of this you would always achieve tremendous speed improvements. In this particular system we found it not to be necessary, and the computer approach is adequate, so we didn't pursue the hardware issue further.

AN INDUSTRIAL EYE THAT RECOGNIZES HOLE POSITIONS IN A WATER PUMP TESTING PROCESS

T. UNO, S. IKEDA, H. UEDA and M. EJIRI

Hitachi Central Research Laboratory, Tokyo, Japan

T. TOKUNAGA

Hitachi Taga Works, Ibaraki, Japan

ABSTRACT

In the production of small water pumps, an automatic machine was requested for water pressure testing of the final products, in which hoses are automatically connected to inlet and outlet flange holes of the pumps. For this purpose, a visual device has been developed to detect the hole positions by means of CCD type TV cameras. This "industrial eye" also includes special electronic circuitry for pattern matching, which utilizes quarter patterns of a circular hole image as templates.

The initial fed-in position of the pumps to be worked on involves some variation caused by the difference of pump types and also the accumulated dimensional errors encountered in prior assembly processes. By equipping mechanical arms with TV cameras, a closed-loop sequential recognition method can be adopted, so that the eye can quickly find the hole position first, and then fine tune the arm position by subsequent measurement with higher resolution.

This industrial eye is now being successfully used in the pressure testing process of a water pump production line in conjunction with manipulator arms as part of the visual feedback loop.

INTRODUCTION

One subject commanding great interest in production automation has been the development of visual devices that can recognize the positions of objects handled in mass production processes. Many studies have been conducted to date on object

References p. 114

recognition and related visual devices [1, 9] and some have been especially successful in automation of production lines [3,6,7,8,9].

Generally speaking, however, the practical application of such devices to actual systems involves many difficulties for several reasons. One such difficulty arises from the fact that the image of an object is easily distorted by the intensity and direction of environmental illumination. Therefore, the recognition algorithm should be tough enough to cope with changes in environmental conditions.

This paper deals with such an "Industrial Eye". The high recognition reliability of the device permitted successful application to the automation of a post-assembly water pump testing process.

Fig. 1. An example of a pump.

OBJECTS AND THEIR IMAGES

A sample pump is shown in Fig. 1. It has an inlet hole and an outlet hole. Another type of hole and the area around it are shown in Fig. 2.

RECOGNIZING HOLE POSITIONS

Fig. 2. An example of a flange hole.

Two solid state TV cameras are used in the device to detect hole positions. The TV images of inlet and outlet flange holes appear as the black circular portion, as shown in Fig. 3.

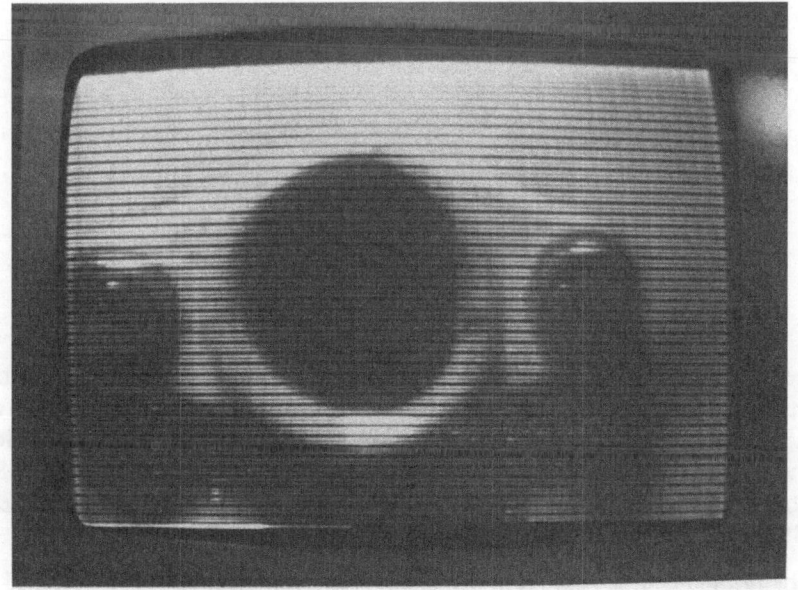

Fig. 3. Image of a flange hole.

Refrences p. 114

It may be thought that detecting a circle from an image is an easy process by image processing. However, this is not necessarily true, because a circle in a visual image is often distorted by the intensity and direction of environmental illumination, e.g. sunlight from a window in our case.

The initial position of the flange holes involves some variation due to pump type differences as well as accumulated dimensional errors caused in prior assembly processes. However, the biggest cause of position variation is the fact that, prior to pressure testing, the pump is set up on a wooden base for subsequent shipment, and the dimensions of the wooden base are hard to control.

The standard flange hole positions for all kinds of pumps are shown in Fig. 4. Their positional tolerances are ±15 mm in the X direction and ±10 mm in the Y direction. A production control computer informs the machine of the type of pump in advance, so that the standard position of flange holes can be determined for the automatic manipulator as its initial position. Accordingly, the TV camera installed on the manipulator arm is registered at the standard X and Y position so that the hole image would be at the center of the image field.

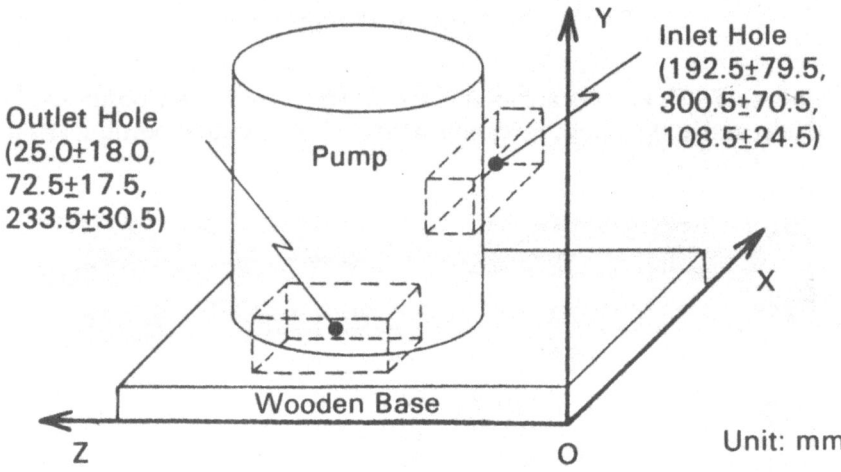

Fig. 4. Standard position of flange holes of all kinds of pumps.

On the other hand, the diameters of the holes are 24 mm, 30 mm or 36 mm depending on the pump type. Therefore, the arm is also adjusted to one of the predetermined standard Z positions, i.e., a position along the direction of the camera's optical axis. This insures that the size of each visual field is set at 100.0 mm (83.3 mm, 66.7 mm) horizontally and 75.0 mm (62.5 mm, 50.0 mm) vertically, when the diameter of the hole is 36 mm (30 mm, 24 mm), as shown in Fig. 5.

The positional accuracy target in this study was 2 mm in the X direction and 1.5 mm in Y direction. This is the tolerance required to permit the manipulator to mechanically fit the hose to the flange hole.

RECOGNIZING HOLE POSITIONS

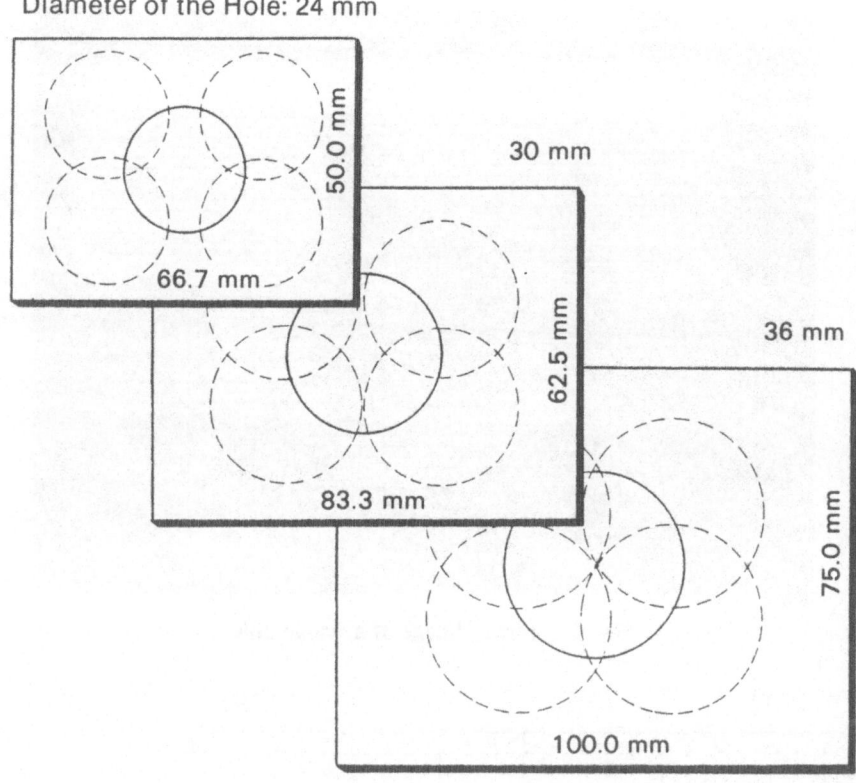

Fig. 5. Size of each visual field.

The TV cameras selected were of the solid state optical sensor type. This was because of the advantage they offer with regard to size, weight and ease of installation on the manipulator hand. These cameras have 100 × 100 picture elements. Their dimensions are 38 mm × 90 mm × 56 mm and their weight is 0.3 kg. They are also equipped with interlace scanning and are of the fixed sensitivity type. An example of the flange hole image from the TV camera has already been shown in Fig. 3, and also in Fig. 6 in binary form.

As there are three standard Z positions, depending on pump type, the image is likely to be out of focus. However, the extent of blur is small, and no influence is observed as the image is adequately thresholded into a binary form.

POSITION RECOGNITION ALGORITHM

Based on the previous conditions, the position recognition algorithm was developed. It consists of two successive steps: (1) macroscopic, and (2) microscopic.

References p. 114

Fig. 6. Binary image of a flange hole.

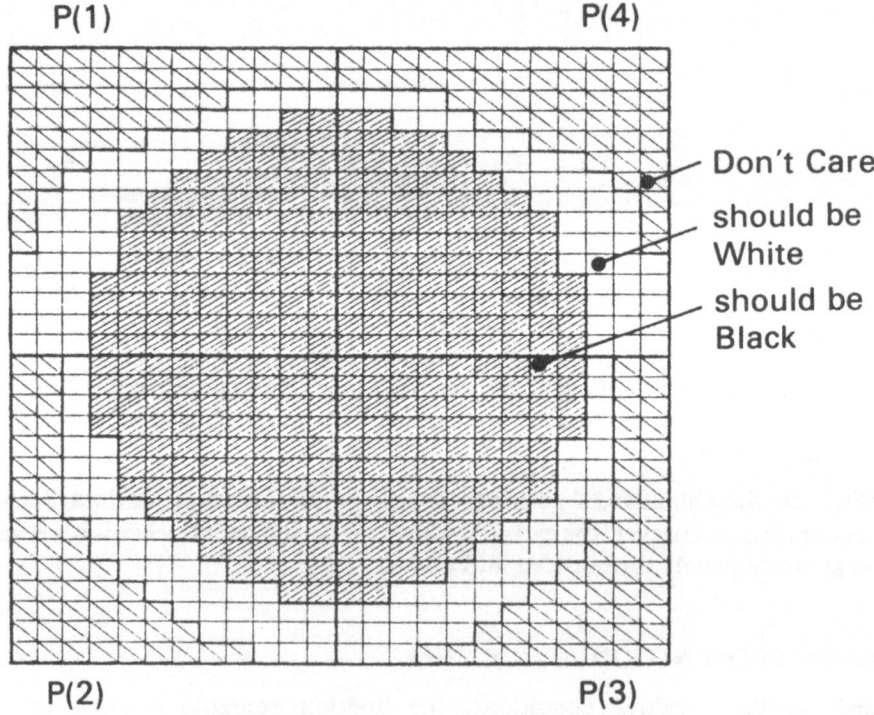

Fig. 7. Templates P(i) for macroscopic processing.

RECOGNIZING HOLE POSITIONS

In the macroscopic step, a binary video signal is sampled every two elements to yield a 50 × 50 element image. In this case, the hole diameter corresponds to 18 elements in the X direction and 24 elements in the Y direction. The difference is simply because the shape of each photosensitive element in this particular camera is designed to have a 4:3 aspect ratio. The macroscopic process is based on pattern matching of this roughly sampled image. Templates for this procedure are illustrated in Fig. 7. Templates are made by dividing the hole image into four pieces. Therefore, this method enables us to reduce the size of image memory for matching to 12 × 15 picture elements, while leaving the size of memory for the templates and their matching circuits at 24 × 30.

Consequently, prior to pattern matching, a preprocessing for template selection is performed on the same roughly sampled image. This reduces by 3/4 the number of matching circuits. In this case, additional templates, hereinafter referred to as sub-templates, are provided as shown in Fig. 8. In these sub-templates, the number of picture elements for which matching is performed is drastically reduced to 11 for each sub-template.

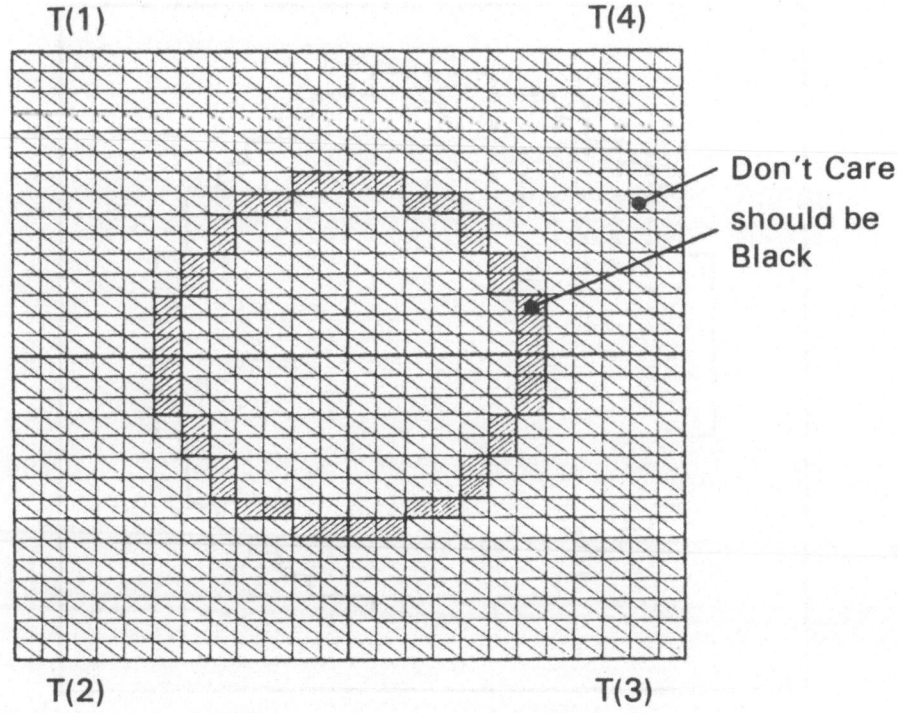

Fig. 8. Sub-templates T(i) for macroscopic processing.

References p. 114

Four sub-templates T(1), T(2), T(3), and T(4) are scanned in parallel over the input pattern. The matching result flag Q(i) is given as follows:

$$R(i) = \begin{cases} 1, & \text{when } S(i) \geq S_{th} \\ 0, & \text{otherwise} \end{cases}$$

where $i = 1, 2, \ldots 4$

$$Q(i) = R(i) \wedge (\wedge_j \overline{R(j)})$$

where $j = 1, 2, \ldots, 4$ and $j \neq i$

where S(i) is the number of matched elements with T(i) and S_{th} is appropriate threshold value, i.e., 10 (only one picture element is allowed to differ out of 11) or 11 (no element is allowed to differ). This equation means that, when the input pattern is matched with the i-th sub-template R(i) and not matched with other sub-templates R(j), Q(i) becomes "1" and the i-th template should be used in the subsequent step. Examples of Q(1) superimposed on a schematic hole image are shown in Fig. 9. Dots on the figure show that, at these points of scan, Q(1) has the logical value 1. Thus, at these points, pattern P(1) is automatically selected for pattern matching.

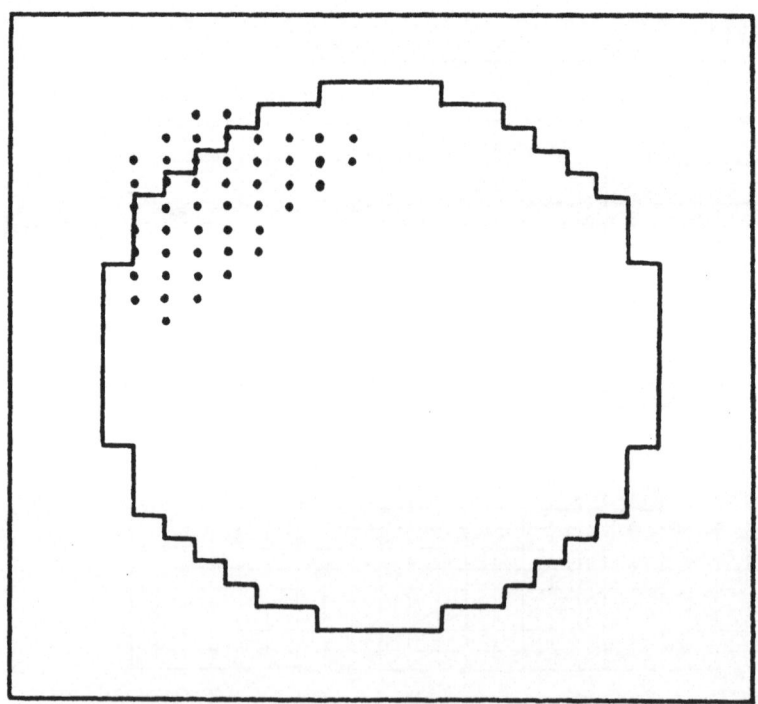

Fig. 9. Examples of Q(1) by macroscopic processing.

RECOGNIZING HOLE POSITIONS

The template P(i) selected is then tried, to determine whether the portion of the input pattern is matched. As mentioned already, the size of each template is 12 × 15 picture elements, and its standard pattern is equal to each quarter of the hole to be detected. The matching result N(i) is given as follows;

$$N(i) = \sum_k O(k) \oplus P(i,k)$$

where P(i,k) is the logic state of the k-th element of the i-th template pattern, O(k) is the logic state of the k-th element of the input pattern, and "k" is the ordinal element number of a 12 × 15 pattern. Therefore, the sum in this case means the number of picture elements whose logic value is 1.

If the input partial pattern is nearly equal to the template pattern, the matching result N(i) will be nearly equal to zero. An example of the result is illustrated in Fig. 10, where larger dots denote more precise matching N(i) value. The processing for five templates T(1), T(2), T(3), T(4) and dynamically selected P(i) is performed within an image sampling period for each inputted image portion to calculate the unlikelihood value N(i). By combining the neighborhood picture elements into one group for each i, the element position having the minimum N(i) value is found for each group. After the scanning of one visual field, several minimum points are determined. From these minimum points, the most probable combination of four (or at least three) minimum

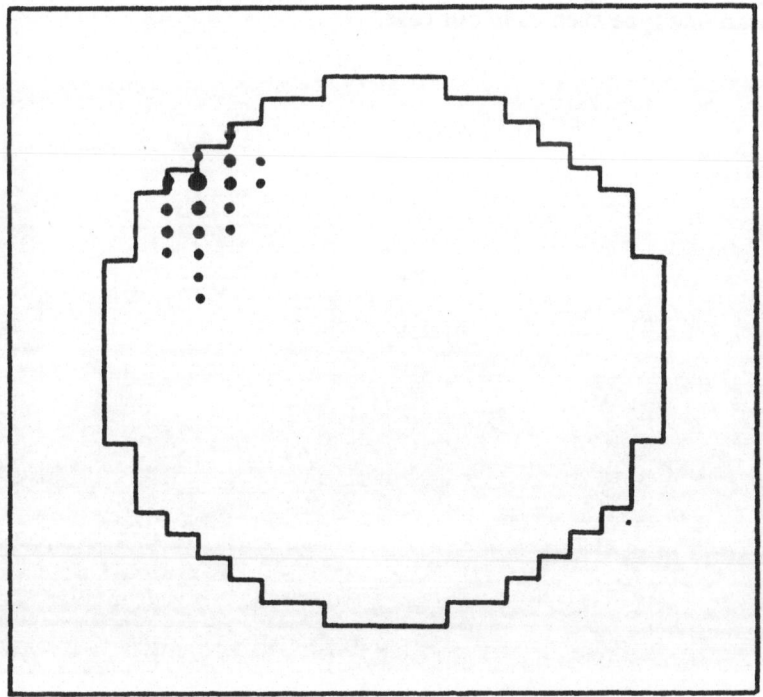

Fig. 10. Examples of N(i) by macroscopic processing.

References p. 114

points is then selected, so that these points have geometrical reasonability determined by the hole image dimension. If the result is satisfactory, the average position of these minimum points is calculated. This value corresponds to the hole position.

Microscopic processing detects hole position more precisely. For this purpose, four rectangular domains are set around the circle as shown in Fig. 11, depending on the results of the prior step. The X and Y position of the circle are determined by calculating the black areas in the domains with the same resolution as the original image. For example, when area M_O and M_1 are detected at each domain as shown in Fig. 12, the precise X coordinate becomes:

$$X = X_O + \frac{M_1 - M_O}{B}$$

Precise diameter D becomes:

$$D = 2A + \frac{M_1 + M_O}{B}$$

where A and B are the geometric dimension of the domains as shown in the figure.

In general, this area measurement method is advantageous in circuitry. This is because it does not need a two dimensional pattern memory, unlike other more traditional, pattern matching methods. This is especially true when the camera is the interlace scanning type such as in our case.

Fig. 11. Four rectangular domains set around circle (microscopic processing).

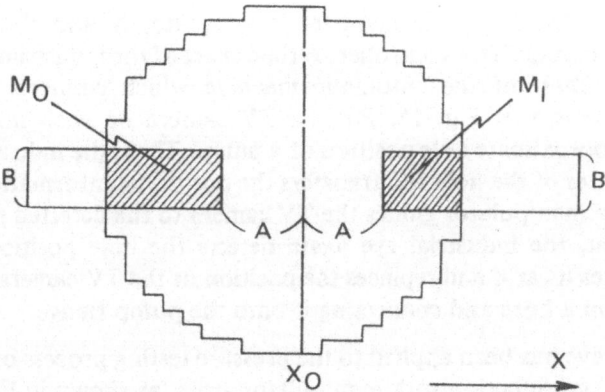

Fig. 12. Two rectangular domains for X position detection (microscopic processing).

SYSTEM CONFIGURATION

An industrial eye for position detection based on these methods has been constructed, as shown in Fig. 13. It consists of two TV cameras with CCD type solid state optical sensors mounted on two arms, an image selector switch, an integration circuit for adaptive thresholding, a pattern matching circuit for template selection, and a pattern matching circuit for the selected template. It also includes a circuit for calculating the black area in four rectangular domains, as well as for calculating X and Y coordinates and the precise diameter of the circle. In the pattern matching circuit, memory adequate to store 14 scan lines is used to store a partial image in real-time mode.

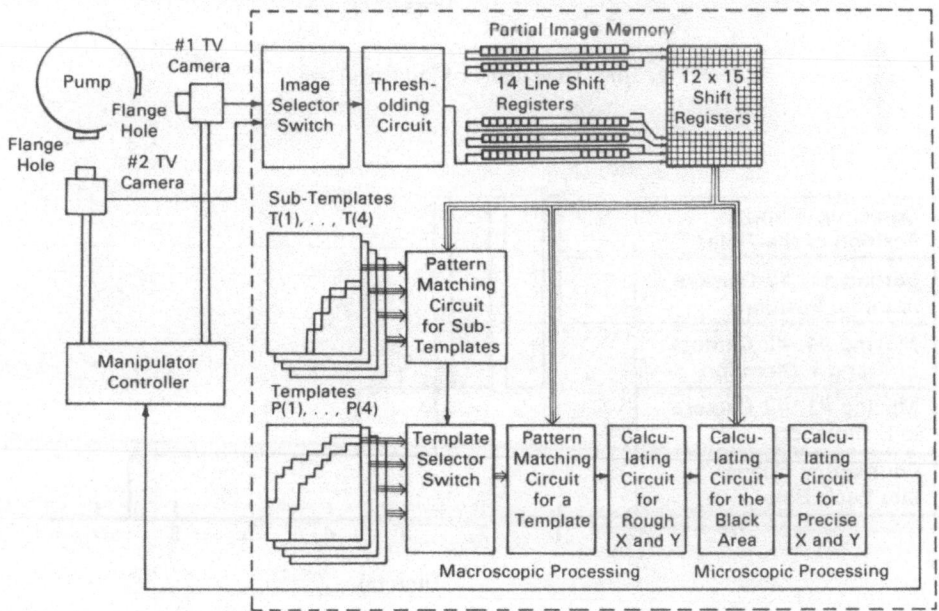

Fig. 13. Block diagram of an industrial eye for position detection.

References p. 114

After verifying that the circle diameter is adequate, X and Y coordinates are transferred to the manipulator controller. A time chart of the industrial eye is shown in Fig. 14. A time chart of the automatic machine which connects hoses with its manipulators is shown in Fig. 15. First, a TV camera on each manipulator arm approaches the approximate hole position of a pump. Then, the industrial eye detects X and Y coordinates of the hole and transfers the positional information to the manipulator. Next, the manipulator guides the TV camera to the detected precise position of the hole. Then, the industrial eye again detects the hole position. Finally, the manipulator moves its arm and replaces the position of the TV camera with a built-in tool for picking up a hose and connecting it onto the pump flange.

This industrial eye has been applied to the pressure testing process of a water pump production line in conjunction with manipulator arms, as shown in Fig. 16.

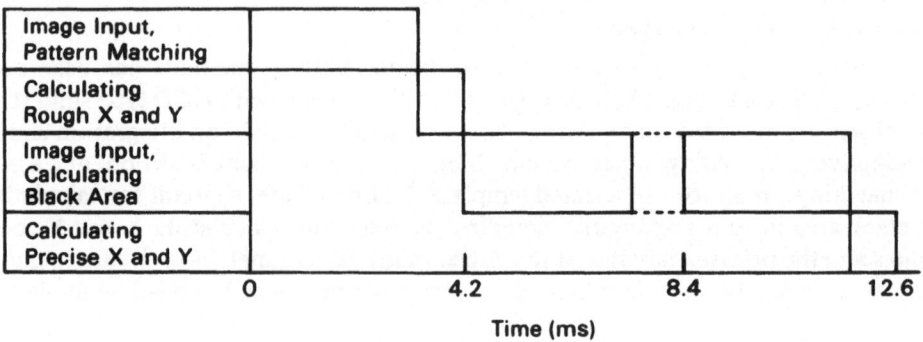

Fig. 14. Time chart of industrial eye.

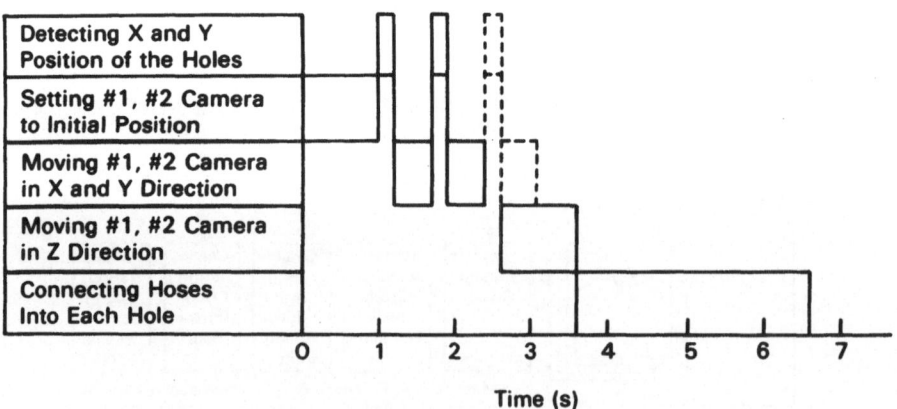

Fig. 15. Time chart of automatic machine for connecting hoses.

Fig. 16. Industrial eye applied to pressure testing process of a water pump production line.

CONCLUSION

An industrial eye for hole position detection is developed by using closed-loop sequential recognition techniques. The recognition method includes a macroscopic and a microscopic step.

The macroscopic step consists of two typical functions. One selects the suitable template to be matched by using sub-templates whose effective number of picture elements is extremely small. The other performs a pattern matching operation by using the selected template and finds the most probable combination of best-matched minimum points.

The microscopic step finds a more precise hole position by measuring the black areas within four rectangular domains generated automatically from the data obtained by the preceding macroscopic step.

The result is a reliable industrial eye that is now being used successfully in the pressure testing process of a water pump production line in conjunction with manipulators as a part of the visual feedback loop.

References p. 114

ACKNOWLEDGEMENT

The authors wish to thank Yutaka Nishiyama, the then General Manager, Toshio Tsubaki, Department manager of Equipment Developing Department in Taga Works, Hitachi Ltd., for their insights and invaluable arrangements. Thanks are also due to Dr. Hiroshi Watanabe, General Manager, Dr. Jun Kawasaki, the then Department Manager, of Central Research Laboratory, Hitachi Ltd., as well as many others whose help and encouragement contributed to the realization of this system.

REFERENCES

1. Olsztyn, J.T., Rossol, L., Dewar, R., Lewis, N. An application of computer vision to a simulated assembly task. Proc. 1st Int. Jt. Conf. on Pattern Recognition (1973), 505-513.
2. Ejiri, M., et al. A prototype intelligent robot that assembles objects from plan drawings. *IEEE Trans. on Computers* C-21, 2 (1972), 161-170.
3. Ejiri, M., et al. A process for detecting defects in complicated patterns. *Computer Graphics and Image Processing* 2, (1973), 326-339.
4. Hale, J.A.G., et al. Control of a PCB drilling machine by visual feedback. Proc. 4th Int. Jt. Conf. on Artificial Intelligence (1975), 775.
5. Saraga, P., et al. An experimental visually controlled pick and place machine for industry. Proc. 3rd Int. Jt. Conf. on Pattern Recognition (1976), 17-21.
6. Baird, M.L. An application of computer vision to automated IC chip manufacture. Proc. 3rd Int. Jt. Conf. on Pattern Recognition (1976), 3-7.
7. Uno, T., et al. A method of real-time recognition of moving objects and its application. *Pattern Recognition* 8 (1976), 201-208.
8. Kashioka, S., et al. A transistor wire-bonding system utilizing multiple local pattern matching techniques. *IEEE Trans. on Systems, Man, and Cybernetics* SMC-6, 8 (1976), 562-570.
9. Mese, M., et al. An automatic position recognition technique for LSI assembly. Proc. 5th Int. Jt. Conf. on Artificial Intelligence (1977), 685-693.

DISCUSSION

D. E. Whitney *(Charles Stark Draper Laboratory)*

You have two arms, one for each eye. You then have two more arms, one for each hose? All together four arms?

Uno

No. This system has two arms and two TV cameras.

J. M. Tenenbaum *(SRI International)*

Where is the hose? It is not easy to see from your slide. You can see the camera, but you couldn't see where the hose was on that same arm.

Ejiri

The system has only two arms. One is for the inlet hole and the other is for the outlet hole. One camera is attached to one arm and another is attached to another arm. Once the camera looks at the hole then the arm switches to the other function by locating its hand to the hole. The hoses are mounted on a circular table and rotate along with the pumps. At the first station the manipulator first grasps the hose and then looks at the hole and connects it. And after connection, the pump and the hose go to the next station. The pressure testing is started there and is performed with subsequent stations. And finally it goes around and the hose is automatically disconnected.

C. A. Rosen (SRI International)

Could you tell me how much rotation of the assembly you can stand for maintaining this operation? Is the assembly jigged to within a few degrees? What if the wooden base is a little twisted? How rugged is the algorithm?

Ejiri

Our tolerances are specified by the x, y and z positions. This means a 3 or 4 degree rotation is possible *(See Fig. 4)*.

P. M. Will (IBM Research Laboratories)

The window looked like two diameters of the hole.

D. Nitzan (SRI International)

What was the labor savings?

Ejiri

I don't have complete figures, but I think two or three people were eliminated by this system. It is cost effective and it is now working. Usually the factory doesn't want to invest much money for uneconomical projects. I think they are happy with it.

Uno

We save about three men by just automating the hole connection, not counting the inspection task.

J. L. Nevins (Charles Stark Draper Laboratories)

Did you do an economic analysis before you installed this?

Ejiri

Actually we didn't do that. But the factory people did and they found that it is economical to make this kind of a system.

J. M. Tenabaum *(SRI International)*

Do you do any verification when you insert the hose to make sure that it has gone into the hole before you turn on the water?

Ejiri

This system is so accurate it is not necessary at this moment to verify that the hose is in. So far, it is difficult to state the error rate of the machine. Let's say it is 100 percent reliable.

Whitney

How much time between one water pump and the next water pump?

Uno

Contact time is about 30 seconds, and three minutes for total testing.

Whitney

A follow up question. Does the same robot also take the hose off at the end of the test?

Uno

No. It is done by other simple mechanisms located at the final station.

Whitney

And the test lasts three minutes?

Uno

Yes. But, dwelling time in each station is about 11 seconds. This is the maximum time allowed for hose connection. Our system completes it within 6.6 seconds. And, one half of that time is seeing.

A. D. Gara *(General Motors Research Laboratories)*

What type of camera were you using?

Uno

The camera was CCD type and was made by Fairchild.

Ejiri

We bought it three or four years ago and at that time it was the only camera that was available. It has 100 by 100 picture elements and was the lightest camera in the world at that time.

APAS: ADAPTABLE PROGRAMMABLE ASSEMBLY SYSTEM

R. G. ABRAHAM

Westinghouse Research and Development Center, Pittsburgh, Pennsylvania

ABSTRACT

In a study performed for the National Science Foundation, Westinghouse concluded that the successful application of adaptable programmable assembly technology could lead to 3 to 1 productivity improvements in batch assembly operations. Careful analysis of assembly requirements led to a conceptual design of a complete pilot line to assemble a representative product line, small motors. This adaptable programmable assembly system consists of a mix of fixed sequence and servo-controlled robots, programmable fixtures and parts presenters, end effectors and tools which are universal for each work station, vision and other sensory systems, mini-computers and microcomputers, special equipment and people. Low cost, microprocessor-based, binary image processing vision systems are used at each work station to ensure good parts quality and to verify that previous assembly operations are performed properly, thereby minimizing system downtime.

INTRODUCTION

Approximately one year ago, Westinghouse completed a Phase 1 research program "Programmable Assembly Research Technology Transfer to Industry".* The objectives of this program were to determine the feasibility of utilizing adaptable programmable assembly system (APAS) research results to increase productivity of batch

---

This program was jointly funded by Westinghouse Electric Corporation and the National Science Foundation; the NSF grant was ISP 76-24164. Any opinions, findings and conclusions or recommendations expressed in this publication are those of the author and do not necessarily reflect the views of the National Science Foundation.

References p. 136

assembly operations, and to develop an experiment that would provide industry with the necessary data to accelerate technological transfer and successful utilization.

We conducted an extensive state-of-the-art review and identified the technology voids and the developments in APAS technology that appear ready for practical application [1, 2, 3]. It was concluded that no one in the world has a complete APAS system and no quantitative economic data on productivity improvement, product quality improvement, and inventory reduction due to the utilization of APAS exists. Furthermore, an integration of all subsystems into a complete pilot APAS system is necessary, as is a practical proof-of-concept experiment using large numbers of parts for a product line with many styles.

We then performed a systematic analysis of 60 product lines, all of which are assembled in relatively small batches, and selected outdoor lighting, compressors and small motors as those most suitable for programmable automatic assembly. Preliminary concepts as to how manipulator arms, programmable parts presenters and fixtures, end effectors and tools, and computer and sensory systems should be configured led to the selection of small motors as the best economic candidate and the one most representative of batch manufacturing [4].

By carefully observing the end bell subassembly (Fig. 1) and final assembly (Fig. 2) operations at the small motor plant, we were able to prepare detailed assembly operations flow charts that documented current manual assembly methods. A study of the 450 different motor styles, which are manufactured with an average batch size of 600 and an average of 13 style changes per day, led to the conclusion that there are eight significant differences in styles (Fig. 3) in terms of assembly requirements. After expanding operations flow charts to include all eight cases, we analyzed the inspection, recognition, parts presentation and insertion operations and identified requirements and alternative APAS solutions (Tables 1, 2, 3).

We conducted vision system experiments at SRI International [5] and insertion experiments using the remote center compliance (RCC) device at The Charles Stark Draper Laboratories [6] using small motor parts. The parts feeding equipment at the University of Massachusetts was observed and analyzed with small motor parts in mind.

Working as a team with these three research organizations, we then developed alternative APAS configurations for end bell subassembly [6] and small motor final assembly [7]. Product design changes were necessary to simplify the more difficult wiring operations. A formal evaluation of configuration alternatives using the weighted criteria of economics, time, performance, risk, utilization of APAS research technology, potential for industry utilization and human resources resulted in the selection of an assembly center concept for end bell subassembly and in-line sequential approach for small motor final assembly [8].

The purpose of this paper is to describe in detail the pilot adaptable programmable assembly system configuration and the conceptual designs of key subsystems.

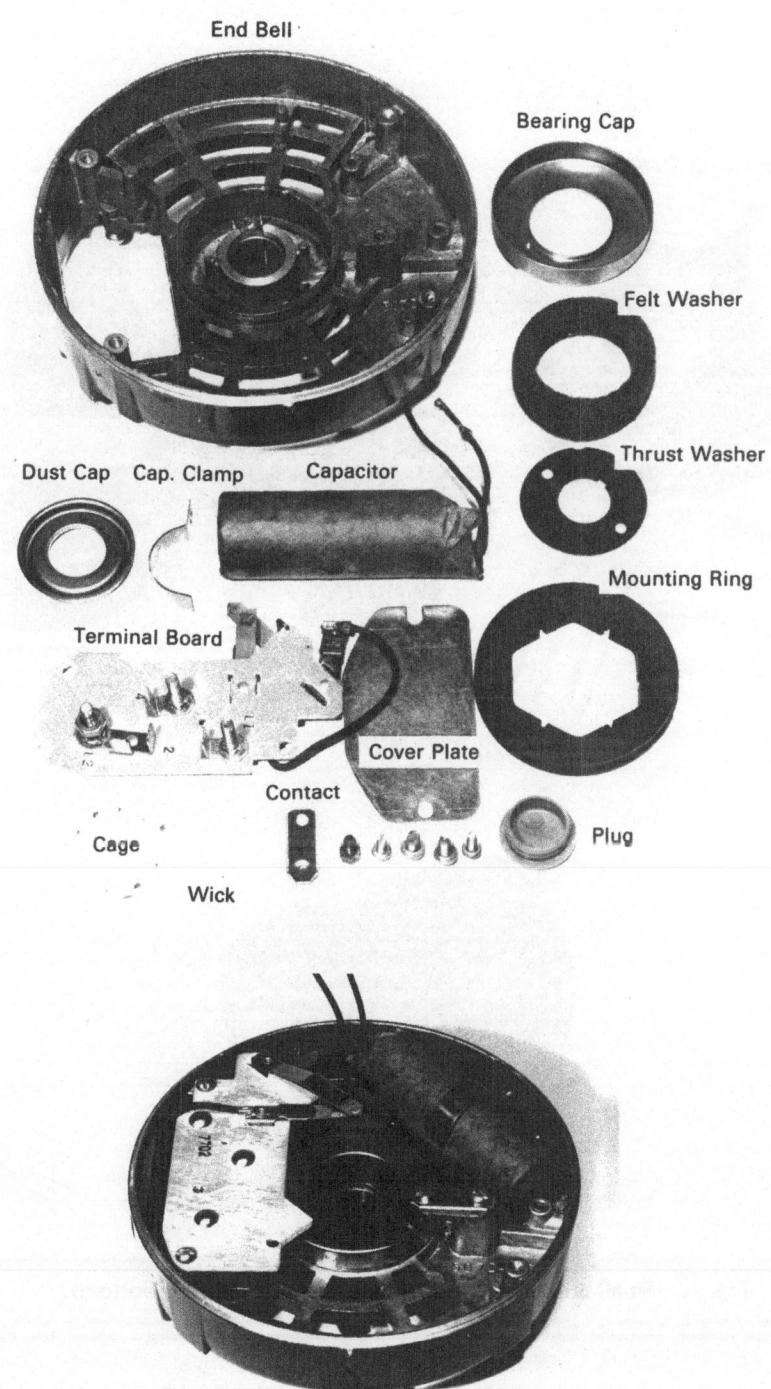

Fig. 1. Top end bell disassembled (top) and assembled (bottom).

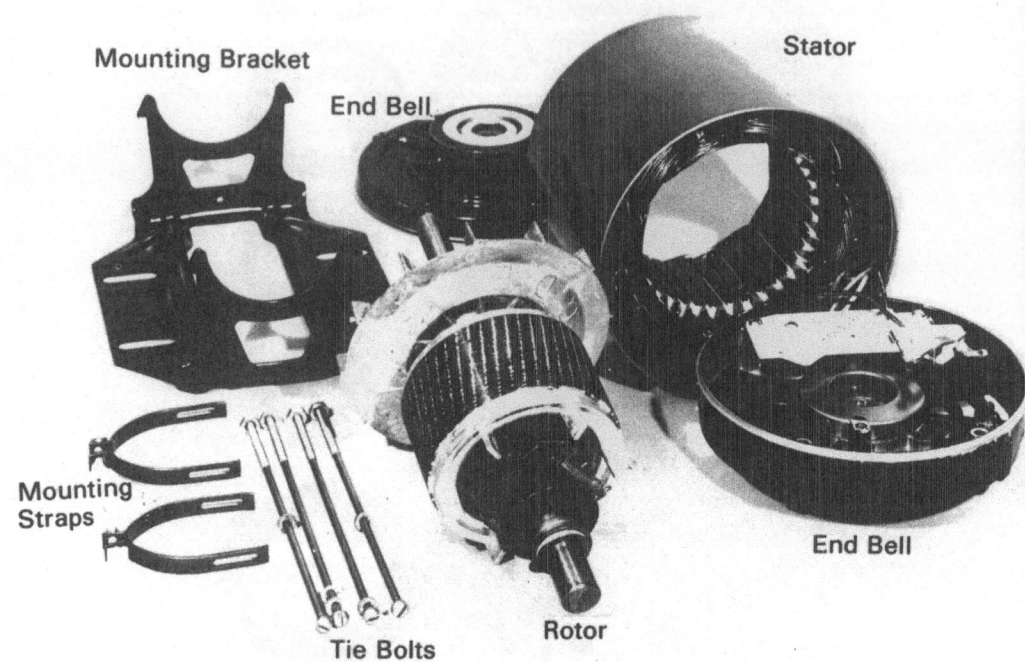

Fig. 2. Small motor disassembled (top) and assembled (bottom).

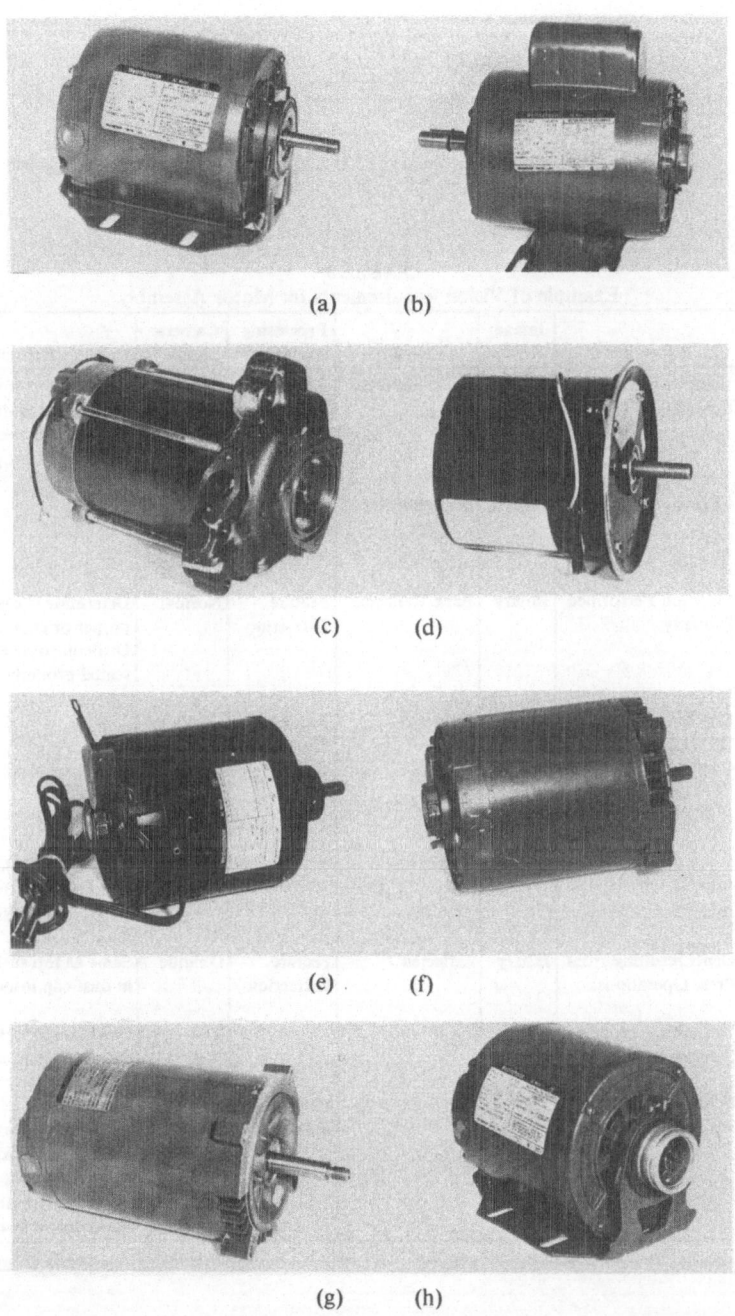

Fig. 3. Classes of motors. (a) General purpose — sleeve bearing, resilient mount. (b) General purpose — capacitor start, ball bearing, rigid mount. (c) Gas pump. (d) Oil burner. (e) Sump pump. (f) Industrial. (g) Jet Pump (centrifugal). (h) Carbonator (vending machines).

References p. 136

TABLE 1
Example of Vision Requirements for Motor Assembly

Part or Operation	Inspection	Image Rq'd.	Lighting	Processing Technique	Camera Angle	Comments
Wick Insertion	Wick Properly Positioned	Binary	Back/Reflected	Extraction Extraction	Normal	Could detect if wick protrudes into shaft hole. Difficult to determine if wick seated properly.
Plastic Cage	Cage has three legs	Binary	Back/Reflected	Image Comparison	Oblique	
Cage Insertion	Insertion Performed Properly	Binary	Back/Reflected	Feature Extraction	Normal	Determine if cage is in proper orientation. Difficult to determine if seated properly.
Rotor	Check for Presence Of Weights On Centrifugal Switch.	Binary	Reflected	Feature Extraction	Oblique	
Stator						The only check that appears possible at this time would be for blockage of bolt holes. Wiring checks would be impossible.
Dust Cap Pressing	Dents resulting from Press Operations.	Binary	Reflected	Feature Extraction	Oblique	Same as top surface dents in dust cap inspection.
Permawick Dispenser	Permawick Properly Dispensed.					Need samples of improperly dispensed permawick.
Flat Washer Insertion	Misaligned flat Washer	Binary	Reflected/Back	Feature Extraction	Normal	Measure area of shaft hole - camera must be centered over hole. If reflected light - measure area of sleeve bearing seen or center of symmetry of sleeve bearing seen.

TABLE 2
Example of Parts Presentation Requirements for Motor Assembly

Parts for Presentation	Method Received	Supplier	Presentation Methods	Comments
Permawick	Bulk 5 gal. containers	Outside vendor	Special dispensing equipment.	Requires special tooling to match end bells.
Bearing cap	Random orientation in cartons	Punch press section or plating section	Bowl feeder. Magazine (off-line loading). Manually loaded parts tray. Belt feeder.	
Felt washer	Random orientation in cartons or strips	Outside vendor	Bowl feeder. Magazine (off-line loading). Manually loaded parts tray. Special presentation equipment.	
Thrust washer	Random orientation in cartons	Punch press section	Magazine (off-line loaded). Bowl feeder. Manually loaded parts tray.	
Mounting ring	Random orientation in cartons	Outside vendor	Magazine. Bowl feeder. Manually loaded parts tray. Belt feeder.	

TABLE 3
Example of Parts Mating Requirements for Motor Assembly

Task Definition	Application	Difficulties	Mating Methods	Comments
I. Insertion				
A. Small diameters, close tolerances	Rotor shaft to bearings	Requires close alignment	1. Force feedback servoing. 2. Passive accommodation, R.C.C. in wrist. 3. Floating fixture 4. Force or position sensing.	R.C.C. wrist appears best with chamfer on parts. No's 1 and 4 appear to be too slow. Floating fixture may have to be lockable.
B. Small diameters, loose tolerances	Permawick retaining cage	None	Rigid.	Gripper location position should be consistent.
C. Large diameter, close tolerance	End bells to stator	Requires reasonable alignment	Component alignment guides on fixture. Part design (taper on end bells and stator).	Part design may be sufficient if locating accuracy is within ±.010.
D. Large diameter, loose tolerance	Baffle to end bell	None	Rigid.	Gripper locating position should be consistent.
E. Flexible parts	Wick, felt washer	Rigidity of parts, part gripping.	Part design (guides). Component alignment guides on gripper or fixture.	Felt washer gripper may present problems. Wick may require a combination of methods.

References p. 136

PILOT SYSTEM DESCRIPTION

The APAS pilot system configuration (Fig. 4) consists of an end bell subassembly center with two high speed, accurate, fixed sequence arms surrounded by parts magazines and two servo-controlled arms at an outlet station (Fig. 5) and a final assembly system (Fig. 6) which is a sequential, power and free, buffered transfer system with programmable stations.

Arms A1 and A2 compose an assembly center for end bells. End bell parts distributed about the periphery of arm A1 are brought to an assembly position at the point where the radii of arms A1 and A2 intersect. Arm A2 holds two end bells during the entire assembly sequence, picking two end bells from orienting escapement devices on the buffer conveyors. Vision is extensively used to inspect parts being picked by arms A1 and A2. The cameras are positioned on arms situated on a rotary post, and one microcomputer processes the video data from both. Vision is also used to orient end bells and inspect dust caps after they have been pressed onto the end bells. Force feedback is used at this assembly center to determine if the pressing of parts onto end bells has been performed correctly.

At the output of the end bell assembly center (Fig. 5), arm A2 transfers two end bells onto special fixtures which maintain the orientation of the end bells. Arms A3 and A4 add special devices, such as baffle plates, toggle switches, capacitors, and reset switches, to the end bells. Arm A4 does the majority of parts acquisition while arm A3 does screwing and fastening tasks. Besides being used to inspect parts, a vision system will properly orient toggle switches. An X-Y table is used at this station to move the end bell fixture into postion to assist arm A3 in performing its tasks. End bells are transferred onto work pallets by arm A4 and a special shuttle device transfers the pallet onto the power-free conveyor which feeds the motor assembly stations.

Arm A5 is the first station in final assembly (Fig. 6). Here the stator and rotor are placed on the bottom end bell and the top end bell is inserted onto the rotor and stator. Vision is used to determine the orientation of a stator in its parts presenter. Visual servoing is used to align the stator during insertion onto the bottom end bell. Passive compliance is used in a wrist-mounted Remote Center Compliance device for all insertion operations, and visual inspection is used on rotors.

Arm A6 inserts the bolts into the partially assembled motor, torques-down the tie bolts and inserts an oil flinger and dust cap onto the top end bell. Torque feedback is used at this station to inspect the motor after tie bolts have been torqued. Programmable parts presenters feed parts to this arm. Vision is used to check the alignment of tie bolt holes and to inspect parts.

Arm A7 inserts the assembled motor onto a frame. Visual servoing is used to mate the motor mounting rings with the supports of the frame. Once inserted onto the frame, it is tied down with spring straps. External capacitors and special switches may be added to the motor at this station. Inspection of all parts and orientation determination of the frame are performed visually. Completed motors are transferred onto a

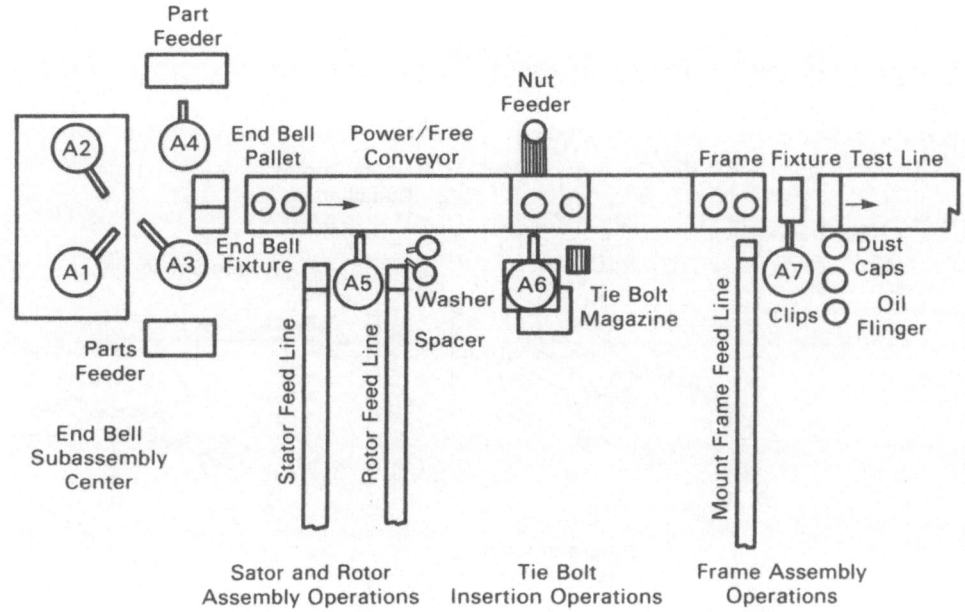

Fig. 4. Pilot line configuration.

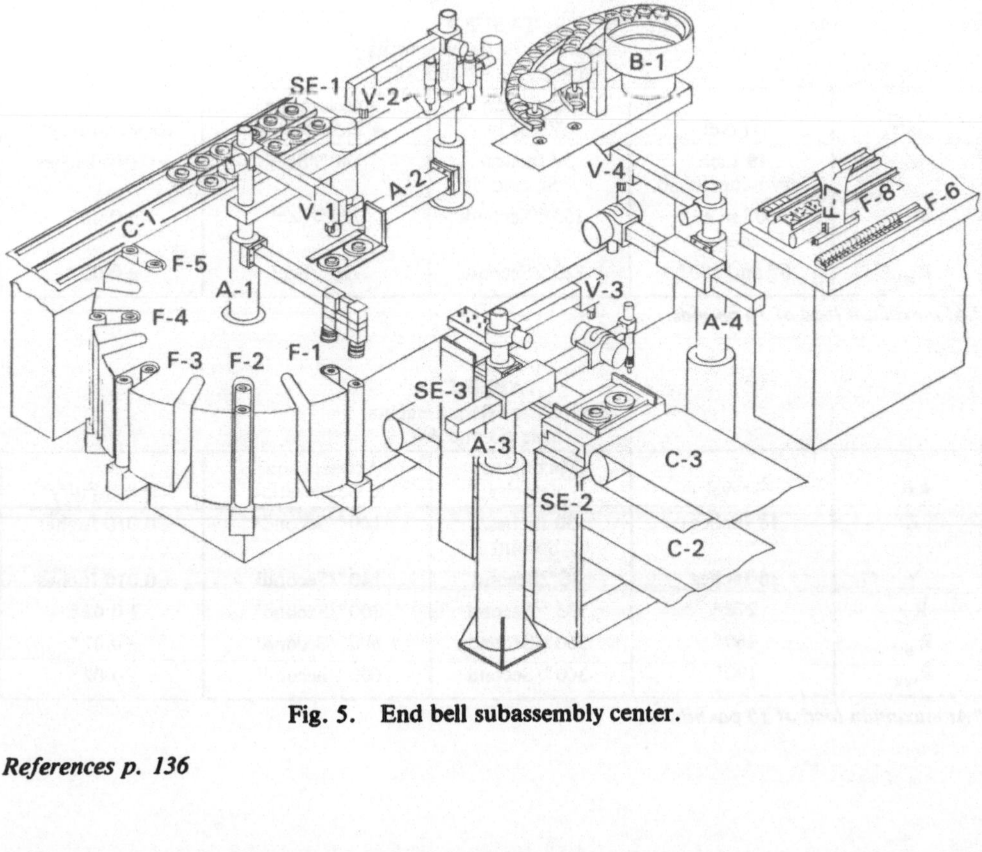

Fig. 5. End bell subassembly center.

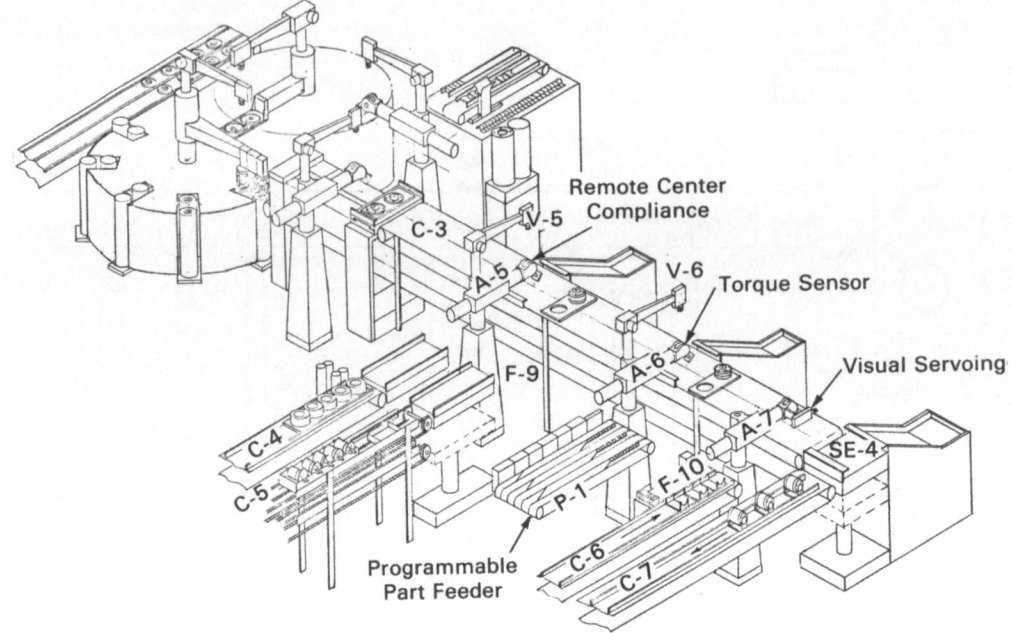

Fig. 6. Final assembly system.

TABLE 4
Manipulator Requirements
Arms A1, A2 and A6

Axis	Travel	Maximum* Velocity	Acceleration* & Deceleration	Repeatability*
Z	18 Inch Servo-controlled	30 Inches/Second	140"/Sec^2	±0.010 Inches
R_Z	360° in 45° Steps	360°/Second	720°/Sec^2	±0.02°
R_{wy}	0° and 180°	720°/Second	720°/Sec^2	±0.02°

*At maximum load of 10 pounds.

TABLE 5
Manipulator Requirements
Arms A3 and A4

Axis	Travel	Maximum* Velocity	Acceleration* & Deceleration	Repeatability*
Z	18 Inches	30 Inches/Second	140"/Second^2	±0.010 Inches
Y	36 Inches	36"/Second	140"/Second^2	±0.010 Inches
R_Z	270°	135°/Second	300°/Second^2	±0.02°
R_{wy}	360°	360°/Second	600°/Second^2	±0.02°
R_{wx}	190°	360°/Second	600°/Second^2	±0.02°

*At maximum load of 10 pounds.

buffer conveyor and work pallets are returned to the end bell subassembly station via a return conveyor.

In this system, human operators will be used for material handling, equipment operations, and maintenance. Material handlers will be used to load, off-line, magazines and pallets, changeover equipment, and for freeing jammed parts presenters. Operators will supervise the running of the system, load local computers with the correct path control and sensory system programs for the styles specified in a production schedule, and set-up experiments and equipment. Maintenance personnel will be responsible for maintaining the equipment based on a preventive maintenance schedule and for restoring equipment which fails during pilot production.

Manipulator Arms — The APAS pilot system contains seven manipulator arms. To overcome a technology void identified in the state-of-the art review work, namely the lack of high speed, low cost, extremely repeatable arms, we designed the modular arm shown in Fig. 7. Desired performance capabilities are presented in Tables 4 and 5 for a fixed sequence and a servo-controlled arm. Cost is minimized, since this concept enables one to use only the degrees of freedom required for the particular batch assembly application. Careful analysis of the assembly requirements of numerous product lines defined the degrees of freedom required as two linear axes Z and Y and four rotary axes R_Z, R_{WZ}, R_{WY} and R_{WX}.

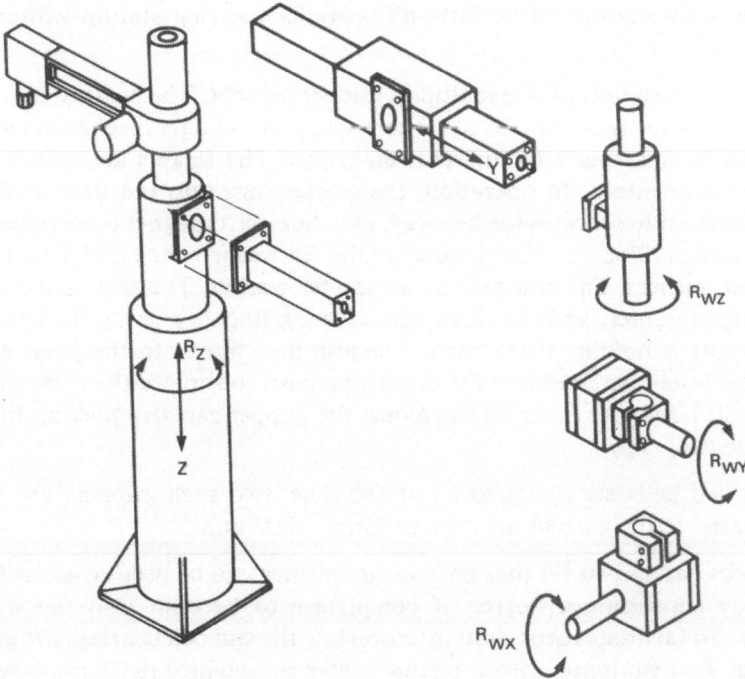

Fig. 7. Modular assembly arm.

References p. 136

The end bell subassembly arms A1 and A2 and one final assembly arm A6, requires only the Z, R_Z and R_{wy} axes. High speed and repeatability are achieved by using fixed sequence R_Z and R_{wy} axes. For these axes, the actuators work between mechanical stops. The R_Z travel is a full 360° in intervals of 45° to any of 8 fixed points. R_{wy} has two fixed points, 0° and 180.° The Z axis must be servo-controlled over 18 inches of travel in order to perform the assembly operations and accommodate differences in styles.

End bell assembly center arms A3 and A4 are five degree of freedom, servo-controlled arms with the desired performance capabilities given in Table 5. The requirements for final assembly arms A5 and A7 are quite similar except that the maximum load is 20 pounds and 40 pounds respectively.

The modular arm concept was developed to address a technology void, and the desired performance requirements were specified to clarify what is actually needed for batch assembly applications. To accomplish the pilot evaluation program, we may have to modify fixed sequence arms, such as those supplied by Autoplace, to achieve arms A1, A2 and A6 and other commercially available arms such as the PUMA arm, soon to be supplied by Unimation, for the remainder of the arms.

End Effectors and Tools — A completely universal and effector capable of performing all motor assembly operations is impractical. The approach selected for the pilot APAS system is to use seven end effectors or grippers (one for each of the arms) which are versatile enough to perform all operations at that station without a tool change.

Fig. 8 shows an example of one multipart gripper concept. The bearing cap, dust cap and rubber mount are picked up by a self-centering, three-finger mechanism which can accommodate circular parts of different diameters. The fingers are operated by an electrical rotary solenoid. In operation, the gripper dips into the steel washer parts presenter. The mandrel locates the key way, and the electromagnet is energized to pick up the steel washer. The arm then indexes to the felt washer stack and dips down into the pile of felt washers. The pins pick up the top felt washer. The arm then moves over to the bearing presenter, and the three self-centering fingers pick up the bearing cap. Now the gripper is holding three parts. The arm then moves to the press anvil, the electromagnet is de-energized and the depositing push rods place the three parts onto the press anvil. Using the three fingers alone, the gripper can also pick up the rubber mount on the dust cap.

Since two end bells are operated on at one time, two such grippers are mounted together, spaced at the correct distance to form a dual unit.

Draper Labs has shown [9] that passive compliance can be used to assist insertion operations by permitting a degree of compliance to be built into the wrist of a manipulator. To facilitate rotor shaft insertion into the end bell bearing, the gripper of arm A5 (Fig. 9) is equipped with a remote center compliance (RCC) unit and force sensor at the wrist joint, where all the service and electrical connections are made. In operation, this gripper picks and places a stator onto the top end bell using the pair of

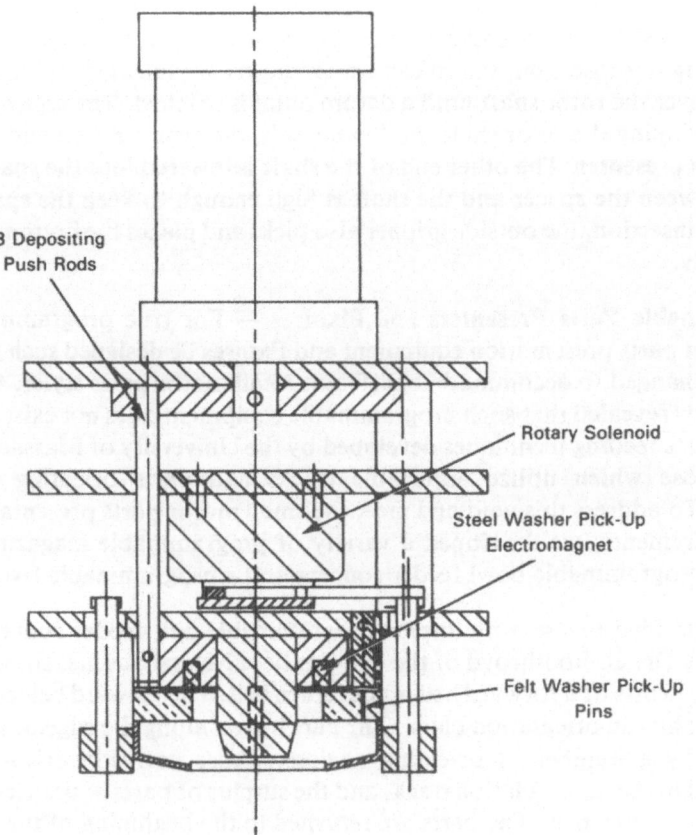

Fig. 8. Gripper for arm A1.

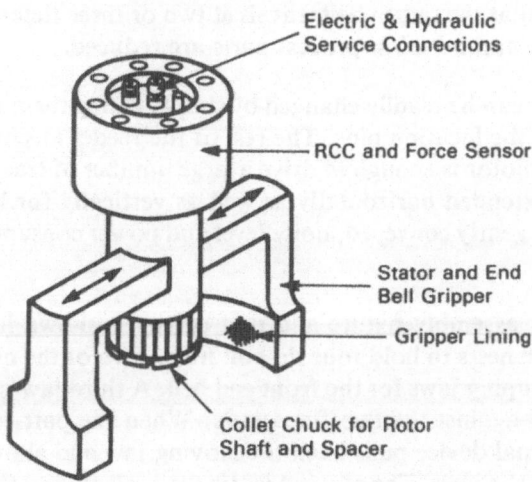

Fig. 9. Arm A5 end bell, rotor and stator assembly gripper.

References p. 136

outside grippers. Then, the collet chuck picks up a spacer and inserts it onto the rotor shaft. During this insertion, the collect chuck fingers are released and the collet chuck is inserted over the rotor shaft until a datum point is reached. The collet chuck is then actuated, gripping the rotor shaft. Following this, the arm is raised and moved to the spacer parts presenter. The other end of the shaft is lowered into the spacer hole. The friction between the spacer and the shaft is high enough to keep the spacer in place. After rotor insertion, the outside gripper also picks and places the bottom end bell into the assembly.

Programmable Parts Presenters and Fixtures — For true programmability, it is essential that parts presentation equipment and fixtures be designed such that they can be quickly changed to accommodate different small motor parts styles. Our state-of-the-art review revealed that such programmable equipment does not exist, but some of the small parts feeding techniques developed by the University of Massachusetts, particularly those which utilize adjustable tracks and parts orienting devices, are promising. To address this void and meet the small motor parts presentation and fixturing requirements, we developed a variety of programmable magazine designs, a high speed programmable bowl feeder concept and a programmable fixture.

We also decided to use a multi-part programmable belt feeder concept originally developed by Dr. G. Boothroyd of the University of Massachusetts, shown in Fig. 10. Basically it consists of a forward belt and a return belt. The forward belt conveys a part along until it hits an orientation blade. The part travels along the edge of the blade and is oriented by a number of orienting devices. Those correctly oriented parts are accumulated in the accumulation track, and the surplus of parts or unoriented parts tip over onto the return belt. The parts are returned to the beginning of the forward belt and another chance of tipping over occurs. At the beginning of the forward belt, a blade diverts the parts to the far side of the belt, so that they will have a fair chance of traveling through the entire length of the orientation blade. The pulley or gearing system is arranged such that the return belt travels at two or three times the rate of the forward belt, so that the number of in-process parts are reduced.

The orientation blades can be readily changed by simply lifting them up and placing another set of blades on the location pins. The rest of the feeder mechanisms remain unchanged. One geared motor is enough to drive a large number of tracks. This versatile belt feeder can be extended horizontally as well as vertically for handling more parts. Since the parts are gently conveyed, noise level and power consumption are very low.

A typical small motor assembly fixture and part carrier is shown in Fig. 11. The fixture has four magnetic nests to hold four tie-bolt nuts. Two of the nest blocks also serve as location and clamping jaws for the front end bell. A third jaw is spring loaded to push the top end bell against the two fixed jaws. When the part carrier is in the loading station, an external device pulls back the moving jaw and allows the top end bell to be deposited onto the jaws. The moving jaw is then released so that the end bell is clamped by the spring pressure. The bottom end bell is simply registered by four

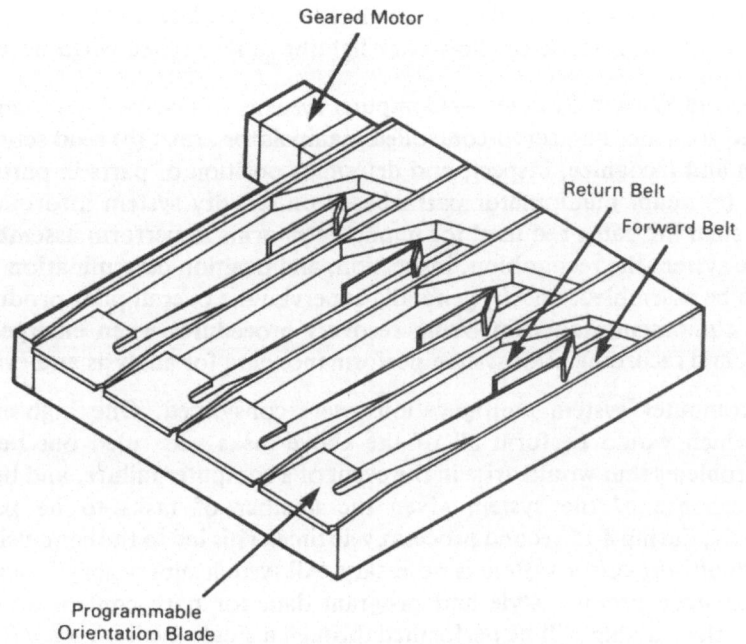

Fig. 10. Multitrack belt feeder.

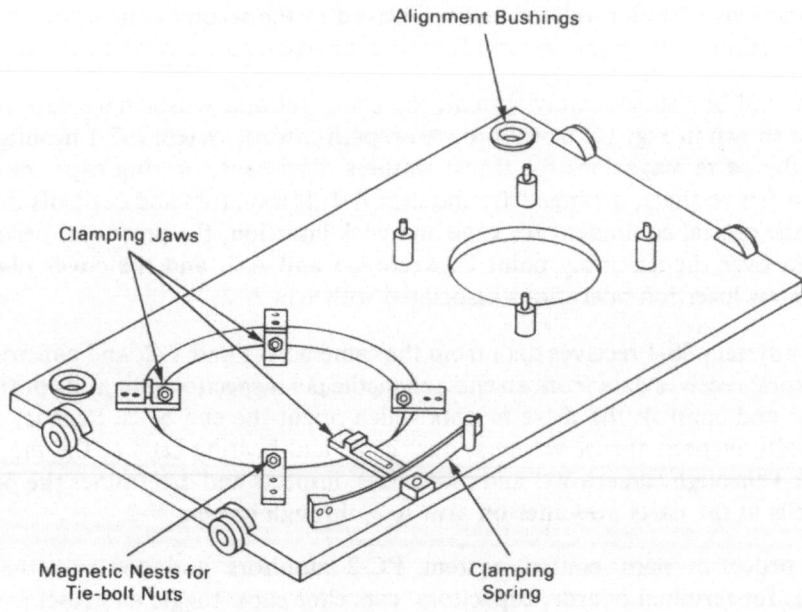

Fig. 11. Programmable fixture for end bells.

References p. 136

location pins. Gravity alone should be enough to keep it in place. Below the end bells, a large circular cut-out is made to allow back lighting to be applied when necessary.

Computer and Sensory Systems — Computer systems will be used to (a) control the path of fixed sequence and servo-controlled manipulator arms; (b) read sensor-based information and recognize, inspect, and determine position of parts in parts feeding equipment; (c) adapt manipulator paths based on sensory system information; (d) teach the system the paths required for manipulator arms to perform assembly tasks; (e) teach the system the recognition, inspection, and position determination tasks for each part to be assembled; and (f) generally supervise the overall pilot production by monitoring equipment status, invoking recovery procedures when emergency situations arise, and recording pilot system performance data for analysis and evaluation.

Several computer system configurations were considered. One high-end mini-computer which would perform all of the above tasks was ruled out because of reliability problems that would arise in the event of a computer failure, and because of speed of response of the system given the number of tasks to be performed asychronously, during a 15 second process cycle time. This led to the conclusion that a distributed multi-processor system is necessary. All system supervisory functions, the data base to store product style and program data for path control and sensory systems, and the training will be performed through a single medium scale (64K word of main memory) minicomputer.

Timing analysis during the studies on potential configurations to assemble motors indicated the need to overlap path control (with servoed arms) and, at least, the processing functions of vision-related tasks. This led to the second conclusion that path control functions and sensory system functions be separated in local microprocessors.

For the end bell subassembly system, the computer and sensor hardware requirements are shown in Fig. 12. Local processor/path control system PC-1 monitors and controls the parts magazines for thrust washers, dust caps, bearing caps, mounting rings, and felt washers, arranged around arm A-1. It monitors and controls the actuation of the special equipment for cage and wick insertion, the press and permawick equipment over the assembly point between A-1 and A-2, and the cover plate and ground screw insertion operations associated with arm A-2.

Sensory system SS-1 receives data from the cameras V-1 and V-2, and controls their drive motors; receives data from an end bell tactile pin inspection unit and controls it's actuators; and controls the drive motors which orient the end bells. Sensory system SS-1 visually inspects thrust washers, dust caps, and bearing caps at the magazines around A-1 through camera V-1 and recognizes, inspects and determines the position of end bells at the parts presenter on arm A-2, through camera V-2.

Local processor/path control system PC-2 monitors and controls the parts magazines for terminal boards, capacitors, capacitor caps, toggle and reset switches, and finger springs arranged around arm A-4. Hex nuts and washers for toggle switch fastening and screws for terminal boards are arranged around arm A-3 and are also

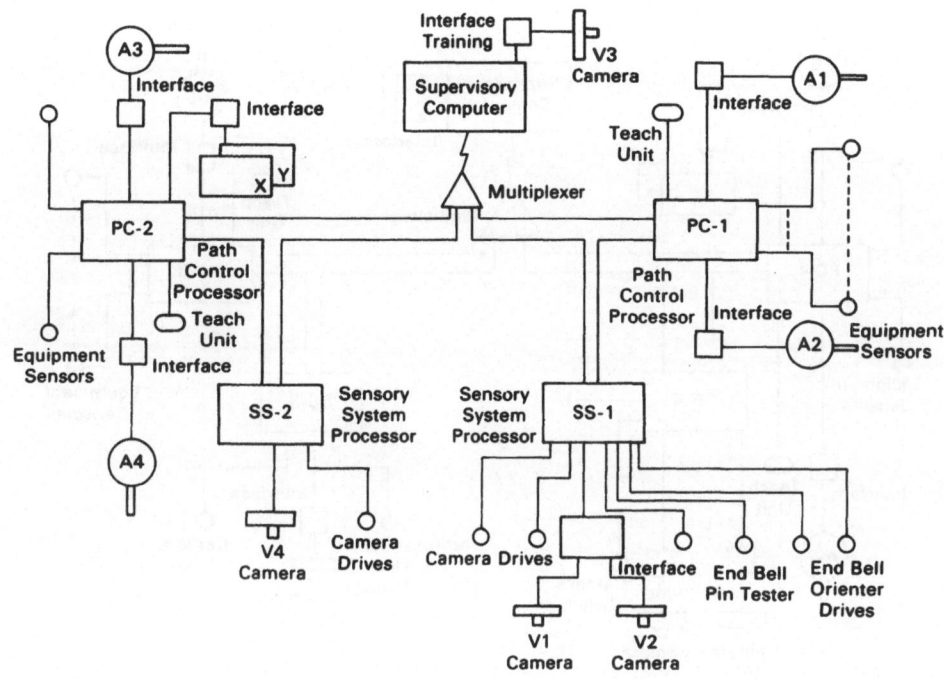

Fig. 12. End bell subassembly: Computer and sensor hardware system configuration.

controlled through PC-2. Finally, this processor controls the X-Y positioning table on which the top and bottom end bells reside. Sensory system SS-2 through camera V-4, inspects terminal boards, baffle plate, and finger springs and determines the position of the arm of the toggle switch.

Vision camera V-3 is used for visual-related training tasks. Path control training in the pilot system will be performed via the arm at the appropriate assembly station. In an actual production system, an additional modular arm would be used off-line to perform path control training.

Fig. 13 shows the computer and sensor hardware configuration for motor final assembly. Local supervisory/path control computer PC-3, in addition to controlling arm A-5, monitors and controls the positioning of the work pallet on the main transfer conveyor, the indexing of the chain conveyors presenting rotors and stators, and the spacer and washer parts presenters. Sensory system SS-13 performs visual inspection of the stator and rotor, visually determines the position of stator bolt-holes, and assists in visual servoing the top-end bell into the correct orientation on the rotor and stator. A force sensor is used to test the centrifugal switch on the rotor.

References p. 136

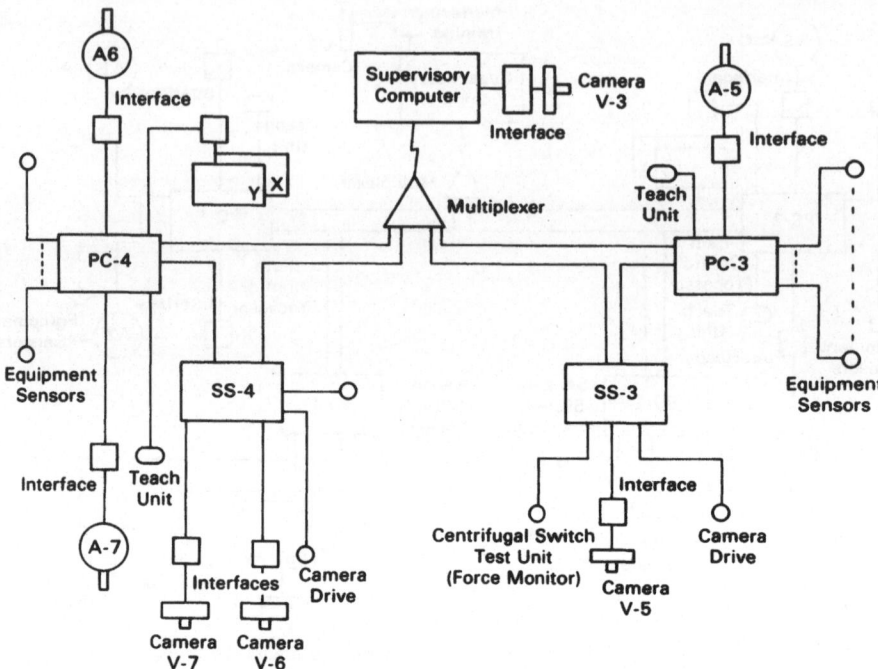

Fig. 13. Motor final assembly computer and sensor hardware system configuration.

Local supervisory/path control computer PC-4 controls arms A6 and A7 and an X-Y positioning base on which A-6 sits. For the A-6 station, the equipment controlled will be that for positioning the work pallet, indexing the tie-bolt magazine, the tie-bolt nut and washer magazines, and the nut-nest equipment. For arm A-7, the equipment controlled will consist of the mounting bracket parts feeder and fixture and the mounting clip magazine arrangement.

Sensory system SS-4 visually inspects the orientation of the motor assembly at the A-6 station for alignment of tie-bolt holes through camera V-6. Camera V-7 at station A-7 will be used in visual servoing of the motor onto the mounting frame. A torque monitor on the wrist of arm A-6 will be used to inspect the rotor for correct rotation before and after tie-bolt insertion operation.

Fig. 14 shows the hardware requirements for binary visual information processing. A 200 × 200 matrix camera is used as the sensing element. For several of the pilot line assembly stations, this camera will be mounted on a fixed arm attached to a rotary post. Parts feeders will be arranged around the circumference of circle swept out by the camera. Back-lighting and reflected lighting will be used to determine the state of parts.

The threshold logic unit receives analog signals from the camera and converts them to binary signals for input to the microprocessor. The threshold voltage for binary

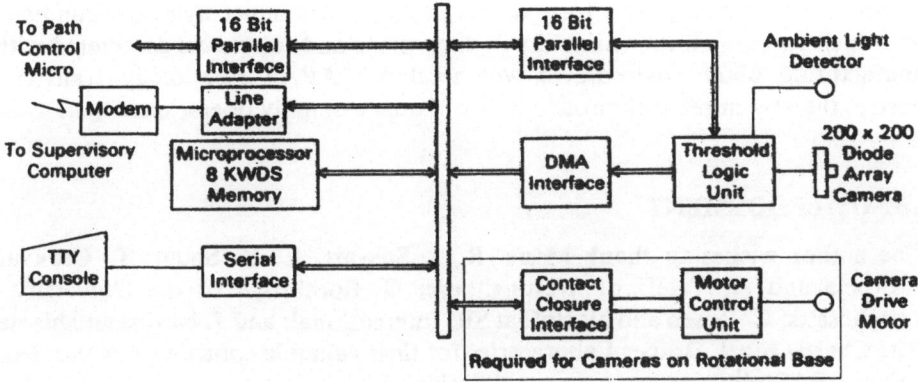

Fig. 14. General vision system hardware configuration.

conversion is supplied to the unit by the microprocessor through a parallel interface. Where several parts are being viewed by one camera, one threshold value will be required per part. The microprocessor will supply the appropriate value to the unit when the camera has been moved over a part.

The microprocessor performs, in software, the connectivity analysis of binary images input via the DMA channel, calculates the features required to determine the state of a part, compares the calculated feature values with a model of expected part feature values, and makes an inspection, recognition, or position determination decision based on this comparison of features. The control of any reject mechanisms on parts feeders will be performed by this processor. Positioning of the rotary post on which the camera sits is also controlled by the processor.

When used to provide part position information to manipulator arm processors, or where the manipulator arm will eject rejected parts, appropriate information will be sent to the arm computer. Any position information transmitted will be in reference to the field of view of the camera; the arm computer will translate this information into offsets referenced to the arms' position in space.

CONCLUSION

It has been shown that the technology base exists for successful design of a complete integrated adaptable programmable assembly system. Novel modular manipulator arms, microcomputer-based binary image processing vision systems, programmable parts presenters and fixtures, and end effectors that are universal for the set of operations at individual stations have been conceived. We are in a position to proceed with pilot system construction and software development and then commence a proof-of-concept experimental program to objectively evaluate technical and economic performance of APAS. A successful cost-effective demonstration that the system can

References p. 136

accomplish timely changeover for a variety of part and product styles and can handle parts which contain all the variability in dimensions and tolerances associated with a manufacturing plant environment will accelerate APAS technology transfer to industry, thereby improving productivity of batch assembly operations.

ACKNOWLEDGEMENT

The author wishes to thank Mssrs. R. J. Stewart, L. Y. Shum, T. Csakvary, J. Korpela and A. Taleff of Westinghouse; G. Boothroyd of the University of Massachusetts; C. Rosen and his staff at SRI International; and J. Nevins and his staff at The Charles Stark Draper Laboratories for their valuable contributions that led to the final APAS pilot system conceptual design.

REFERENCES

1. Abraham, R. G., Stewart, R. J. S., and Shum, L. Y. State-of-the-art in adaptable-programmable assembly systems. Soc. Manuf. Eng. Rep. MSR 77-16, 1977.
2. Abraham, R. G., Stewart, R. J. S., and Shum, L. Y. *State-of-the-art in adaptable-programmable assembly systems.* International Fluidics Services, Ltd., Bedford, Eng. 1977.
3. Abraham, R. G., Stewart, R. J. S., and Shum, L. Y. State-of-the-art in adaptable-programmable assembly systems. Soc. Manuf. Eng. Tech. Paper MS 77-757, 1977.
4. Abraham, R. G. Programmable automation of batch assembly operations. *The Industrial Robot* 4,3 (1977), 119-131.
5. Abraham, R. G., et al. 3rd bi-monthy report - programmable assembly research technology transfer to industry. Westinghouse R&D Rep. 77-6G1-APPAAS-R2, March 31, 1977.
6. Abraham, R. G., et al. 4th bi-monthly report - programmable assembly research technology transfer to industry. Westinghouse R&D Rep. 77-6G1-APAAS-R4, June 10, 1977.
7. Abraham, R. G., et al. 5th bi-monthly report - programmable assembly research technology transfer to industry. Westinghouse R&D Rep. 77-6G1-APAAS-R5, August 25, 1977.
8. Abraham, R. G., et al. Final report - programmable assembly research technology transfer to industry. Westinghouse R&D Rep. 77-6G1-APAAS-R6, October 31, 1977.
9. Drake, S. G. Using compliance in lieu of sensory feedback for automatic assembly. D.Sc. Th., MIT, the Charles Stark Draper Lab. Rep. T-657, Sep., 1977.

DISCUSSION

J. Albus *(National Bureau of Standards)*

What is the status of your manipulator arm?

Abraham

We don't intend on building this particular modular arm concept. That would divert too many funds away from putting together the complete pilot system and conducting the evaluation. Our intent in designing that manipulator arm was to encourage supplier firms to do something about that particular void. What we plan to do in the next phase is to take the PUMA arm and utilize it for some of the final assembly stations and we will modify something like the Autoplace arm to get the kind of performance that we specified for the fixed sequence arms. We feel the key is more of the systems integration and demonstration of the economics and technology at this point in time. We are going to accelerate the transfer.

E. Fredkin *(Massachusetts Institute of Technology)*

How did you determine the speed and capacity of the microcomputer?

Abraham

You may have noticed the difference in the two slides. One processor had 8K of memory and the other had 16K. What we've actually done is to design the overall software system requirements and looked into storage as well, and distributed the various tasks, including supervisory, to the micros.

Fredkin

But you don't actually have a running program?

Abraham

We are trying to stress the utilization of these simpler image processing methods and go with the simple algorithms because we are concerned with execution times. We have a PDP 11/03 in our lab now and some of these basic routines that I've mentioned here have been programmed. Before the end of the year, on a limited set of small motor parts, we will be conducting experiments to verify the lighting and the execution times before we go ahead with the full pilot system.

C. A. Rosen *(SRI International)*

The hardest thing those 11/03's have to do is the image processing. We have done a lot of that work at SRI and have put very sophisticated programs into an LSI/11.

J. K. Dixon *(Naval Research Laboratory)*

When does your schedule call for General Electric to begin selling motors that have been robot assembled?

Abraham

Anyone who worked fairly close with us during our first phase of effort saw that we were quite open not only with the results of our effort but also with much of the data. Two smaller motor manufacturing firms have corresponded with us and asked for the detailed NSF reports, and we provided them to these firms. Frankly, we feel that in this particular field a lot of cooperation is required if we are going to effect the technology transfer. So I'd rather cooperate and be successful in this transfer. I do think that the team that does the work does get some natural advantage in lead time in making any kind of future production systems.

J. L. Mundy *(General Electric R&D Center)*

I'm not connected with Abraham's group so I can be objective with the following question. I was wondering what you saw as the greatest technological advantage of your modular arm configuration over purchasing axes individually from a vendor?

Abraham

If a firm has robots with modular degrees of freedom, which when put together would meet the kinds of performance requirements specified here, then we would certainly use them. We're not interested in building such a robot.

D. E. Whitney *(Charles Stark Draper Laboratory)*

Are you planning to assemble any motors that are of the sleeve bearing type? And if so, when you do the torque test and the motor does not spin freely, which I understand does happen sometimes, what are you planning to do?

Abraham

At this point we haven't gotten into the detailed design considerations. I can't answer your question now.

J. McCarthy *(Stanford University)*

What is the state of commitment to actually do this project?

Abraham

There was a National Science Foundation Board meeting on Friday (September 22, 1978), which is the final step in the whole process, and I have no indication as yet as to what the outcome is. We would expect to know one way or the other shortly *(NSF approved the project)*.

M. L. Minsky *(Massachusetts Institute of Technology)*

Why do you want to make these modular arms? What fraction of the system cost would it save, or how much more would it cost if all the arms had six degrees of freedom? The reason I ask is that this doesn't seem to me like a very programmable system. It can make a certain variety of rather concentric things and I admit there are a million motors made every now and then.

But I noticed all sorts of little things like when you change over jobs you're going to put little blades of various kinds on feeders and someone has to do that. One extra degree of freedom, however, might be a tenth of one percent of the cost of the system and for all I know the robot could reach out and perform the operation. So you've made your system very unprogrammable because if there is a slight change in that job that doesn't resemble one of your six types of motors, then you're down.

Abraham

No, I would disagree on the aspect of programability.

Minsky

What fraction of the cost would it be? It would be a very small part of the system.

Abraham

The robot arms themselves are a small part of the system. At the time we did the work most of the arms were costing from $50,000 - $60,000 and didn't meet the repeatability requirements. Our attempt was to reduce some of the mechanical costs because the electronic costs had been going down dramatically all the time. We estimate the cost for each degree of freedom to be $5,000.

Minsky

But if the concept is successful the arm costs should come down also.

Abraham

You had two points to your question. One related to the degree of programability and the other with the arms themselves. In the case of the motor application there are considerable variations. As the Draper Laboratories work has shown, there are many other products with z axis assembly requirements. If we can handle all of these assembly operations that we've defined for small motors, then the class of other products for which this approach would apply represents a significant step on the learning curve. So we have a number of generic types of inspection and assembly operations already addressed that will handle many other product lines.

The issue of change over of blades in feeder mechanisms and the requirement for a man to do that is a practical consideration at this point in time. Based on what we feel the state of the art is, if a man can get in there and quickly change the blades and if that wins out in a tradeoff analysis over having some robot arm do it, then we're going to use the man. The key is the economics in the tradeoff analysis.

Minsky

But it sounds as if you already did that analysis.

Abraham

Yes we did, and this is the result of the analysis that I've described in terms of all of these concepts.

J. F. Engelberger *(Unimation)*

I want to put Marvin's (Minsky) mind to rest that this question is going to be answered. Because there will at least be one approach that will use only sophisticated arms. So we will know as soon as NSF blesses you and Unimation blesses us.

PUMA: PROGRAMMABLE UNIVERSAL MACHINE FOR ASSEMBLY

R. C. BEECHER

General Motors Manufacturing Development, Warren, Michigan

ABSTRACT

The General Motors PUMA development is an approach to the mechanization of relatively low volume automobile sub-assemblies. PUMA is a combination of robots, parts feeders, special purpose machines, and humans working in concert on an indexing conveyor or turntable.

DEFINITION

PUMA is a Programmable Universal Machine for Assembly, planned as a combination of robots, fixed automation, transfer devices, parts feeders, and critically important human beings. It may be constructed as a straight indexing line, a rotary index, or any configuration which is convenient and applicable to a given product family and/or location (Fig. 1).

HISTORY

PUMA was conceived in mid-1975. For some time, General Motors Manufacturing Development had been considering the use of robots in assembly. When one speaks of assembly in the automotive business, one thinks of engines, wheels, hoods, fenders, auto bodies, gas tanks, seats, all of which are relatively large components. Consideration was given to the assembly of automobiles, but the final assembly of an automobile presented some very formidable challenges. The assembly of large component parts on continuously moving lines, complex model mixes, a multitude of part numbers, and highly complex assembly tasks are examples of these challenges. A systematic review of all of auto related General Motors products revealed that over 90% of the parts in an automobile weigh less than 1.4 kilograms, or three pounds.

Fig. 1. PUMA robot inserting light bulbs in an instrument cluster.

Cataloging and analyzing the tasks required to assemble these smaller components made assembly by robots appear feasible.

The concept of a stand-alone robot which would assemble a complete product from start to finish was rejected for several reasons. In many applications, the supplying of parts to the machine would be extremely complicated, and in order for such a costly machine to be feasible, high production speeds would be needed. Also, because only one robot would do the complete assembly, downtime would cause lost production and standby units or manual assembly lines, needed to take up the slack, would waste floor space.

Robots were considered which would perform assigned, repetitive tasks, much as humans do. After analysis of robots with varying degrees of freedom, a robot with five degrees of freedom was found to be satisfactory. Since all robots would be the same, they would be interchangeable, and could be applied to a wide variety of tasks, as dictated by demand.

There were tasks of which robots were not, at that time, capable. Some assembly operations required sight, some very subtle tactile feedback and, worse yet, some required the use of judgment. It was felt that waiting for vision and tactile systems to be perfected and made economical would not make assembly with robots feasible for many years. The decision was made to retain humans in the proposed system where they could make the greatest contribution, and to act as supervisors for the machines. Where special-purpose machines could perform a task better or cheaper than robots, they would be included. The combination becomes a "Programmable Universal Machine for Assembly" — a PUMA. It is envisioned as a system which will be made up of humans and machines, each doing what it does best. If one of the robots breaks down, it could be replaced, very quickly, by another robot or human operator. As planned, a system of this type could be implemented on a production floor with a minimum of disruption.

The program for each robot will be carefully planned and recorded on a magnetic disc in a methods type laboratory. In this way, each robot will use the most efficient method to perform its task. Because Manufacturing Development was interested in completing this development as soon as possible and keep everything as simple as possible, it was decided to use state-of-the-art technology. The robot was to be capable of adaptive control when it became available, but first applications would utilize dead reckoning and accurate fixturing.

With these basic findings and some preliminary cost studies, it was proposed to buy, build, or have built, a prototype manipulator. With this prototype, the concepts could be proved and feasibility demonstrated. Also, performance specifications for a production model could be determined.

Two robots were purchased which would suit our developmental needs, the ASEA 6 kilogram model and a Unimate 5000. Within a short time, demonstrations were available which proved that assembly with robots was indeed feasible.

Working with a wide variety of products, Manufacturing Development drew up detailed performance specifications for the "ideal" assembly robot (see Appendix). These specifications were then sent for bid to a number of potential of suppliers. Unimation, Incorporated was the successful bidder. The first prototype robot was delivered to General Motors on May 1, 1978.

CURRENT STATUS

Since May 1 extensive repeatability and durability tests have been made, and planning for the first installation of the production models was begun.

In order to prove the reliability and durability of the robot, a variety of installations is planned. The first will be an assembly application. There will also be some stand-alone installations where the robot will be loading and unloading other machines.

The most difficult assembly problem continues to be parts feeding and orienting. Parts feeding and orienting is still something of a "black art". Whether or not programmable assembly devices succeed could very well be determined by the success or failure of parts feeding and orienting mechanisms.

An assembly system that consists of robots and human beings can be changed over from assembling one product to an entirely different product in a matter of a few minutes. However, it may take several hours to retool a line with a type of dedicated parts feeding and orienting mechanisms with which we are familiar today.

When using any machine that has no adaptive capability, it is essential that parts be presented to that machine in precisely the correct location and orientation.

The initial PUMA applications will use dead reckoning and accurate fixturing. After these are successfully underway, attention will be increasingly directed toward refinement of the system and the addition of adaptive control.

The robot is now capable of accepting adaptive control. Tactile and visual feedback, acoustics, lasers, and a variety of proximity sensors are being considered.

CONCLUSION

Potential applications for the PUMA system cover a broad spectrum of products. PUMA can be used to assemble such components as instrument clusters, heater and air conditioning controls, wheel brake cylinders, power seat transmissions, and many more.

Each new application brings with it a unique combination of challenges. PUMA is not the answer for all assembly problems, any more than any other assembly system. However, it does have its place as a highly flexible, economical system for the assembly of untold thousands of products.

APPENDIX

FUNCTIONAL PERFORMANCE REQUIREMENTS

PUMA SYSTEM

General—The PUMA system is a programmable machine for light assembly, utilizing easily programmable manipulators with an indexing conveyor system.

The manipulator is to be used to assemble small, lightweight, (2.5 kg. max.) components.

The manipulator will perform its tasks sequentially with other manipulators, humans and/or fixed machinery. It will have the capability to use various small tools such as power screw drivers, nut runners, etc.

The PUMA system as presently conceived consists of:

- Manipulators and controls
- Indexing conveying system
- Parts feeders and orienters

The reference to a conveying and parts feeding system is merely to provide an overall description of the PUMA system.

The manipulator and controls portion of PUMA will consist of the following:

1. *Mechanical System*—Includes the manipulators and transmission which converts the motion of the prime mover into the required action.

2. *Drive and Measuring System*—Includes the method of driving the manipulator arm plus a system for determining and regulating the arm's position.

3. *Control System*—Functions to store information (memory) and act on input, output and interlock signals from peripheral equipment to control the entire system. It also functions to control the drive and measuring system of the manipulator. The teach unit is considered to be an integral part of the control system.

MANIPULATOR CONSTRUCTION REQUIREMENTS

MECHANICAL SYSTEM

Configuration—The manipulator is to be small and lightweight, but rugged enough to withstand the rigors of an automotive production environment.

The basic dimensions shown in Fig. 2 should not be exceeded. The arm configuration may be either articulating or cylindrical coordinate type. The primary considerations given to the arm design shall be reliability, ease of maintenance, and cost. Five degrees of freedom are the minimum required.

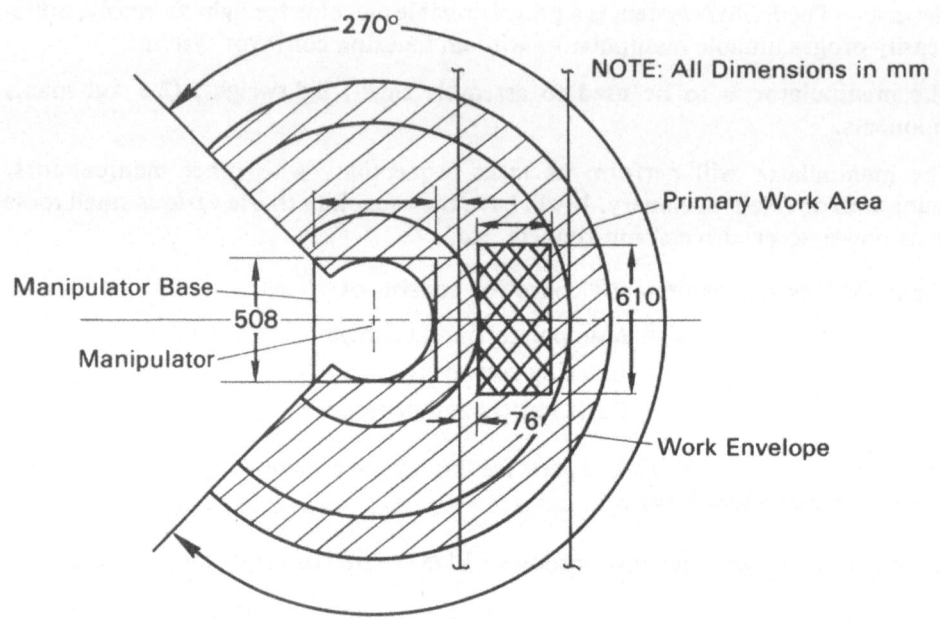
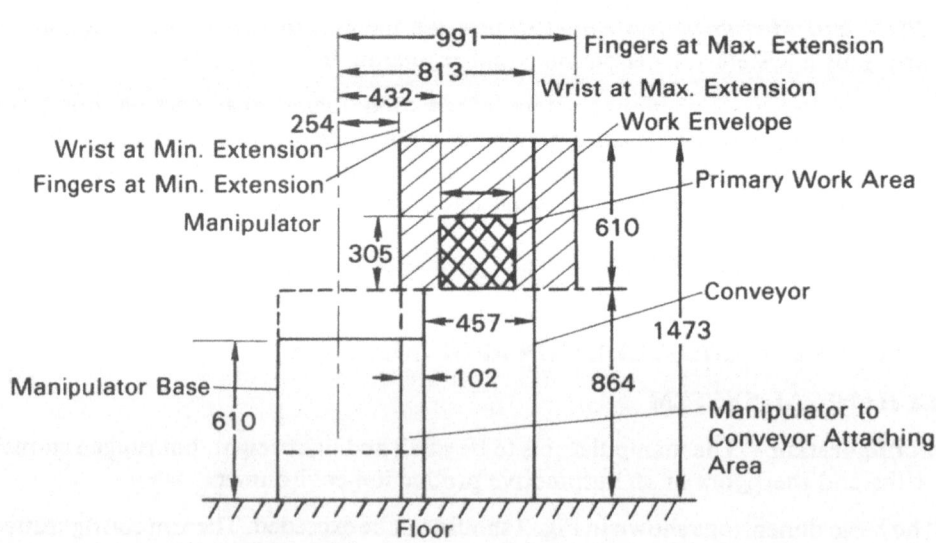

Fig. 2. PUMA manipulator specifications.

Capabilities

Axis	Range	Slew Rate
Arm — horizontal travel	See Fig. 1 (559 mm)	1000 mm/sec.
Arm — vertical travel	See Fig. 1 (610 mm)	500 mm/sec.
Arm — swing	270°	135°/sec.
Wrist — bend	180°	360°/sec.
Wrist — rotate	360°	360°/sec.

Note: The acceleration and deceleration of the tool and/or part center should be at least 1G with transient motion critically damped.

Repeatability — Repeatability requires is ± .10 mm at the end point with all five degrees of freedom having been exercised through their maximum range at maximum speed and carrying the maximum weight.

Mounting — The manipulator must have provisions for mounting to the side of a conveying system and capable of being readily removed and/or repositioned. When repositioned to the conveying system the manipulator must have a provision for easily and accurately locating the unit to its previous position (shot pins, fixed keeper, etc.).

Weight Capacity — The total weight handling capacity of the manipulator (including gripper, fingers, tools or parts) should be a minimum of 3.5 kg.

DRIVE AND MEASURING SYSTEM

Drive System — The preferred drive system for each major axis of movement is electric motor.

The drive system must provide for as fast an acceleration and deceleration time as is practical. The system must also be capable of applying the force (approximately 6 kg.) necessary to use a small power screw driver for driving self tapping screws into a light gauge sheet metal. The applied force must be within the design limits to prevent damaging the manipulator (see collision or impact under Control System.)

It is preferred that the drive system be integral with the manipulator base however, it may be remotely located (approximately 8 meters).

Measuring System — The measuring system for determining and regulating the arm's position can be optical, electrical, mechanical or any combination required to maintain accuracy and reliability.

CONTROL SYSTEM

Control Computer — The control system shall be computer based and capable of operating in a production environment. Each manipulator in the system must have its own controls and run independently of the others. It is preferred that the control computer be integral with the manipulator, however, it may be remotely located (approximately 8 meters).

Memory — Memory storage capability for four 32-step programs must be resident in each manipulator.

Repeater Function — The control computer must be capable of conditional branching (repeating portions of the program).

Input-Output — Each manipulator's control system must accept a minimum of six optically isolated 110V-AC and six TTL inputs and outputs from peripheral equipment.

Collision Protection — In the event of collision or impact, the manipulator shall be protected to minimize damage to the unit or other entity. A method shall also be provided to stop the conveyor or other equipment.

Power Failure — In the event of power failure, the manipulator arm must be arrested or controlled rather than free falling onto the conveyor or peripheral equipment. Provision must be made to protect the integrity of a working program and/or stored information in the event of brief power outages.

TEACHING

General — As envisioned, a manipulator identical and interchangeable with the production units will be used at a remote location for programming. The programming or teach unit must be small, lightweight and portable, capable of being used with any of the manipulators in the system. Cost, ease of use and effectiveness should be considered in the design of the unit. Some methods of teaching to be considered, but not excluding others, are joy stick, lead through and push button.

The taught program will then be transferred to the production units, which will repeat the transferred program.

Production Units — Provision should be made to allow minor adjustments and/or corrections to be made on the production floor to the taught program by *authorized personnel only*.

Teaching Requirements — Software, teach console and displays may be integral with the control computer but the functions and devices may also be part of a remote or portable programming support system. In any case the programming system must provide the following capabilities:
- Program storage and loading (tape cassette preferred)
- Selectable point to point or continuous path control with speed control and accurate positioning.
- A correction function to allow alteration, cancellation and/or additions to an existing portion of a program.

- The ability to select a speed slower than the normal operating speed for each axis.
- Movement of the manipulator in predetermined minute incremental distances along each axis.
- Addition or deletion of instructions between steps already taught without re-entering or reteaching the original program.

DISCUSSION

D. Nitzan *(SRI International)*

Could you elaborate more on your study as to how long you thought adaptive control would be applicable to your system. You implied it would take generations.

Beecher

When we first got into this, which goes back to 1975, we were relative neophytes at this whole business. In our opinion, adaptive control such as tactile and visual sensing, was a long way off. It has come on much faster than we had thought. However, we have a problem we must recognize, that of acceptance within our divisions. They are real world people who aren't sure the robot is a good thing in the first place, much less one that can sit there and do all these great things, such as feeling around and seeing. So we're in no hurry to apply adaptive control on the production floor. We want to make sure that this thing proves out first. In the meantime we'll be spending a lot of time with adaptive control in the lab. It's not a long way off. It's here today but it has not yet gained acceptance in our situation.

C. A. Rosen *(SRI International)*

Do you have any compliance at all in the wrist?

Beecher

No.

Rosen

You assume dead reckoning to be sufficiently accurate?

Beecher

That's right. We are however looking at the Remote Compliance Center (RCC) and have one in the laboratory.

H. Freeman *(Rennselaer Polytechnic Institute)*

In the application that you showed it appears that the tolerance wasn't very great. Now you just mentioned that there is no compliance. Are there any tight tolerance uses, such as fitting a bolt very precisely, into a hole?

Beecher

"Very precisely" is a relative thing. We are talking here about repeatability of roughly 0.002 inches in this prototype system. The vendor is still doing a lot of work on it. I used the expression, "initially we are trying to pick the cherries." We are not going after those applications that are going to stretch this to its limits. Where we do have precise fits, obviously we would be looking at something like the RCC.

D. L. Waltz *(University of Illinois)*

Do you have any specifications for how long these arms are to operate without requiring servicing?

Beecher

Forever. If we could get the same reliability out of this arm that we see in today's production arms we would be very pleased.

J. A. Feldman *(University of Rochester)*

Could you outline for us what you set out as specifications and what fundamental things you felt you had to have out of the PUMA arm.

Beecher

Fundamentally, we were talking about something that was roughly the size of the human arm, because we wanted to put it into the same size work station that a human being would operate in. We wanted it to be able to move no faster than human beings were capable of, but in fact it does, and we like that safety margin. We wanted it to be able to handle up to seven pounds including the tool. The PUMA meets these requirements. Essentially, we talked about something that would operate on a product smaller than a breadbox. We analyzed, as many people have, the basic assembly tasks, and just committed them to writing. A copy of the specification will appear in the proceedings.

Nitzan

Continuing along the philosophy that you need a general purpose manipulator for a variety of tasks, why didn't you go all the way and chose six degrees of freedom?

Beecher

We just couldn't justify the sixth degree.

Nitzan

On what basis?

Beecher

We analyzed the assembly tasks that were required.

Nitzan

But in the future you might change your mind.

Beecher

Oh yes, a lot of things might happen. I know that Unimation is working on six degrees. So we feel safe from that standpoint. As far as we were concerned we could see no immediate need or foreseeable need for the sixth degree. I know that sounds like heresy to some people.

G. J. VanderBrug *(National Bureau of Standards)*

You spoke of using this in a system with people. What are your immediate plans for safety?

Beecher

We have had a project underway for a couple of years to make people safe around these things. And we're concerned not only with the PUMA but with all other robots. We've looked at a number of different devices for stopping the machine dead in it's tracks if it gets too close to a human being. We don't have the answer to that yet. In the meantime, safety of the first production installation will be up to the particular division. They may very well build a cage around the thing.

R. Davis *(Chesebrough-Ponds)*

Have you established a firm target date for implementation in the factory?

Beecher

I personally have. We're shooting for the first device operating on the production floor in 1979.

M. L. Minsky *(Massachusetts Institute of Technology)*

In order to put the light bulb into the hole you have to be able to get to the hole from a certain direction and then rotate. You can get away with that if you pick up the light bulb in the right orientation, but it seems to me that in general you need a sixth degree of freedom. Didn't you say some of those holes were at different angles?

Beecher

No. What I meant was that the bayonet feature was located at a different angle. The key slot was at a different angle, not the bulb. The bulbs all came in straight up and down.

J. F. Engelberger *(Unimation)*

For those who are worried about the elegance factor on the sixth degree, the last revolute articulation in the PUMA has room in the arm and it is designed to carry six. It will be available with six.

J. L. Nevins *(Charles Stark Draper Laboratories)*

It is worth pointing out, parenthetically, that the task illustrated is a four degree of freedom task.

Minsky

If the bulbs are parallel when you start.

Nevins

They were. If they were not then a fifth degree of freedom would be used to pick them off a slanted tray.

Freeman

Were all the bulbs aligned by the bayonet base as they were fed to the robot?

Beecher

Yes.

PROGRAMMABLE ASSEMBLY SYSTEM

M. SALMON and A. d'AURIA*

Olivetti, Torino, Italy

ABSTRACT

The Olivetti SIGMA robot is reviewed and its industrial applications are listed. Timing, tooling, and economic data are given. Both positive and negative aspects of the robot approach are examined and compared with those of alternative solutions. The primary conclusion drawn is that lack of flexible feeding devices causes the main limits of programmable assembly. Future activities will concentrate on upgrading present robot programmable assembly systems to incorporate flexible feeding systems, active sensors and universal grippers for greater flexibility in batch production.

INTRODUCTION

In order to automate its assembly lines, Olivetti began a project to develop an automatic assembly programmable system. In 1974, the project proceeded to a point which enabled Olivetti to construct the "SIGMA" Robot (Fig. 1) and utilize it in the Olivetti assembly lines.

The basic machine is fitted with two arms each of which has three degrees of freedom, straight movements X, Y, Z controlled by a single control unit consisting of a specific electronic hardware and a minicomputer. The modularity of the design makes it easy to produce more complex systems with up to 8 degrees of freedom or more simple systems with only two degree of freedom in a single arm. The machine is constructed of welded sheet metal and consists of two guide rails along which move one to four trolleys each holding an arm, similar to overhead traveling cranes.

In this way, up to four arms share the guide rail of the longitudinal axis known as the X axis. Each of the overhead cranes has an X-shaped part which runs along the trolley itself (direction Y). A tubular element which forms the arm itself runs in the X-shaped

*Presented by Enzo Lamonaca
References p. 163

Fig. 1. SIGMA robot with two arms.

guide (direction Z). The volume served by the two arms is therefore a parallelepiped whose horizontal dimensions are the same for all the machines of the series. A maximum linear dimension of 30 cm. is normally available for each arm.

The basic structure can be altered either by removing or adding parts in order to meet the specific requirements of the various machines. Thus, in some cases, only one arm is fitted and in other cases, the Z movement is completely eliminated. Each arm may have more than one gripper to allow it to simultaneously perform the same operation on several parts.

PROGRAMMABLE ASSEMBLY SYSTEM

The SIGMA can assemble in several directions but, for this feature, additional rotations are required on the arms. Since a maximum number of 8 controlled axes can be handled, when 4 arms (at maximum configuration) are utilized only 2 axes per arm can be controlled.

This robot has been described many times in detail and shown at many exhibitions (Milan, Paris, Hannover, Moscow, Zurich, Stuttgart). The main specifications of the SIGMA are listed in Table 1. SIGMA is an acronym for: SISTEMA INTEGRATO GENERICO MANIPOLAZIONE AUTOMATICA i.e., GENERIC INTEGRATED SYSTEM FOR AUTOMATIC HANDLING. After two years of using the SIGMA at Olivetti, it was decided to market the Robot outside Olivetti. Since January 1977 a new subsidiary: OSAI - Olivetti Sistemi Automazione Industriale, has been established for producing and selling the SIGMA.

TABLE 1
SIGMA Robot Specification

Overall dimensions	1950 × 1200 × 2600 mm.
Working area	1010 × 400 × 400 mm.
Total Weight	1500 kg.
Maximum Payload	10 kg.
Number of arms	1 to 4
Number of controlled degree of freedom	Up to 8 step motors
Arm speed	30 m/second
Force feedback: Maximum detectable force Lowest detectable force Force discrimination	2 kg. 50 gr. 16 gr.
Position accuracy	0.05 mm.
Data input through:	Joy stick for position data and/or TTY for SIGLA language instructions

Principle SIGLA Language Functions
- Edit the program
- Motor control
- Force reading
- Branch on force or logical conditions
- Move until touch
- Change of arm movement
- Parallel arm programming
- Reference system positioning

References p. 163

So far, nearly 50 systems are in operation of which 20 are at Olivetti and 30 are in Italy and abroad (France, West Germany, East Germany, USSR, U.K., Spain, Japan). This paper describes the experience gained by Olivetti in this field.

APPLICATION

The application fields covered by the SIGMA Robots differ very much from each other. Some examples are shown in Figs. 2 and 3. Listed in Table 2 are the specifications of the main applications which also shows the main limits and features resulting from the assembly works carried out on the SIGMA.

The number of different parts to be assembled by one robot is limited by the space available for the feeders and, mainly, by the need of specific part-oriented grippers. Normally, it is impossible to grip different parts with the same gripper. Therefore, even if more than one gripper is mounted on each arm, the total number of parts that can be assembled simultaneously cannot be higher than 10.

The number of equal parts to be assembled is limited only by the memory which can be extended indefinitely (the maximum number of points the memory can store is about 5,000). Therefore, this does not constitute a limitation to the practical applications. The number of different parts with equal shapes, however is limited by the feed mechanism of the parts to be assembled. This is a typical case in electronic assembly.

Although practically an unlimited number of permutations on the same part can be performed, in practice, 3 to 6 different permutations is all that can be handled. Two installations have a dozen different configurations. These special cases correspond to families of products of similar shape (for instance, groups of pistons, rods, pins) in which the relevant tools and grippers are the same for all different part diameters and lengths. Obviously, the number of different configurations is unlimited when the components are similar.

TABLE 2
Major SIGMA Applications

	Customer	Type of Business	Name of Product	Number of Arms	Number of Differ. Parts	Total Number of Parts	Time to Produce 1 Group	Time to Assemble 1 Part
1.	Bassani	Electr. Switch.	Domestic Socket	3	5	6	3″9	0″66
2.	Elsag	N.C. Machines	IC. Board	2	46	100	200″	2″
3.	Necchi	Sewing Mach.	Compressor Group	2	3	3	4″	1″33
4.	Piaggio	Motor Vehicles	Dumper Valve	2	9	9	15″	1″5
5.	Alfa Romeo	Vehicles	Gear Box	2	3	15	60″	4″
6.	Petercen	Electr. Supplier	Relay Box	3	4	10	4″5	0″45
7.	Riello	Thermotechnic	Cable Box	2	4	55	36″	0″65
8.	Copreci 1°	Electr. Supplier	Safety Group	3	6	6	3″	0″5
9.	Copreci 2°	Electr. Supplier	Therm. Group	2	8	10	6″	0″6
10.	Arman	Vehicles Supp.	Windscreen Wiper (6 Types)	3	7	5	6″	0″85
11.	Paris Rhone	Vehicles Supp.	Coupling	3	4	5	5″	1″

PROGRAMMABLE ASSEMBLY SYSTEM

Fig. 2. Keyboard assembly.

Fig. 3. Two arms operating on a casting.

References p. 163

The SIGMA operating speed depends on the system configuration (number of arms) and on the use of multifinger grippers. Fig. 4 shows how several hands are mounted on the same arm to improve operating speed. The "basic" time is the time needed for an elementary "pick and place" cycle; this time is about 3 sec. Therefore, systems having 3 arms and 2 grippers per arm all performing the same tasks, are capable of handling one part an average of every 0.5 seconds and therefore of assembling a group consisting of 6 parts takes 3 seconds.

It is difficult to describe in general terms throughput comparisions between the SIGMA and alternative solutions, because of the multiple arm - multiple grippers per arm design of the SIGMA robot. However, some comparison concepts have been ascertained as follows:

Fig. 4. Multiple grippers on two arms operate on five parts at the same time to increase thruput.

SIGMA vs. Man: The SIGMA productivity is about 8-10 times the manual one. Indeed, the manual-assembling speed is, in general, about 4-5 sec/part. In practice, due to the peculiarities of each product, it is difficult to always have well-balanced systems and, as reported in Table 2, the actual operating speeds are lower than the one indicated above and overall results show the ratio of robot speed to man speed to fall between 4 and 6 to 1.

SIGMA vs. Automatic Machines: In respect to the automatic machines, not only speed but also quality and reliability have to be taken into account. In general, automatic machines perform much shorter cycle times and reach a greater productivity, but the lack of flexibility is one of the major drawbacks the SIGMA should overcome.

SIGMA'S STRONG POINTS

The above-figures can only give a quick outlook on the SIGMA optimal utilization field. Indeed, the innovative characteristics of this system allow a qualitative new utilization as compared to the conventional, manual and mechanical methods. Besides the flexibility, which is limited, the main feature of the grippers is the presence of force sensors, continuously controlling the correct performance of the assembly job, as shown in Fig. 5. This ability to detect and report failures offers a significant improvement in the assembly reliability. The SIGMA is therefore able to work with defective parts resulting from errors in earlier assembly steps, thus avoiding breakdowns or production stops which usually occur with automatic machines.

Also included among the SIGMA strong points is the system "logic" response, that is, the system capability of performing measurements and checks during the assembly operation. This is shown in Fig. 6 and outlined in Fig. 7 wherein, besides assembling, the system also ensures a constant clearance between rod, pin and piston.

SIGMA'S WEAK POINTS

In addition to the above-mentioned limits, the main difficulties in using the industrial robot in the assembly work originate from the lack of tooling flexibility and constraints imposed by the traditional production environment.

So far, in our SIGMA, there is a limited capability of operating on different groups with the same tooling and, in practice, assembly works are only possible with a limited variability of dimensions within the same family. For instance, in the case outlined in Fig. 7, a dozen different rod-piston groups can be assembled, since they are nearly equally shaped and the same gripper can handle pistons of any diameter. On the other hand, it is not possible to assemble, with the same tooling, different groups: the feeders, the grippers, and the tools are in fact studied and designed to a specific application and therefore it is necessary to completely change them to change production.

This change is technically feasible and can take place fast enough to meet the production requirements, but it is not economically feasible due to the extra cost of retooling. Indeed, to date, according to our experience, the specific tooling of a SIGMA

References p. 163

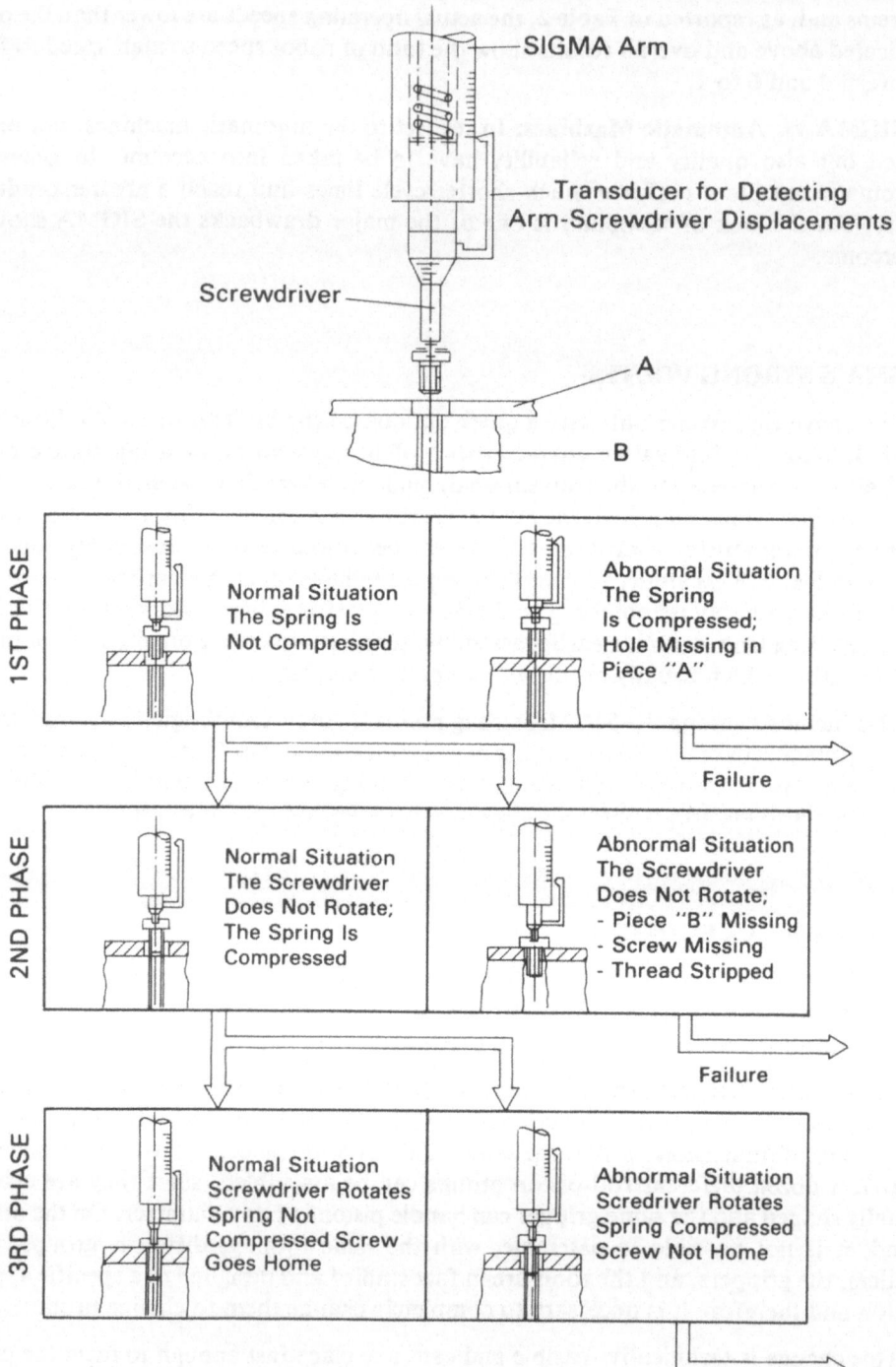

Fig. 5. Tranducer detects failures in assembly by detecting pressures on spring in the gripper.

PROGRAMMABLE ASSEMBLY SYSTEM

Fig. 6. SIGMA robot with checking fixture.

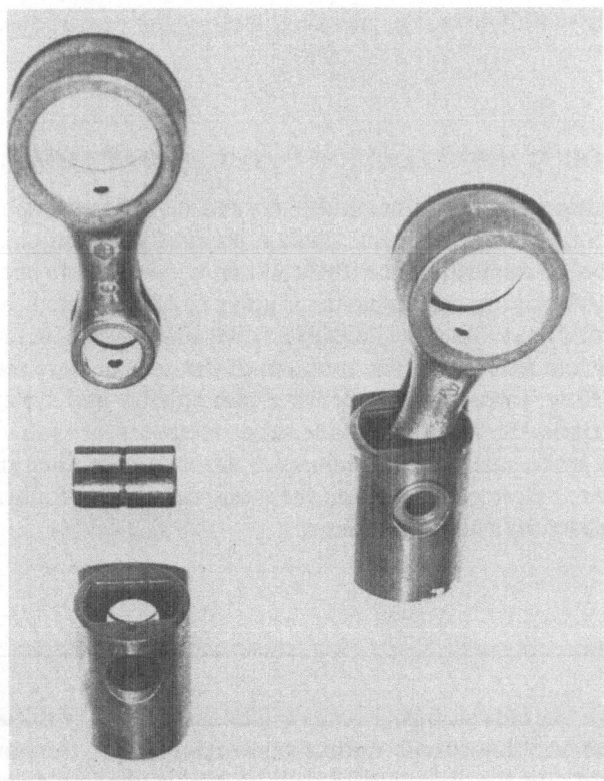

Fig. 7. A robot can assemble up to one dozen different connecting rod-piston groups since they are equally shaped and the gripper can handle pistons of any diameter.

References p. 163

amounts to about 30%-50% of the total cost: retooling therefore correspondingly increases the total cost and delays the pay-back time.

The automatic gripper-changer, now a SIGMA optional, is just an aid and not a final remedy for the solution of flexibility problems, since the extra costs for feeders and tooling still exist. This lack of flexibility limits an industrial robot to parts in a single group or family of groups, thereby excluding their applications from batch manufacturing, seasonal manufacturing and, in general, from low production volumes.

Not withstanding these problems, the industrial robot offers considerable progress over the conventional assembling system. As a matter of fact:

- Complete re-tooling of the system is possible at fractional times and costs in respect to the conventional "hardware" automation, wherein no recovery at all is usually possible.
- It is possible to work on families of products provided that a morphologic similarity exists among the parts.
- The "logic" capabilities of the computerized industrial robot allow a real quality jump in the assembly-gauging, checking, and setting operations.

INDUSTRIAL ROBOT INSTALLATION IN THE MANUFACTURING PROCESS

So far, the attention of the Robot researchers and manufacturers has been concentrated in implementing the robot itself. Now it seems from many indications (IBM, Vicarm, Unimation, in addition to the SIGMA) that a new line of robot models, better matching the manufacturing requirements, is going to be marketed. Nevertheless, the industrial problems of installing the industrial robot in the manufacturing process has not received sufficient attention. The problem of the proper part feeding, materials and information flow, inter-relations between part quality and system productivity, the roles, the functions, the work load of the robot operators, are some of the big problems to be solved when installing an industrial robot in a production environment. These are prbblems that, in practice, limit the industrial robot productivity in a worse way than those caused by robot limitations.

CONCLUSIONS

From the above considerations, it can be concluded that robitized assembly is now a concrete fact and that robots outline a new approach in the world of industrial assembly. But, until now the objective for real flexible system implementation has failed. By comparison, it can be said that a greater flexibility has certainly been reached in the field of NC machine tools.

To overcome this limit, the R & D directions to be followed are preferably:

- Flexible part feeding systems and use of active sensors,
- Universal gripper systems capable of handling different parts,
- Assembly layouts which minimize the change of grippers and/or the specific feeder components.

These are not easy since, so far, and no solution has been taken on a scientific or mathematical basis, but only on an experimental cut-and-try basis which is seldom successful. The industrial robot installation in the manufacturing process presents a big gap between the few theoretical studies and the industrial world. The introduction of automatic "intelligent" machines requires a complete revision of production organization and attention from all management levels.

Today programmable assembly is an industrial reality and, in spite of considerable limits, its use is profitable. The electronic technologies, the active sensors, and mainly a scientific study of production methods can and must lead to great improvements and allow industrial robot assembly tasks to attain levels of utilization much greater than those presently available.

REFERENCES

1. d'Auria, A., Salmon, M. An integrated general purpose system for automatic automation. Proc. 5th Int. Symp. on Industrial Robots. (1975), 185-202.
2. Salmon, M. Applazione di un robot di montaggio ad una produzione con elevato volume orario. Ing. Meccanica (1976).
3. Salmon, M., and Buronzo, A. SIGMA: un sistema di automazione multiscopeo XXIII Rassegna Int. Flett. (1976).
4. Holligum, J. No hands make lighter work of your assembly. *The Engineer* 242, 6264 (1976), 34-35.
5. Salmon, M. Advance in automatic assembly: the SIGMA system. TM International, (1977).
6. Salmon, M. SIGLA - the Olivetti SIGMA robot programming language. Proc. 8th Int. Symp. on Industrial Robots. (1978).

DISCUSSION

C. A. Rosen *(SRI International)*

Regarding the assembly job you showed with the switch box, how long would it take to make the special fixtures, handling tools, and parts presenters if you started today?

Lamonaca

It is a matter of months. Three to five months plus debugging. A minimum of six months for a very easy job.

Rosen

That's a crucial place to try to cut down.

Lamonaca

I agree.

R. L. Paul *(Purdue University)*

The robot cannot do the job without a lot of special tooling. So one has to spend a great deal of money and engineering to put the tooling in to get this dramatic productivity increase. If you put that much effort and tooling into the people, what would the productivity increase be if you had people doing the work with this new higher level of tooling and work organization?

Lamonaca

It is a question more of reducing the cost of robots, of tooling, and reducing lead time for the production of tooling.

R. G. Abraham *(Westinghouse R & D Center)*

In the Return On Investment (ROI) analysis, you not only consider the total system cost, I'll say system instead of robot because of all these components we talked about earlier, but you have to work in other expense, including the tooling design. Then when you get into the savings portion of the equation you make a comparison of both direct and indirect operational and labor costs with the previous system. There is much more to ROI analysis than quoting productivity figures.

Paul

I suspect the robot has nothing to do with the productivity increase.

P. M. Will *(IBM Research Laboratories)*

That's the important point. I've never seen anything published on precisely that point.

B. Chern *(National Science Foundation)*

When you redesign the product you save the most. And in most cases you can not justify automatic assembly.

J. F. Engelberger *(Unimation)*

I think there are a number of things that don't lend themselves too readily to hard automation. One of the problems we have here is we are looking at little parts that one could have done with hard automation. That is, pancake assembly. But there are many things that we have looked at in automotive plants that just couldn't be done with hard automation because the part sizes vary too much, the quantities are too small, there is necessity for hand-to-hand coordination and so on. So what we do is look at the individual, and at least a time study, and we have a standard to work with. Then we say, all right, take that man out of the loop and put robots into his place. This is a critical issue. If you don't get the speed out of this machine your economics are finished.

J. A. Feldman *(University of Rochester)*

Continuing on Lou's (Richard L. Paul) point, what happens is like new math. We saw good results when there were highly motivated people involved. But ten years later when people aren't much interested, payoff decreases. I think we are seeing the same thing here.

G. G. Dodd *(General Motors Research Laboratories)*

Along the same line, at Stanford University there work in the medical field, in the Mycin project, to put into a computer system the information about infectious diseases and how to diagnose them. I've been told by the people working on this project that the great contribution has not been the artificial intelligence techniques but it is finally getting the medical doctors to sit down and categorize how they diagnose infectious diseases. They are now getting more requests for that catalog of information than they are for the program.

Rosen

But suppose we do make a robot assembly system and in the process we develop a beautiful set of jigs and what-nots which people can then use. The point is, something is being done. I think one of the results of Abraham's study is that we now know how to make motors a lot better than before.

Chern

A philosophical point that may be useful. There is always an effort to bring in the idea that somehow robots are useful, and can be used in industry to improve productivity. What worries me is that there are two approaches to robot utilization. If you want to use very interesting complex six or seven degree of freedom arms, then you look for a way to use them. That is one approach. But if we look from the other end of the spectrum and say, "I really want you to design systems for doing things," then somehow there must be a relationship to fundamental operations.

R. R. Bajcsy *(University of Pennsylvania)*

There might be some jobs that nobody indeed wants to do. I can give you a long story of a working woman who has had a heart transplant and needs a good house robot.

J. Albus *(National Bureau of Standards)*

This whole subject of productivity depends upon what the cost of labor is.

B. H. McCormick *(University of Illinois - Chicago Circle)*

Maybe we have many of the wrong ideas of what we call assembly.

Lamonaca

It is true that the productivity of the robot depends upon the hourly costs, of course. But the costs you have for producing robots and for using robots may differ. You also have to consider that some jobs are uninteresting.

R. B. McGhee *(Ohio State University)*

I'm surprised nobody has mentioned product quality. I thought that this was one of the main reasons for replacing man on assembly lines with robots, so that cars built on Mondays are the same as those produced during the rest of the week.

SESSION III
FUTURE VISION SYSTEMS

Session Chairman
H. G. BARROW

SRI International
Menlo Park, California

COMPUTER ARCHITECTURES FOR VISION

D. R. REDDY AND R. W. HON

Carnegie-Mellon University, Pittsburgh, Pennsylvania

ABSTRACT

It is estimated that systems capable of executing 1 to 100 billion operations per second will be required to achieve near-real time performance in image analysis tasks. In this paper, we present three aspects of architecture for realizing economical solutions to this problem. At the Processor-Memory-Switch (PMS) level, we consider parallel and pipeline architectures using multiprocessor and array processor systems. At the Instruction-Set-Processor (ISP) level, we consider the firmware implementation of special instruction sets for image processing. At the processor design level, we consider the design of complex Arithmetic-Logic-Units (ALU) for execution of complex image operations. Given these alternative aspects, we present hardware and software design considerations affecting the overall architecture. Together, they appear to hold the promise of several orders of magnitude improvement in speed over general purpose systems.

INTRODUCTION

Computational problems that arise in vision may have many special properties that are difficult to take advantage of using general-purpose processors. As a result, it has been uneconomical to undertake studies in vision involving large and complex images. Several speakers at this symposium have repeated the theme that "vision is hard." How hard is it really? What factors contribute to the difficulties? How can we overcome these difficulties? The answers to these and other such questions are difficult to come by. There have been almost no quantitative measurements on how slow and expensive "vision" really is. In this paper we will attempt to quantify the computational requirements of some typical vision problems and show how different

References pp. 184-185

computer architectures might help in alleviating the apparently insurmountable computational complexity of the vision problems.

The bane of computer vision is human vision. Humans appear to perform visual perception tasks so easily and effortlessly that we expect instantaneous, flawless performance from a computer vision system. This requirement, in turn, leads to memory capacity, bandwith, and processing requirements that are beyond the capabilities of most general-purpose computer systems available today. For example, a real-time video interface to a simple black and white TV camera generates about 10 million pixels of data every second — a rate of 60 to 80 million bits per second. Most current general-purpose systems cannot accept such high data rates nor do they have the primary memory capacity to store the digitized images. Even if we solve these problems of memory capacity and bandwith, there still remains the vexing problem of how to process data which is arriving at a rate of a pixel every 100 nanoseconds. To give you some idea of how much data that is, consider taking all the numbers that arrive in a second, placing them on a conveyor belt one after another, having a robot pick one up every second, working non-stop, seven days a week, 24 hours a day. It would take the robot approximately four months to complete the task.

Other problems that make vision a challenging and difficult area of Artificial Intelligence (A.I.) arise because noise, error, and uncertainty pervade all levels of the image interpretation process. In unoriented images, shadows, highlights, and occlusions may cause the expected features of objects to be missing. Variability due to noise and object characteristics leads to errors in detection. Incomplete and/or inaccurate representation of knowledge sources leads to more errors. Thus, vision systems must accept the inevitability of errors and handle them in a graceful manner.

The acquisition and representation of (knowledge about) models of objects is particularly difficult in vision systems. Consider, for example, the problem of representing a model of a steering knuckle in a vision system in such a way that it can be used for matching and recognizing objects on a conveyor belt. A human faced with a similar problem appears to perform the task of "model acquisition" of objects with little or no effort and does so in no more than a few hundred milliseconds of total observation time. Unfortunately, we have neither the concepts nor the vocabulary of representation needed to facilitate similar model acquisition by machines. This problem is ubiquitous at all levels of the subconscious, informal, commmon-sense problem solving activity of humans while performing sensory and motor tasks. Unlike algorithmic knowledge (as in Gaussian elimination or payroll computation) or formal knowledge (e.g. knowledge acquired from books on molecular biology or physics), effective use of informal knowledge raises severe problems in knowledge identification, acquisition, representation, and utilization. Each of these problem areas has implications to AI beyond the vision problem which is of immediate interest to us.

How hard is the vision problem? How much computational power is required to interpret a typical image? Preliminary estimates of required processing power seem to range from 1 to 100 billion instructions per second. The specific requirement will, of course, depend on several factors: the image size, the response time, and the computational complexity of the operations to be performed.

Estimating the computational complexity of image interpretation algorithms is difficult because there are no well understood algorithms for performing the entire task. However, given that several well understood components of low-level vision such as smoothing, differencing, histogram calculation, correlation, color mapping, and registration do exist, one can attempt to establish some lower bounds for the required computational power.

Establishing a lower bound for the computation required, even for a simple operation such as differencing (commonly used in edge detection algorithms), raises difficulties. Consider, for example, the problem of accessing the density (gray scale) value of a pixel in an image. Normally images are represented as arrays and one, naturally, assumes that a pixel value can be accessed in 4 to 6 operations (or instructions executed) on a general-purpose computer. This simplistic assumption, however, ignores the fact that most real images are too large to fit in primary memory. So, some form of virtual addressing and paging from secondary memory is required. For instance, on a PDP-11 processor, one of our algorithms executes 24 instructions to access the value of a pixel in a subimage paged from the disk. This is for accessing a pixel from a page already in core and does not include times for page-fault handling or disk accessing. Thus, it appears that a differencing operation using a 3×3 subimage would require 200 to 300 operations per pixel (possibly more in high level language implementation).

Extracting all the low level features of the image required for subsequent processing requires several thousand operations per pixel. Ohlander's [13] segmentation program needed in excess of 10,000 operations per pixel. Careful algorithm analysis plus the use of planning and abstraction reduces this significantly [15]. But given that this covers only a subset of the low level operations eventually required in a total integrated system, it appears that a computational effort in the range of 1,000 to 10,000 operations per pixel is not an unrealistic estimate.

The Landsat earth observation satellite generates about 200 images (approximately 10^{11} bits) per day using the multispectral scanner (MSS). Each MSS scene, covering 185KM x 185KM, contains about 8 million pixels (an approximately 2500 x 3300 array). If we assume 10,000 operations per pixel and desire a response time of less than a second, then we require computational power in excess of 80 billion instructions per second (BIP) to interpret these images. If the response time can be slower, say about two minutes, than a one BIP machine would be adequate. Even so, the fact remains that a 1-BIP processor is still three orders or magnitude faster than commercially available systems. Does this mean that the vision problem is too hard to be attempted at present or is there some hope?

Reddy and Newell [17] show that speed-ups of several orders of magnitude are possible in AI systems provided certain conditions are satisfied. If a system consists of a hierarchy of levels, where the levels are relatively independent, then improvements

References pp. 184-185

at various levels combine multiplicatively resulting in a large total speed-up. In AI systems usually there are many levels of potential optimization, for example, device technology, computer architecture, software architecture, system organization, algorithm analysis, knowledge sources, and heuristics.

Table 1 shows the estimates by Reddy and Newell [17] of the potential for improved performance achievable at various levels. Note that by far the largest improvement is expected in the areas of technology and architecture. We group these two levels together because they are not entirely independent and because the present trends in very large scale integrated circuit technology will permit new and interesting architectural organizations which would be radically different from the conventional Von Neumann machine.

TABLE I
Expected Speedup Due to Optimizations at Various Levels.
Improvements at these Levels Combine Multiplicatively.

Speedup	Level
10^3-10^5	Architecture and Technology
2-10	Software
2-10	System Organization
10-100	Algorithm Analysis
10-1000	Knowledge Sources
10-1000	Heuristics
2-5	Program Optimization

In the remainder of this paper we will examine various aspects of the design space of computer architecture and consider task-oriented computer architecture concepts that might lead to significant improvement in performance. The three main aspects we will consider are: custom processing elements for vision, tailored instruction sets for vision applications, and parallel computer architectures. The last of these three is perhaps the most widely considered alternative in systems to date, but we believe that the other two are equally promising.

CUSTOM PROCESSING ELEMENTS FOR VISION

As we progress from the era of large scale integrated circuits (LSI) to the era of very large scale integrated circuits (VLSI) with an accompanying 100 fold increase in the number of transistors per chip, we can expect significant advances in the types of computational elements that will be economically feasible. To capitalize on this opportunity, we need to examine the underlying concepts that have been used in the past to perform various arithmetic and logical functions. Conventional minicomputer systems usually use a single ALU (arithmetic-logical unit) to perform simple arithmetic operations on scalers. In larger systems the scalers can be in fixed or floating-point representation. Very few other types of ALU's have been considered except for a few ultra-large (and expensive) computers such as the STAR-100 [16]. At the same time, designers of special purpose systems have seen the need for implementing special ALU's tailored task, for example, a butterfly multiply ALU found in FFT processors. In this section, we will examine various attempts at custom ALU's in three broad categories: algorithm driven architectures, memory driven architectures, and specialized functional units.

Algorithm Driven Architectures — Given the need for systems that can perform image and other signal processing functions inexpensively, some system designers have proceeded to develop fast, low-cost processing elements which can be added to a general-purpose system. These add-on processors capitalize on the fact that their designs need not deal with issues prompted by a need for I/O controllers, priority interrupts, large addresses, wide data paths, and flexible instruction sets. Instead they concentrate on the fast execution of arithmetic functions under local program control. Such processor architectures can be said to be "algorithm (or task) driven". Two examples of successful architectures of this type are the Floating Point Systems FP-120B processor, widely used in signal processing applications, and the Control Data Corporation Flexible Processor.

A more recent example of this type of architecture is SPARC [7], a system being developed jointly by CDC and CMU. Fig. 1 shows a functional diagram of SPARC. SPARC is intended for use as an auxiliary processor to a host computer, probably a conventional minicomputer. A high speed memory holds instructions which are to be executed by a set of functional units; each instruction may initiate an operation on four different units. Operands are routed to and results taken from the functional units by a crossbar switch whose interconnections are controlled by a field in the instruction. High throughput is achieved by pipelining operations, for example an inner product can be calculated by fetching operands from the memory data memories, multiplying the individual terms and passing them to an adder which accumulates the sum. Each functional unit is pipelined internally so that operations may be initiated every clock cycle (20 ns).

The instruction memory is large enough (4K words) to allow complex operations to be performed. By retaining the flexibility of a programmable machine while providing a high processing rate, SPARC bridges the gap between general-purpose machines and special purpose hardware devices.

The key feature of most algorithm driven architectures is that they relieve the user of the "designer's prerogative." Rather than having to put up with whatever operations the designer has chosen to provide, the user can define his own operations. Although these add-on processors are like other general-purpose processors (without the fancy I/O structure, etc.) their primary use is as flexible, programmable ALU's for fast execution of complex operations.

Memory Driven Architectures — There are many operations in image processing where the main bottleneck is not the speed with which some operation can be performed but rather the speed with which data can be accessed from memory. Hence

References pp. 184-185

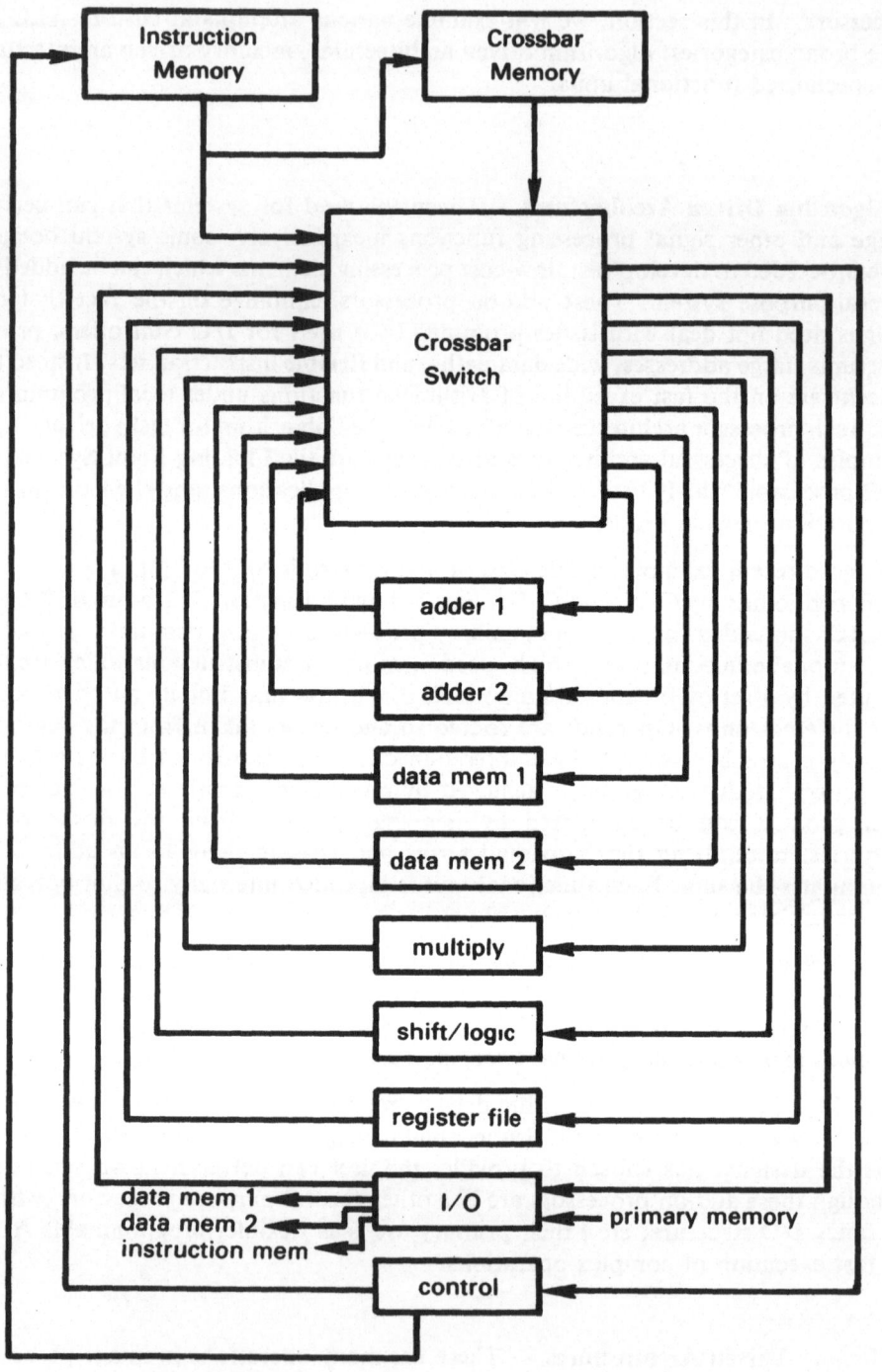

Fig. 1. Structure of SPARC Processor element (CMU/CDC joint effort).

the concern is to minimize the number of accesses to memory and to minimize the bookkeeping and indexing operations required to access data from a complex data structure. The key technique used in such cases is to stream the data in an ordered fashion and to arrange the arithmetic units in such a way that all the relevant operations on a given data can be performed without requiring additional accesses to the main memory. Such architectures can be said to be "memory driven". Unlike conventional memory structures, which provide data only on request, in memory driven architectures the memory delivers a data item each cycle in some predefined order, leaving it up to the processing element to do whatever it wishes with the data.

Several systems have effectively used the concept of memory driven architectures. Most blood cell analyzers [14], the dynamic spatial reconstructor developed by Gilbert et al. [23] at the Mayo Clinic for computer aided tomography, the Sobel edge detection chip developed at USC and Hughes Aircraft [12], and the multi-resolution image chip developed by CMU and Texas Instruments provide some examples of memory driven architectures.

Fig. 2 shows the structure of the programmable ALU multi-resolution image chip set. The chip depends on the fact that if enough on-chip buffer memory can be provided, many previously impossible functions can be realized using the memory driven approach. The task is to perform some program selected operation on the entire

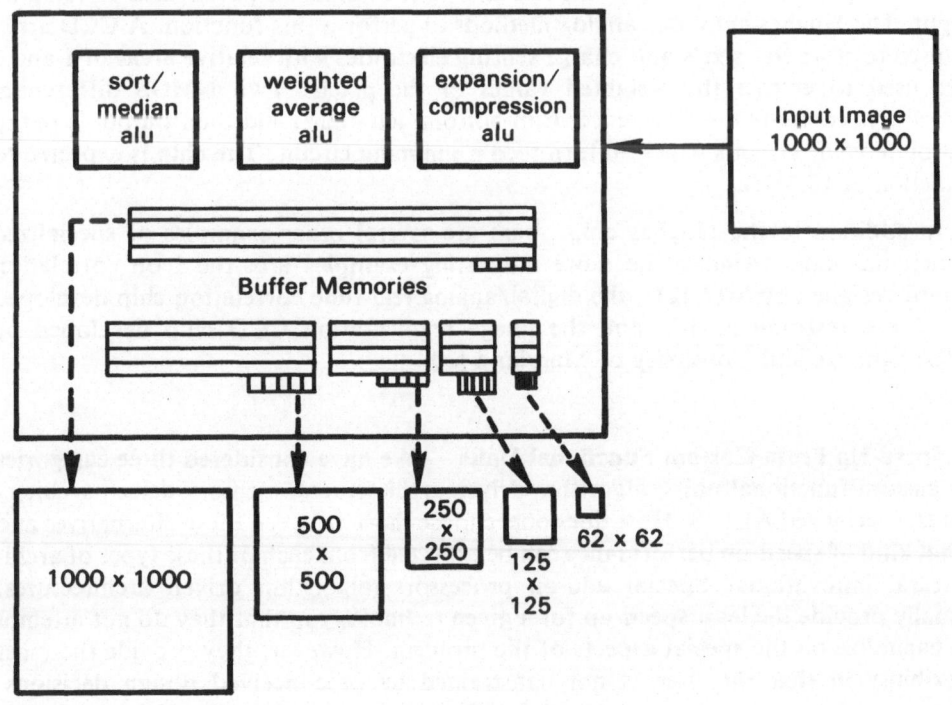

Fig. 2. Structure of CMU/TI STL chip set.

References pp. 184-185

image or to produce reduced resolution versions of the image in one pass. The initial design, currently under consideration, permits operations such as median calculation or a convolution with programmable coefficients on 5 x 5 subimages. Multiple buffers containing 4 rows of the original image and reduced resolution versions of the image permit computation of all the lower resolution versions in a single pass. This architecture is expected to operate at 10MHz pixel processing rate (a task impossible on most general purpose systems).

Specialized Functional Units — When a particular operation is used often, it becomes desirable to design a hardware unit specifically tailored for the function. Edge detection, correlation, masking, and convolution operations are commonly used in image processing applications. A specialized unit requires tailoring of the arithmetic operations for a given function without attempting to be flexible enough for use in other similar operations. This extreme step is meaningful only if the operation is a basic function used over and over again. Fig. 3 shows the block diagram of the USC/Hughes Sobel chip [12]. The Sobel operator provides a measure of the gradient in a 3 x 3 neighborhood centered on the pixel at "e". The value of the Sobel function, S, is found by subtracting the weighted sum (i.e., lower left pixel gray scale value + twice lower center + lower right) of the bottom row of pixels from a similar weighted sum of the top row, taking the absolute value of this difference, adding it to a similar quantity derived from the left and right columns of pixels, and dividing by eight. The Hughes chip uses analog methods to perform this function. A CCD array is used to store the pixels and charge sensing electrodes with relative areas of 1 and 2 are used to extract the weighted values of the pixels. Two NMOS differential amplifiers compute the differences (top-bottom, left-right) and their output is fed to absolute value circuits which in turn feed a summing circuit. The chip is expected to function at 10 MHz.

In addition to the Hughes chip, there are several other examples of specialized functional units. Among the more interesting examples are: the Icon correlation board designed by MIT [21], the digital/analog real-time correlation chip developed by Texas Instruments [4], and the image segmentation CCD chip developed by Westinghouse and University of Maryland [20].

Speed-Up From Custom Functional Units — We have considered three categories of custom functional units; algorithm driven architectures, memory driven architectures, specialized ALU'S. How does one choose those between these alternatives and what kind of speed-up performance can be expected from each of these types of architectural innovations? Special add-on processors (algorithm driven architectures) usually provide the least speed-up for a given technology in that they do not attempt to capitalize on the special aspects of the problem. However, they provide the most flexibility in that the user is not constrained by preconceived design decisions. Memory driven architectures preempt decisions concerning the order in which data is accessed, but usually provide some user programmability (or selectability) of the

COMPUTER ARCHITECTURES FOR VISION

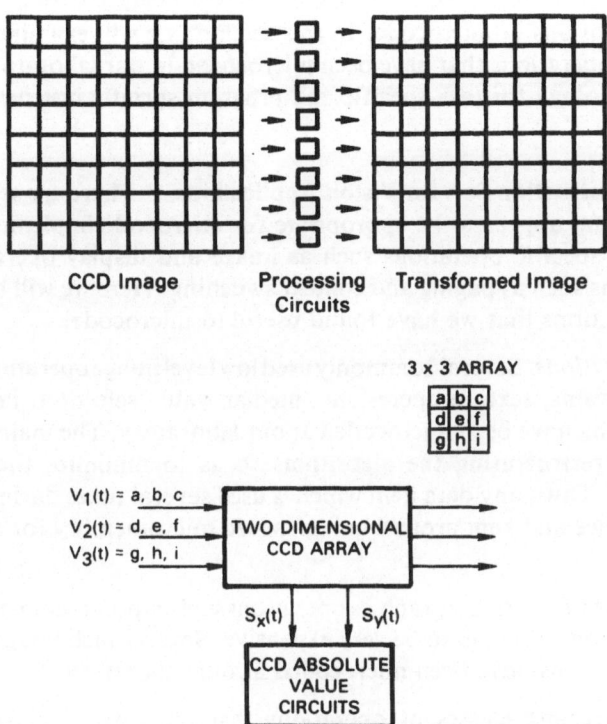

Fig. 3. Structure of USC/HUGHES CCD chip (From Nudd, et al).

function to be performed. Specialized ALU's have the least flexibility in that both the function and memory accesses are entirely predefined, but usually provide the largest speed-up.

Speed-up factors of 10 to 1,000 have been observed in various custom implementations of processing elements. It is reasonable to assume that similar speed-up can be achieved in most computation intensive problems where cost-effective performance is essential. The exact degree of speed-up achieved in the implementation of a given algorithm depends on several factors: locality of memory references, memory access time and bandwidth, number of independent functional units usable for parallel computation at micro-unit level, and the speed of the integrated circuit technology of implementation.

CUSTOM INSTRUCTION SETS FOR VISION

With the increasing availability of processors with writable control stores, the most direct option is to tailor the instruction sets by defining the commonly used functions

References pp. 184-185

as primitive operations in microcode. In this section, we will examine several of the general types of operations that have been microcoded in our laboratory, and suggest a general methodology for task specific and program specific instruction set design.

Examples of Instruction Sets for Vision Applications — There are several classes of instruction sets that appear to be appropriate for microcode implementation. These range from task specific operations such as image and display operations to more general operations such as paging and context switching. Here we will briefly mention a few of the functions that we have found useful to microcode.

1. Image Operations. Several commonly used low level image operations such as edge detection, histograms, texture operations, median value selection, image reduction and image zooming have been microcoded at our laboratory. The main effort in most cases went into restructuring the algorithms so as to minimize the references to primary memory. Thus, any data item which is used several times during computation is fetched only once and kept around in high speed micromemory for future use, size permitting.

2. Display Operation. Raster graphics operations such as point, drag, reduce, expand (zoom), and window [11] tend to be very expensive. Several such image manipulation and display operations have been microcoded at our laboratory.

The other interesting classes of operations that have been microcoded include access and store operations of complex data structures, routines for paging of large images from secondary memory using virtual memory addressing techniques, and context switching operations such as the saving and restoring of register contents on a stack each time a procedure call is invoked.

The PICAP system [8] combines a microprogrammed picture processor (PICAP) with a special television I/O interface (TIP) which allows PICAP access to a 512 x 640 pixel TV field. Control functions and non-picture related processing are handled by a minicomputer connected to PICAP and TIP. PICAP operates on a basic data unit which is a 64 x 64 pixel (x 4 bits) window; PICAP may position this window arbitrarily in the TV field. Nine memories, each capable of holding one such window, are provided for the storage of intermediate results.

Two main modes of operation, multiply picture processing and neighborhood processing, are provided. In the former, two or more 64 x 64 pixel windows may be operated on to produce a result (for example the difference of two windows). Neighborhood processing produces a result window whose pixels are determined by a function operating on the eight neighbors of the corresponding pixel in the input window. A rich class of logical and arithmetic instructions suited to pattern recognition problems is provided.

Methodology for Task Specific Instructions Set Design — Given that it appears desirable to have tailored instruction sets, how does one go about selecting a set of instructions of maximum utility? The key to task specific or program specific instruction set design appears to be the discovery of the most commonly used operations or sequences of operations in a set of programs and to provide new primitive instructions to emulate these operations. The discovery of commonly used instruction sequences can be accomplished by static analysis of the parse trees [18] to discover common subtrees, and dynamic program execution analysis to detect the inner loops where most of the time is spent. Both types of analyses are currently performed manually or by using interactive performance analysis facilities. It appears that one can develop entirely automatic techniques for program specific instruction set design in the future.

Speed-up From Tailored Instruction Sets — Speed-up factors of 3 to 20 have been observed in programs where specific operations or inner loops have been microcoded. The actual speed achieved depends on several factors: relative speed of the microengine, number of additional registers available to the user in micro-memory, and the number of accesses to primary memory. In general, since half of all the memory accesses are to data it is often assumed that, even if the micro-engine ran infinitely fast, no more than a factor of two speed-up is possible. However, when substantial program locality exists (and where all the local variables are kept in micromemory) substantial speed-up may be possible. In these cases the key limiting factors are: number of references to global data structures, and relative speeds of micromemory and program memory.

MULTI-COMPUTER ARCHITECTURES FOR VISION

In this section we will examine architectural innovations at Processor-Memory-Switch (PMS) level [1]. At this level we are primarily concerned with how processors and memories are organized and interconnected (possibly in parallel) so as to achieve improved performance. Speed-up through parallelism can be achieved at the functional unit level or at the processor level. The former is called micro-parallelism and the latter macro-parallelism. The main difference is that in micro-parallelism specific arithmetic and logical units are designed to work together in parallel while in the macro-parallelism entire processors (and memories) are interconnected to work together. This provides a greater flexibility in the type and the complexity of processes that can be executed in parallel. In this section we will examine alternative interconnection strategies used in different families of multi-computer systems.

Array Processors (tightly coupled systems) — In this case, processors are interconnected to permit instruction-by-instruction interaction. They are arranged as a rectangular array executing the same instruction on different items of data, leading to the term single instruction stream, multiple data stream (SMID) architectures. While such systems provide the highest throughput of all multi-computer families, the class of algorithms they can execute tends to be restricted.

References pp. 184-185

The ILLIAC-III [10], ILLIAC-IV [2], CLIP [3], and the Mpp (Massively Parallel Processor) currently under development at NASA provide examples of tightly coupled system architectures. The CLIP (Cellular Logic Image Processor) System (Fig. 4) consists of a 96 x 96 rectangular array of single bit boolean processors and switching logic; each processor is associated with a particular pixel. Thirty-two bits of image storage are provided at each processor site, allowing, for example, four images of 8 bit pixels to be stored and operated upon.

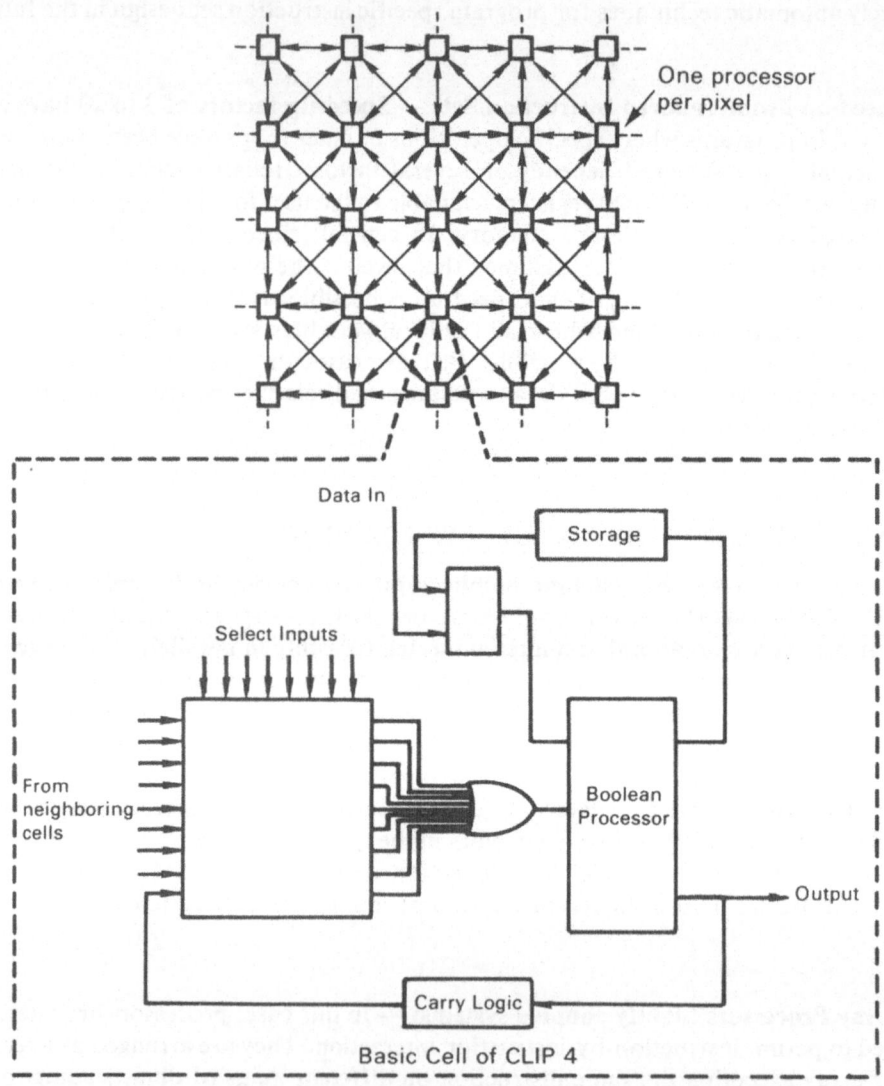

Fig. 4. Structure of CLIP 4 (Duff, et al, 1976).

COMPUTER ARCHITECTURES FOR VISION

Complex operations are performed by sequences of simple instructions (e.g. load, store, perform any of 16 logical functions) which all processors execute simultaneously. Each processor is bidirectionally connected to its eight neighbors and has logic to select which (if any) of its neighbors' outputs are used during any step of a calculation.

Multiprocessors (closely coupled systems) — Multiprocessors are closely coupled in that they permit rapid interaction (in microseconds) and share a uniform address space but permit asynchronous execution of different instruction streams (MIMD — multiple instruction stream, multiple data stream architectures). They are more flexible than SIMD architectures but the interaction overhead can be high.

The C.mmp and Cm* systems developed at CMU provide two interesting examples of multiprocessor systems. C.mmp [22] (Fig. 5) is 16 PDP-11's connected by a crossbar switch to 16 banks of memory. Each processor functions independently and may access any word of memory at the same cost.

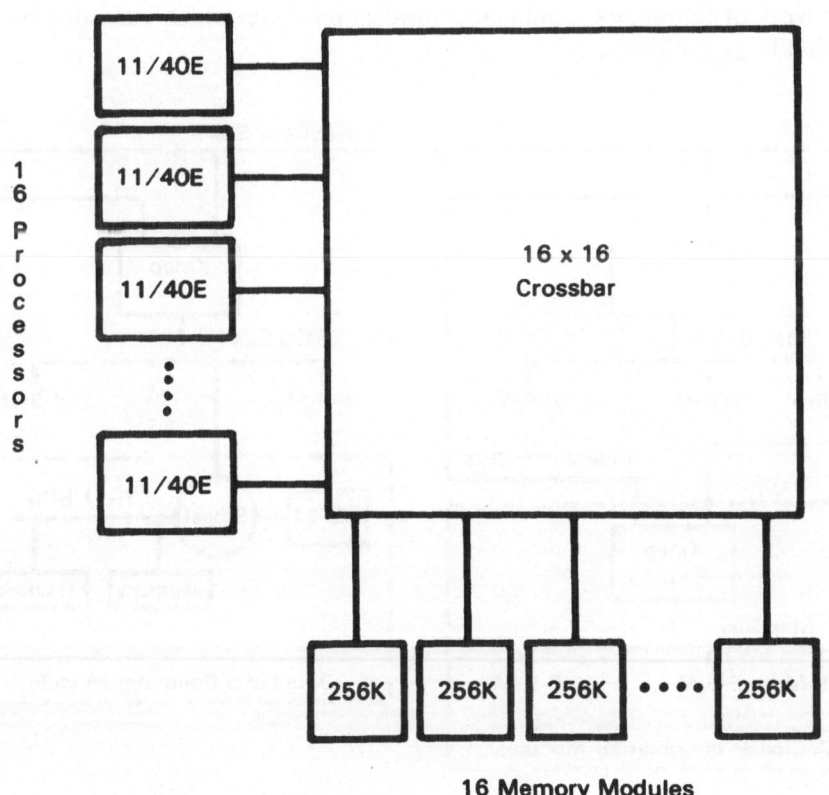

Fig. 5. Structure of C.mmp multiprocessor.

References pp. 184-185

Cm* [19] (Fig. 6) is a system of microprocessors (currently 48 LSI-11s) connected by a packet switching network. Up to 14 computer modules (an LSI-11, local memory, and a local switch) may be connected with a single mapping and routing processor called a Kmap to form a cluster. Each processor functions independently and sees a uniform address space, but memory references have varying time costs depending on if they are to a processor's local storage, to storage at another processor in the same cluster (intra-cluster), or to storage at a remote processor (inter-cluster).

Both C.mmp and Cm* raise questions as to how to best partition vision algorithms and data to provide the best performance. Other difficult issues include the synchronization and communication of processes. These questions are still under investigation.

Networks (loosely coupled systems) — Networks are loosely coupled in that they are usually connected to each other by coaxial cable and do not share a common address space. The interaction rate is also relatively slow — .5 to 5 ms per communication. Although there are several well-known computer networks such as the ARPA net, none of them have been used extensively in cooperative problem solving in vision (or any other complex problem for that matter). The DSYS system [6] is an example of a network of computers running independent processes which may soon be applied to solving large problems.

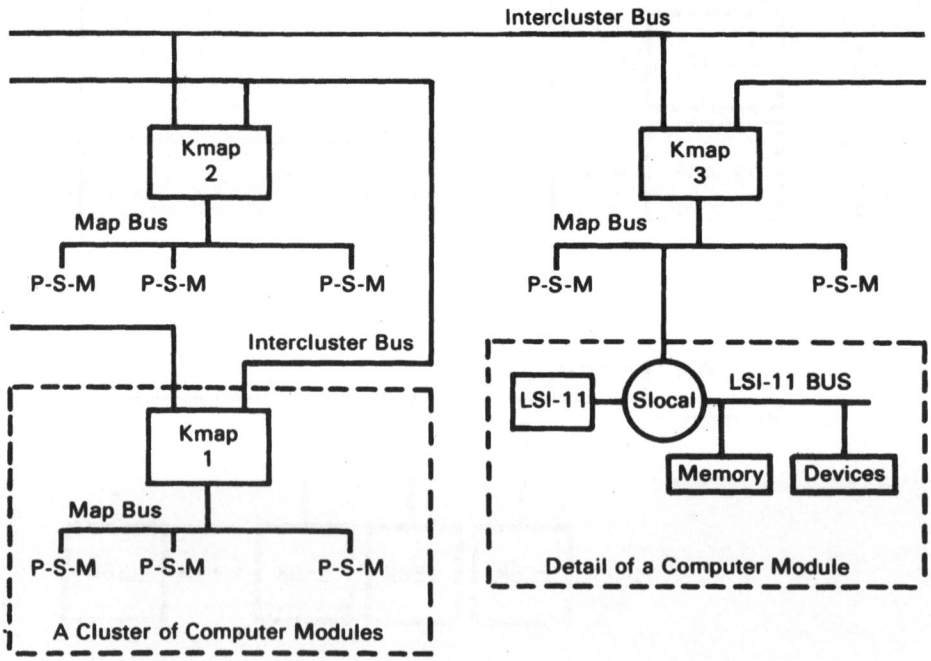

Fig. 6. Structure of CM* multiprocessor (Swan, Fuller and Sierwiorek, 1977)

Speed-up from Multi-Computer Architectures — Speed-up factors of 10 to 10,000 have been achieved in vision applications using multi-computer architectures. The actual speed-up depends on the number of processors, the implicit parallelism of the algorithm, the memory interference resulting from requests from several processors to the same global data, and the secondary memory bandwidth of the various multi-computer architectures. The highest speed-up is often achieved by the least flexible architecture as was the case with specialized functional units. But this speed-up can be achieved only for a restricted set of problems which can be directly mapped onto the architecture. Maximal speed-ups can only be achieved when the algorithms being executed can be decomposed to take advantage of the strengths of the architecture at hand. An algorithm, running on a multiprocessor system of say, 10 processors, where each process passes a block of data (via a pointer) to the next may run slowly on a network of 20 processors. In the loosely coupled system, the overhead of shipping large blocks of data may be more expensive than the actual computations, especially if there is contention for the communication channel. In this case it may be appropriate to code the data, or perhaps decompose the problem entirely differently.

Independently running processes may also require substantial overhead to administrate their cooperation. The cost of inter-process communication can severely limit the effective algorithm decompositions for high performance.

CONCLUSION

This paper is an attempt to synthesize architectural concepts that have been found to be useful in vision systems. It is not intended as a detailed survey of computer architectures for vision. Instead, it provides a framework for architectural specialization under three broad categories: processing elements (PE), instruction set processor (ISP) structure, and processor-memory-switch (PMS) structure. Under each of these categories, we consider various architectural alternatives and indicate what level of improvement (speed-up) might be reasonably expected, given a serious attempt at specialization. Together, they appear to hold the promise of several orders of magnitude improvement over general-purpose systems.

Of the three areas of architectural alternatives considered, the custom processing element design appears to have the highest potential benefit. Continued improvement in LSI technology will make custom PE design simpler and cheaper. Further, the custom PE approach preserves the sequential control structures of uniprocessors systems at the macro level, thereby substantially reducing the burden of algorithm reformulation for parallel processing. In the end, one may have to use a combination of all three alternatives to achieve speed-ups necessary to solve vision problems in real-time.

The problems of vision are characteristic of problems that arise in many other areas involving interpretation of signals. Radar image processing, x-ray tomography, and electron density maps from protein crystallography all have similar problems.

References pp. 184-185

Interestingly, solutions developed on one area will lead to solutions in the other areas also.

As we have observed, vision is indeed hard. Hard from the AI point of view on how to use knowledge to deal with noise, error, and uncertainty; hard from algorithm analysis point of view for parallel decomposition; and hard from the computational effort point of view. The difficulties are only partly understood in ongoing vision research and many new techniques and algorithms will be developed in the future. This means both flexibility and speed are essential for continued advances in the area. If AI research succeeds in solving the vision problem, we can also thank the semiconductor revolution and other advances in computer science for making it possible.

Vision may be hard today, but once the solutions are developed for the key unsolved problems, the cost using vision sensors in industrial automation will be negligible. We hope that industrial designers will plan ahead to make effective use of these advances and eschew costly solutions which will prevent effective use of future technological innovations.

REFERENCES

1. Bell, C. G., and Newell A. *Computer Structures: Readings and Examples*. McGraw-Hill, New York, 1971.
2. Barnes, G. H., Brown, R. M., Kato, M., Kuck, D. J., Slotnick, D. L., and Stokes, R. A. The ILLIAC IV computer. *IEEE Trans. on Computers* C-17,8 (1968), 746-757.
3. Duff, M. J. B. Review of the CLIP image processing system. *Proc. NCC* 47 (1978), 1055-1060.
4. Eversole, W. L. Personal Communication.
5. Eversole, W. L., Mayer, D. J., Frazee, F. B. and Cheek, T. F. Jr., Investigation of VLSI technologies for image processing. *Proc. Image Understanding Workshop* Science Applications SAI-79-814-WA, 1978.
6. Feldman, J. A., Low, J. R., Rovner, P. D. Programming distributed systems. *University of Rochester Computer Science/Engineering Research Review* (1977-78).
7. Juetten, P., Allen, G., Hon, B., Reddy, R. An image processor architecture. *Proc. Image Understanding Workshop* Science Applications SAI-78-656-WA (1977).
8. Kruse, B. Experience with a picture processor in pattern recognition processing. *Proc. NCC* 47 (1978), 1015-1024.
9. Druse, B. A parallel picture processing machine. *IEEE Trans. on Computers* C-22,12 (1973), 1075-1087.
10. McCormick, B. H. The Illinois pattern recognition computer-ILLIAC III. *IEEE Trans. on Computers* EC-12,5 (1963), 791-813.
11. Newman, W., and Sproull, R. *Principles of Interactive Computer Graphics*. Second Edition, McGraw-Hill, New York, 1979.
12. Nudd, G. R. CCDS image processing circuitry. *Proc. Image Understanding Workshop* Applications SAI-78-549-WA, 1977.
13. Ohlander, R. B. Analysis of natural scenes. Ph. D. Thesis, Dept. of Computer Science, Carnegie-Mellon University, Pittsburgh, PA, 1975.
14. Preston, K. Use of the CELL SCAN/GLOPR system in the automatic identifcation of white blood cells. *Bio Medical Engineering* Vol. 7, 226.

15. Price, K. E. Change detection and analysis in multi-spectral images. Ph.D. Thesis, Dept. of Computer Science, Carnegie-Mellon University, Pittsburgh, PA, 1976.
16. Purcell, C. J. The Control Data STAR-100 — performance measurements. *Proc. NCC* 43 (1974), 385-387.
17. Reddy, R., and Newell, A. Multiplicative speed-up of systems. *In Perspectives on Computer Science* (A. K. Jones, ed.) Academic Press, New York, 1977.
18. Saunders, S. E. Compiling customized executable representations and interpreters. Ph.D. Thesis, Dept. of Computer Science, Carnegie-Mellon University, Pittsburgh, PA.
19. Swan, R., Fuller, S., Siewiorek, D. Cm* — a modular, multi-microprocessor. *Proc. NCC* 46 (1977), 637-644.
20. Willett, T. J., and Bluzzer, N. CCD implementation of an image segmentation algorithm. *Proc. Image Understanding Workshop* Science Applications SAI-78-656-WA, 1977.
21. Winston, P. A. MIT progress in understanding images. *Proc. Image Understanding Workshop* Science Applications SAI-79-749-WA, 1978.
22. Wulf, W., Bell, C. G. C.mmp — a multi-mini-processor. *Proc. FJCC* 41 II (1972), 756-777.
23. Gilbert, B. K., Stokma, M. T., James C. E., Hobroc, L. W., Yang, E. S., Ballard, K. C., and Wood, E. H. A real-time hardware system for digital processing of images. *IEEE Trans. on Computers* TC-25 (1976), 1089-1100.

DISCUSSION

R. J. Popplestone *(University of Edinburgh)*

When Carl Hewitt visited us the other day he talked about an architecture which you did not explicitly cover. It is one where you have a lot of processors and each processor has its own memory and processors are connected to each other having some organization, for instance, Boolean n-cube. Have you evaluated this sort of system?

Reddy

No, I have a complete bias against any tightly coupled architectures such as the Boolean n-cube. I do not know how to use them effectively. I do not know how to make a 4th powered failsafe alogrithmic machine. I want an architecture which will continue to run with three processors down today and 17 of them down tomorrow. It is software organization as well as hardware organization. Any tight coupling of this kind is, in the long term, a loser.

J. McCarthy *(Stanford University)*

You studied various kinds of algorithms and how they can be done. Did you look at how pattern matching algorithms can be done?

Reddy

One of the very important problems in vision is how to deal with error and uncertainty and noise. And this is where the AI techniques come in, "When in doubt, sprout." Namely, when you have no other way of deciding when this alternative is right or that alternative is right, you essentially get into search problems. For many of these problems, search dominates signal processing by almost an order of magnitude as images become complex. And it turns out substantial parallelism exists but not the kind of parallelism that can be easily done in a tightly coupled network. You need global information access. There are many ways in which you can have a tightly coupled architecture which also has an orthogonal connection to a data structure machine which has a global data structure so that anybody can access global information as needed. This may in fact lead to alternate solutions. But it turns out, to answer your specific question, we have looked into that from the speed point of view which also has some search problems. We are building a graph-structured machine which has a specialized parallel pipeline network architecture which achieves substantial amounts of parallelism. Here we spend 90% of the time in search and only 10% of the time in signal processing.

B. McCormick *(University of Illinois - Chicago Circle)*

One of the things I think these architectures neglect, is those techniques which allow you to segment the pictures and operate on separate pieces very rapidly and very effectively. You have specialized basically on pixel operations and local operations. The packaging problems seem to me to be the more important problems.

Reddy

Exactly. There are some algorithms which can do region extraction in one pass using a standard tightly coupled machine, but the point you raised is very important. After awhile there are only a certain number of those things you want to do and the problem of utilizing the architecture effectively and doing some type of recursively segmentation becomes very hard and you spend a lot of time trying to fit the solution to this type of architecture.

McCormick

I see a vision machine becoming more and more the inverse of the computer graphics machine.

THREE-DIMENSIONAL COMPUTER VISION

Y. SHIRAI
Electrotechnical Laboratory, Tokyo, Japan

ABSTRACT

There are some difficulties in applying 3-dimensional computer vision to industry. One of them is that a vast amount of computation is required for low level processing. Some special hardware systems are described and one device is shown in more detail. Applications of the hardware are discussed in three examples. Two methods for range data acquisition employing special processors are described. Then our studies for range data processing are introduced. Finally, a complete production system is proposed which makes use of unified geometric models to be shared by CAD, CAM, and visual processing. This approach can solve a second difficulty in applying 3-dimensional computer vision to industry, namely the problem of the extensive programming effort that is required.

INTRODUCTION

More than one decade has passed since research on 3-dimensional computer vision started. The early works aimed at realizing robots with hands, eyes, and intelligence. There have been few such systems actually used in industry because of the following reasons:
1. It is not easy to recognize 3-dimensional objects from 2-dimensional light intensity images.
2. Low level processing is slow because it usually needs much computation.
3. It requires considerable effort to write computer programs.
4. Cost of computer systems is high.

Many practical computer vision systems have cleverly avoided those difficulties. One typical example is a device for inspection of printed circuit boards [1]. The first difficulty is eliminated by looking at 2-dimensional patterns. The second is also

References pp. 203-205

reduced by converting the input patterns into binary patterns. The fourth is solved by constructing special hardware. Besides, the device replaces a human inspector who repeats simple work using only eyes. Its cost performance is much better than a robot which might replace a skillful worker using legs, hands, eyes, etc. (e.g. [2]).

Later more complex works were realized by using computers. In a transistor wire-bonding system [3], the cost of a computer and special hardware is reduced by a hierarchy of a computer, special hardware, and TV cameras.

A computer vision system at GM for locating and inspecting IC chips [4] might be the first system which deals with gray level pictures rather than binary ones. It has demonstrated a comparable performance with a human by processing a 50×50 digitized image within a second.

However, a task which requires a higher resolution may take about a minute to process, say, a 500×500 digitized image. This problem might be solved by employing special hardware for processing gray level pictures. But it usually costs more compared with the one for binary pictures. We need flexible hardware with many standard functions applicable to a variety of image processing problems.

The next section of this paper overviews special hardware for image processing and describes one in detail. We then discuss some applications of the hardware in processing light intensity data. Following sections deal with methods for range data acquisition and the feasibility of hardware for stereo vision, and a review of range data processing focusing on a region method and the use of local constraints placed on vertices. The last section deals with the third problem of programming difficulty mentioned earlier by proposing the use of 3-dimensional geometric models.

HARDWARE FOR IMAGE PROCESSING

There are many special purpose image processors or binary image processors. But there are few flexible systems actually constructed.

In the PIPS (Pattern Information Processing System) project, three companies take different approaches to image processors as shown below:
   a. A computer system with two central processors having large local memories and sharing a large common memory. Fast processing is achieved by firmware programmed for a particular class of image processing operations.
   b. An array processor of 4×4 processing elements with small local memories [5]. Fast processing is achieved by parallel and pipeline processing.
   c. A processor with many special hardware components and a microprocessor [6]. Many local parallel operations are executed by one or by some combination of the hardware components. A microprocessor provides some flexibility.

There are trade-offs in speed, generality and economy. (a) is most general since it can process any kind of image by suitable microprogramming. For a class of operations, (c) is fastest so long as its special hardware can be applied effectively. (b) is in between in both aspects, but it requires sophisticated software to make maximum use of the array processing elements.

Presently (c) seems to be closest to practical applications because we can apply it easily to a fairly wide variety of local operations which usually take a large amount of time. Fig. 1 shows its diagram. Image data is supplied to a basic function module, and the output is addressed to a proper place of an output memory plane. The capabilities of the function modules are as follows:

1. 2-dimensional convolution. Output $G(i,j)$ is a function of the neighboring pixels $F(i+k, j+l)$ as $G(i,j) = \sum_k \sum_l F(i+k, j+l)$, where W_{kl} is a weight matrix.
2. Logical filtering. $G(i,j)$ is determined by the binary pattern of its neighboring pixels.
3. Linear coordinate transformation. This is performed by the address controller.
4. Point mapping.

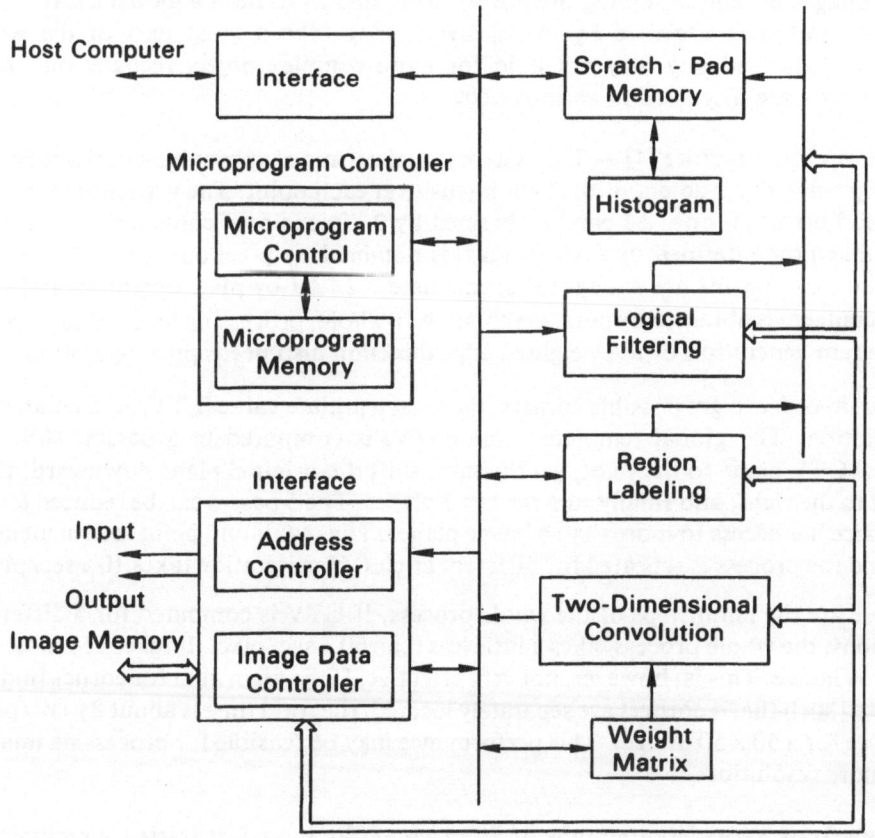

Fig. 1. Block diagram of parallel pattern processor.

References pp. 203-205

5. Region labelling. This is used for labelling connected regions.
6. Histogram making.
7. Pixel operation. Any arithmetic and logical operations can be performed by a microprogram controller. This controller can make use of the other modules.

Processing time is 1 μ sec/pixel for (1) - (6), and depends on the processing complexity for (7). This means, for example, that calculation of 2-dimensional convolution with an 8×8 window over an entire 512×512 image takes about a quarter of a second. The speed is more than a hundred times that of conventional microcomputers.

DISCUSSION OF APPLICATIONS

The image processor (c) described previously was made originally for remote sensing data processing [7]. Certainly it often requires many simple local operations such as coordinate transformation, thresholding, averaging, generating a histogram, etc. Since images in remote sensing are not so structured as to need a global analysis, the local operations performed by the hardware may take a great part of the whole process. This tendency may not hold for more complex object recognition. Some applications are discussed in various fields.

IC Chip Manufacture [4] — The system first determines approximate orientation of the chip using the gradient of the light intensity at each point. The vertical component ΔI and horizontal one ΔJ can be obtained by 2-dimensional convolution. The gradient magnitude defined by $|\Delta I| + |\Delta J|$ is obtained by pixel operation. To get the direction, two passes are enough: first calculate $I/\Delta J$ by pixel operation and then the arctangent is obtained by point mapping. As a whole, processing time is 9μ sec/pixel. Histogram generation of the weighted edge direction also needs pixel operations.

Next, in order to get possible corners, the local template value (LTV) is calculated by convolution. The global template value (GTV) is computed in 5 passes: shift the original GTV plane to the right, get the sum, shift the original plane downward, then shift it to the right, and finally sum the last 3 planes. The 5 passes can be reduced to 4 if the device has access to more than 4 image planes. The maximum point is then memorized and the process is repeated for different angles. One iteration takes 10μ sec/pixel.

These are the main steps of the whole process. If GTV is computed for 5 different directions, the whole process takes a little less than 60μ sec/pixel. It takes 150 msec for a 50×50 image. This is, however, not very effective. If the algorithm for corner finding is revised such that 4 corners are separately located, the total time is about 8μ sec/pixel (20 msec for a 50×50 image). This performance may be feasible for processing images with more resolution.

Analysis of X-ray Photographs of Stomach — We have just started a preliminary study for inspection of x-ray photographs of stomachs to detect cancers. The first task is to determine the stomach region and the various parts of the stomach. This requires the consideration of the variance of the intensity, position, size, shape of the stomach and the overlap with other structures such as bones.

Currently, we are trying the following steps for the initial estimation of the stomach region.
1. According to the histogram of intensity, determine upper and lower bounds of the threshold (the stomach is assumed to be brighter).
2. Calculate the gradients of intensity over the image.
3. Set the threshold to the lowest bound.
4. Generate a binary picture such that pixels containing 1 correspond to those whose intensity is above the threshold and whose gradient is small enough.
5. Generate connected regions in the binary picture and determine the stomach region based on position and area.
6. If the size is too big, increase the threshold and go to (4). Otherwise go to (7).
7. Get the holes in the stomach area. If holes with a large area are found, increase the threshold and go to (4). Otherwise, fill the holes and go to (8).
8. Expand the stomach area twice. The expansion is carried out by assimilating each adjacent pixel into the stomach area if the intensity is above the current threshold and the 3×3 neighbor pixel pattern satisfies a certain criterion.
9. Get the holes again as in (7). If there are large holes, increase the threshold and go to (4). Otherwise fill the holes and go to (10).
10. Get the boundary and find obvious overlaps. If they are found, remove the overlapping regions.

The total processing time for a 256×256 image should require an average of 5 minutes with a medium size computer (we are now using a 256×256 image for (2), and 128×128 for the rest, and it takes about 2 minutes).

We are going to install the hardware described above. It can be straightforwardly applied to steps (1) - (4). In (5), connected regions are labelled by two passes, one is to assign temporary region labels making use of the region labelling modules and the other is a simple point mapping. The areas of the labelled regions are obtained by histogram generation. If the binary picture of the stomach is inverted, and the connected regions are obtained, the holes are found. In (8), the candidate pixels for the expansion can be marked by applying logical filtering. In (10), however, most of the processing can not be executed by the hardware. The boundary has to be traced, the concave boundary segments have to be checked in terms of their curvatures and their spatial relations with one another and the whole stomach.

As a result, the hardware should reduce steps (1) - (9) to about 2 sec. Step (10) will remain at about 20 sec. We need additional hardware such as (a) or (b) described in the previous section to process the remaining parallel algorithms which can not be realized by the hardware discussed in this section.

Desk Scene Analysis — We studied a system for recognition of everyday objects [8]. Since we did not impose many constraints on the scene, complex programs were required to analyze it. Fig. 2 shows the diagram. Main modules are an edge finder, a description maker which segments series of edge points into segments and approximates them by straight lines or curves, and a recognizer. The system, at first, finds the most obvious edges and tries to recognize some objects using these edges, and then more edges are searched for to recognize more objects. As this cycle is repeated, less obvious edges are found. A reference map is used to make the edge finding process efficient. At first it contains approximate gradient values. The edge finder searches only candidate regions suggested by the map. While searching for edge points, the edge finder updates the map for later use. The recognition results are used for modification of the edge description and for specifying the object region so that the edge finder will not search there.

Computation time for locating several objects in a 256 × 256 image stored on the disk is a few minutes. Most of the time (80%) is spent for edge finding while about 15% is spent for generating edge descriptions. Note that the system has many mechanisms to reduce the edge finding time, such as a reference map, feedback from obtained facts, a suitable application of 2-dimensional and 1-dimensional operators, and special routines for handling raw data. It is interesting that calculation of the gradient over the entire image takes more time.

Let us consider the contribution of the hardware. To make best use of it, we should change the strategy for edge finding such that the hardware extracts edge points in the entire image at an early stage. Since three types of edges are used, more steps are required besides calculation of the gradient. Average time is about 15 μ sec/pixel (1 sec/image). Since we need a series of edge points, a serial edge following process is

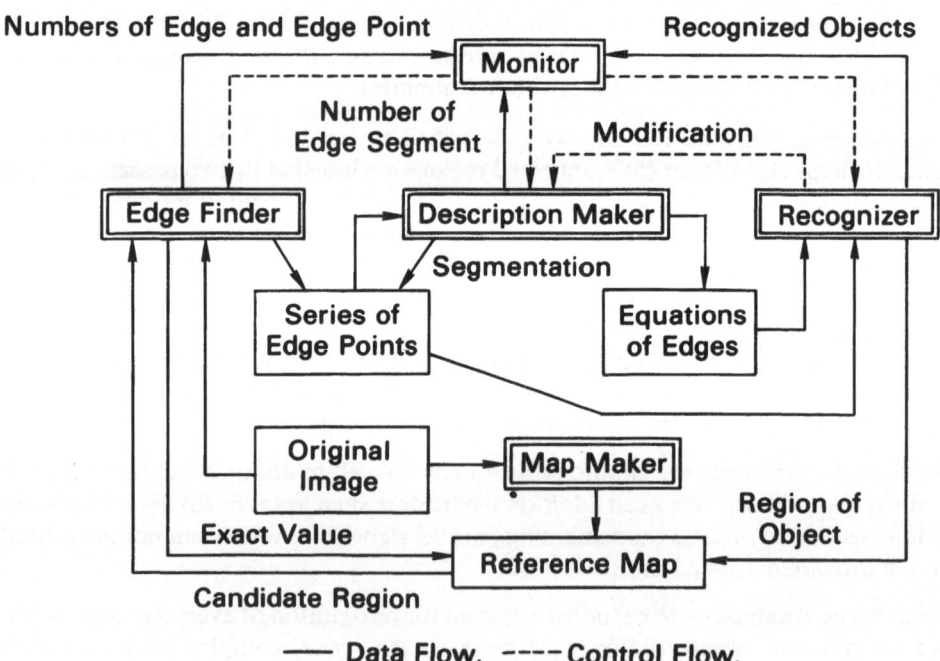

Fig. 2. Block diagram of desk scene analysis.

necessary. Most of the process for making edge descriptions can be, as described previously, realized by a parallel processor other than this one. As a result, this hardware can only halve the total time. The stronger effect, however, might be on the programming complexity. Many mechanisms in the system for efficient image processing can be removed, and the complex control structure can be simplified.

RANGE DATA ACQUISTION

We have various sensory systems to obtain an array of light intensity. On the other hand we have no such standard systems for range data.

Two basically different optical techniques have been used to measure range: triangulation and time of flight. Triangulation is subdivided into passive and active methods.

Both methods are reviewed and discussed in this section.

Stereo Vision — The passive methods take the stereo pair of pictures in two different positions just as binocular stereopsis in animals. We can get the range by finding corresponding points in both pictures. Most of the work related to stereopsis has focused on solving this difficult correspondence problem (e.g. Levine [8] and Genery [9]). Models of human binocular depth perception have been proposed to account for the perceptual phenomena related to it (e.g. Julesz [10], Marr [11] and Sugie [12]).

Given a point in the left image, the corresponding point in the right image is determined according to a certain criterion. Usually the correlation between small neighbor regions of the points is used. Let σ_L^2 and σ_R^2 denote the variances in the neighbor regions in the left and right image respectively and σ_{LR}^2 denote the cross correlation. The correlation C is given by $C = \sigma_{LR}^2 / (\sigma_L^2 \sigma_R^2)^{1/2}$. We can use the sum of squares of the normalized intensities as defined in the following:

$$D_n = \sum_j \sum_i \left\{ \frac{I_L(i,j) - \mu_L}{\sigma_L} - \frac{I_R(i,j) - \mu_R}{\sigma_R} \right\}^2$$

where $I_L(i,j)$ and $I_R(i,j)$ denote the intensities of the left and right images. Since $D_n = 2 - 2C$, maximizing C is equivalent to minimizing D_n.

We conducted an experiment [13] of stereo vision using a stereo adapter attached to a television camera. First we selected points in the left image for which corresponding points are to be found in the right image. Those points (feature points) should have greater difference of intensity in the horizontal direction. The feature points of Fig. 3(a) are shown in Fig. 3(c). For a feature point, the corresponding point lies on a line in the left image if it exists. Assuming that the mean and the variance of both neighbors of the corresponding points are similar, we used the following criterion for economy of computation:

References pp. 203-205

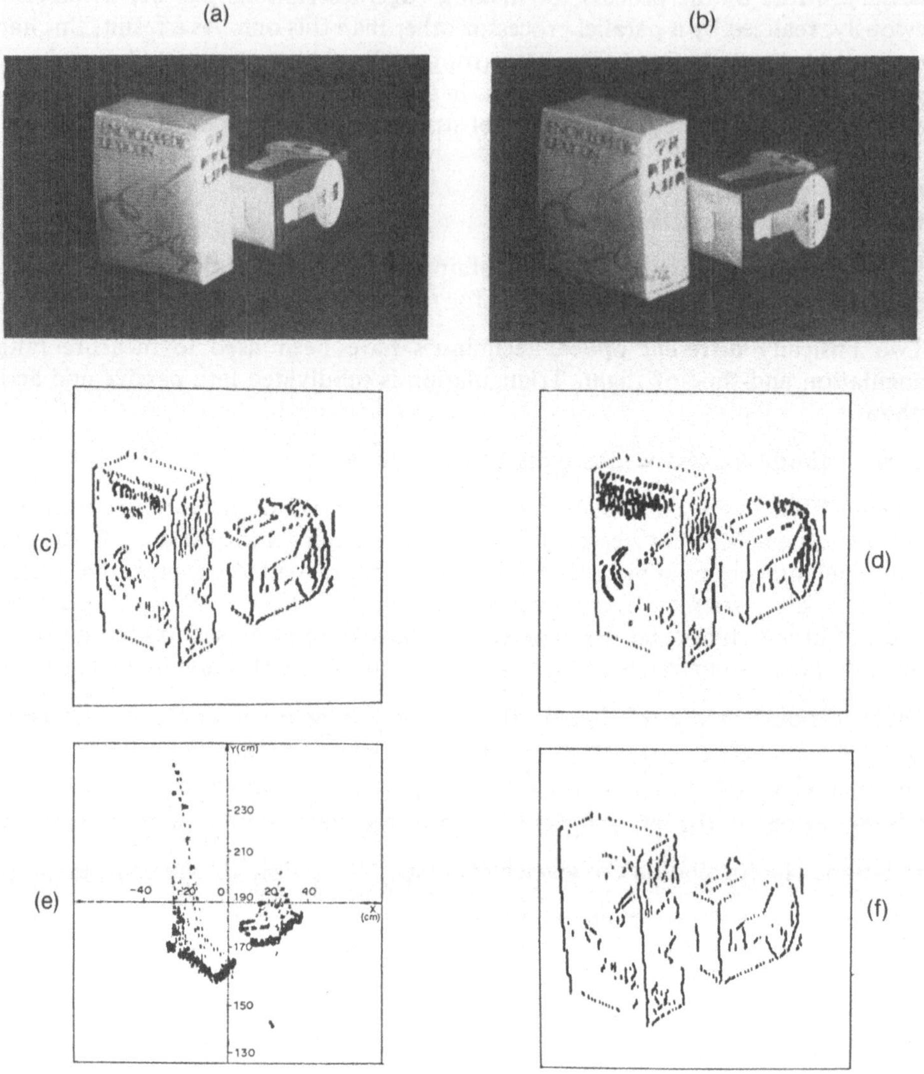

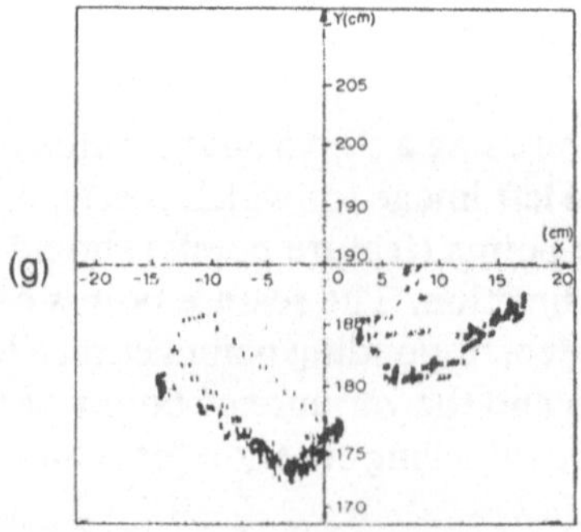

Fig. 3. Example of stereo vision.
(a) Left image.
(b) Right image.
(c) Feature points.
(d) Result of forced matching.
(e) Projection to horizontal plane.
(f) Result of flexible matching.
(g) Projection to horizontal plane.

$$D = \sum_{j=-u}^{u} \sum_{i=-u}^{u} (I_L(i,j)-I_R(i,j))^2/\sigma_L^2 (2u+1)^2$$

where both neighbors are squares of length $(2u+1)$.

If we take the point which minimizes D (with $u=4$), we get the corresponding points shown in Fig. 3(d). Fig. 3(e) shows the same points projected on the horizontal plane. There are many erroneous points, because some feature points which have no corresponding points are forced to be matched. We must determine whether a feature point has a corresponding point or not. Another problem is a mismatch. Let $D(p)$ denote the criterion function along the line in the right image. If u is too small, that is, if regions are compared that are too small, more than one valley of $D(p)$ may be found. To solve these problems, the following algorithm is adopted using 3 thresholds $d_1 < d_2 < d_3$.

Initially let $u=4$ and compute $D(p)$. If $D(p)$ has only one valley such that $D(p_0) < d_1$, then p_0 is the corresponding point. If $\min [D(p)] \geq d_2$, there is no corresponding point. Otherwise for the candidate points such that $D(p) < d_3$ the above process is repeated with $u = u+1$. If no decision is made until $u=9$, there is no corresponding point. The corresponding points and the horizontal projection are shown in Fig. 3(f) and (g) respectively. This method is not only reliable but also efficient because it reduces the number of calculations of $D(p)$ with larger u's. In this case, approximately 1500 feature points are tried, 500 are determined to have no corresponding points, 770 points are matched with $u=4$, 180 points with $u=5$, and the remaining 50 points need larger u's.

The computing time for extracting points is 2 min. and matching requires 25 min. In matching a point, at least $D(p)$ must be calculated at all the candidate points (here, 160 points) with $u=4$. Sometimes more calculation is required with larger u's. Considering these facts, the matching time of 1 sec/point is not very slow. But it is far from a practical application.

Now let us consider the utilization of the hardware again. Feature points are easily extracted as described earlier. The mean and the variance in a small region are obtained quickly by convolution. Storing the image of the small region centered at a feature point in the weight matrix, we can get the convolution $C'(p)$ along a line. For $u=4$, $C'(p)$ is computed in 260 μ sec for 160 p's. If the corresponding point can be determined as the one which maximizes $C'(p)$, the total time for matching 1500 points is only 400 msec., which is reasonable. But in order to simulate the process described above, we must get the peak of $C'(p)$, normalize it with the mean and variance, and make the decision. Sometimes the process must be iterated with a larger u. In the above example, the total time may be about 700 msec. This is not very slow for practical applications. The problem, however, is that the computation time is proportional to the number of feature points.

Other Methods — The active method replaces one eye in stereopsis by a light projector to avoid the correspondence problem. Typical systems project a sheet of light on the scene and get the image by means of a television camera (Shirai [14] and Agin [15]).

We developed special hardware [16] to determine the center position of the stripe images. This hardware accepts digitized video signals as its input, and gets the peak V_p and the postion x_0 in real time. During the horizontal blanking time, it determines the left side and the right side of the peak using a threshold V_t as shown in Fig. 4, and obtains the center point as the midpoint. V_t is determined by two parameters c_1 and c_2 in the following manner:

$$V_t = V_p - 2^{-c_1} V_p - C_2$$

Thus the time required for determining the center positions of a stripe image is exactly the same as the one for scanning a frame.

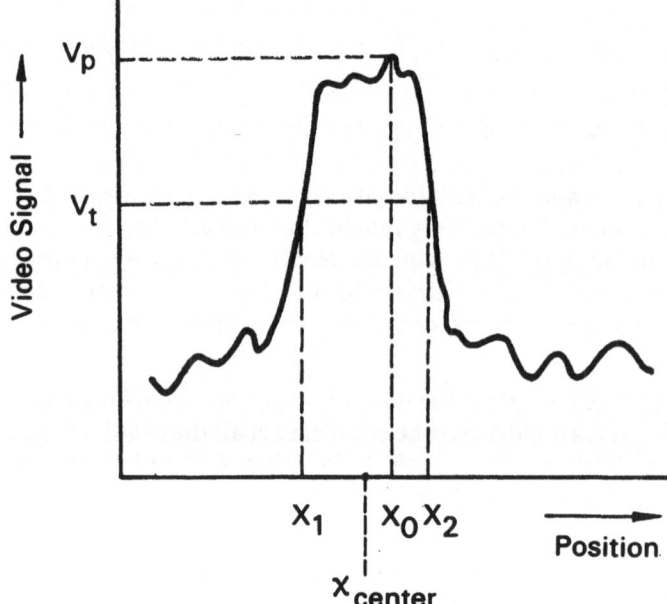

Fig. 4. Determination of center point.

Ishii [17] employed a laser tracker which consists of a laser spot controller and a special image dissector camera to detect the spot position quickly. Its characteristics are similar to those of an image dissector, i.e. it can rapidly get the range of an arbitary point, but takes more time for scanning a complete scene. The current system is slower because it controls a laser spot mechanically.

Accuracy of triangulation methods depends on the distance between two views. Long distances, however, lead to missing data because many points in the scene can not be seen in both views.

The time-of-flight methods eliminate this problem. Johnston [18] got range by using a pulsed laser and measuring the elapsed time of the reflected light. Nitzan [19] made a ranging system by using a continuous wave modulated beam and measuring the phase shift. This system can also get the intensity of the reflected light at the same time and normalize the intensity using the range data.

Although these new systems have considerable potential for the future, there are many problem to be solved for practical applications.

RANGE DATA PROCESSING

Segmentation of Range Data — Segmentation of range data is one of the most important processes in describing 3-D scenes. The method can be divided into "region method" and "edge method" just as in segmentation of light intensity data.

Most of the early systems adopted a way which is most suitable for the input device.

Ishii [17] traced the contour of an object by controlling a laser spot. Shirai [14] segmented the stripe images into piecewise linear lines, and then constructed planes out of the lines.

Popplestone [20] dealt with polyhedra and cylinders in the similar way. Agin [15] fitted quadratic curves to the images of sheets of laser beam, and guessed the axis of the objects.

Sugihara [21] used a typical "edge method" for segmenting range data obtained by a range finder [14]. Input data consist of a series of light stripe images corresponding to the projection angle of the light. He constructed an array of the data whose element $R(i,j)$ represents the horizontal coordinate of the j-th point of the i-th stripe image, and applied to it local operators for detecting the edges of surfaces.

Shirai [23] proposed a region method to represent polyhedra (see Kyura [24] for details). Oshima [25] extended the method to represent curved objects. Fig.5 illustrates the main steps of the process. First, we calculate the 3-D positions of points over a scene from the input array described above. We group the points into small surface elements. These elements are the smallest units used in subsequent processing. Assuming each element to be a plane, we fit a plane equation to it (Fig.5(c)). We merge similar surface elements together into approximately plane regions which are called elementary regions (Fig.5(d)). The elementary regions are classified into three classes: plane, curved and undefined (Fig.5(e)). Finally we try to extend the curved regions by merging adjacent curved or undefined regions (Fig.5(f)). Since the method uses range data explicitly from the initial segmentation and many data points contribute to the segmentation, it is more reliable than conventional edge methods, though it needs more computation.

Interpretation of Range Data by Using Local Constraints — While the local constraints of line types and junction configuration are used for interpretation of line drawings, similar constraints can be used for interpreting raw range data. If we can extract a line from range data, the meaning (label) can be easily found. Taking advantage of this range information, Sugihara [21, 22] used a junction dictionary as a guide

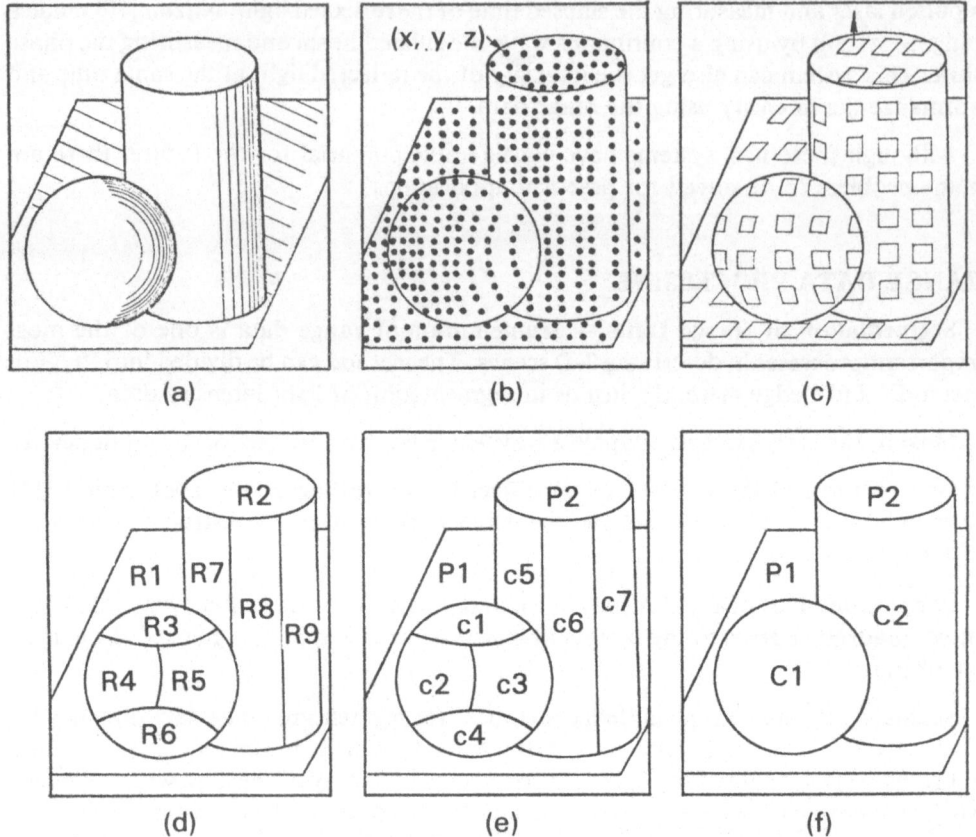

Fig. 5. Conceptual scheme for representation of curved objects.
    (a) Original scene.
    (b) Range data points.
    (c) Surface elements.
    (d) Elementary regions.
    (e) Classified regions.
    (f) Global regions.

in describing 3-D scenes with trihedral curved objects. Since we have range data obtained by triangulation, we can discriminate four types of lines according to physical meanings. Three of them are the same as Huffman's labels and the fourth type represents occluded lines denoted by dotted arrows as shown in Fig.6. The junction dictionary is the directed graph, whose nodes consist of the possible junctions and impossible ones as shown in Fig.7. Each branch of the graph shows that the endnode can be made by adding a new line to the startnode. Curved lines around a junction are approximated by the tangential lines. Applying the local operators described in the previous section, the program first extracts contours (occluding and occluded lines) and locates junctions on them. Starting with those junctions it searches for new

3-D COMPUTER VISION

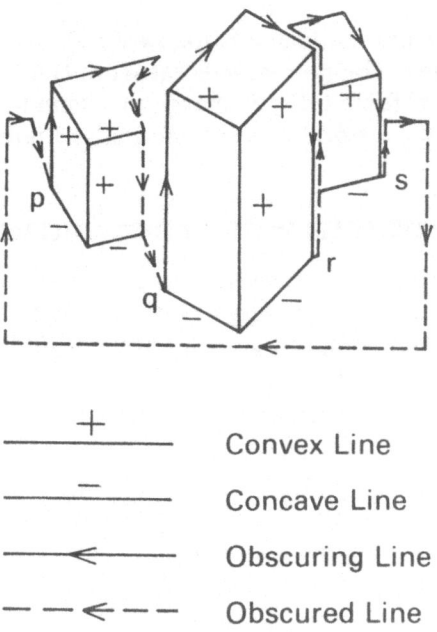

Fig. 6. Labelled line drawing.

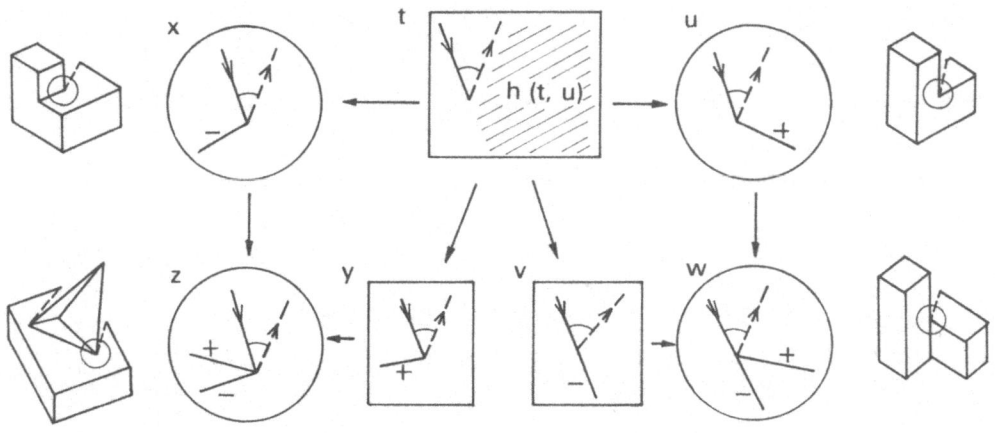

Fig. 7. Part of junction dictionary.

References pp. 203-205

lines in the neighborhood of each junction according to the suggestion by the dictionary. Whenever a new line is found it is followed until it terminates, and then new junctions on it are located. This process is repeated until all junctions become possible and no more lines are found. Fig.8 shows an example of the process. Thus, knowledge about the object world makes scene analysis reliable and efficient.

COMPUTER VISION SYSTEMS WITH GEOMETRIC MODELS

Three-dimensional Models — Generally computer vision systems need object models to interpret a scene. The simplest way of generating object models is to store the 3-dimensional data as in computer graphics. Roberts [26] and Falk [27], for exam-

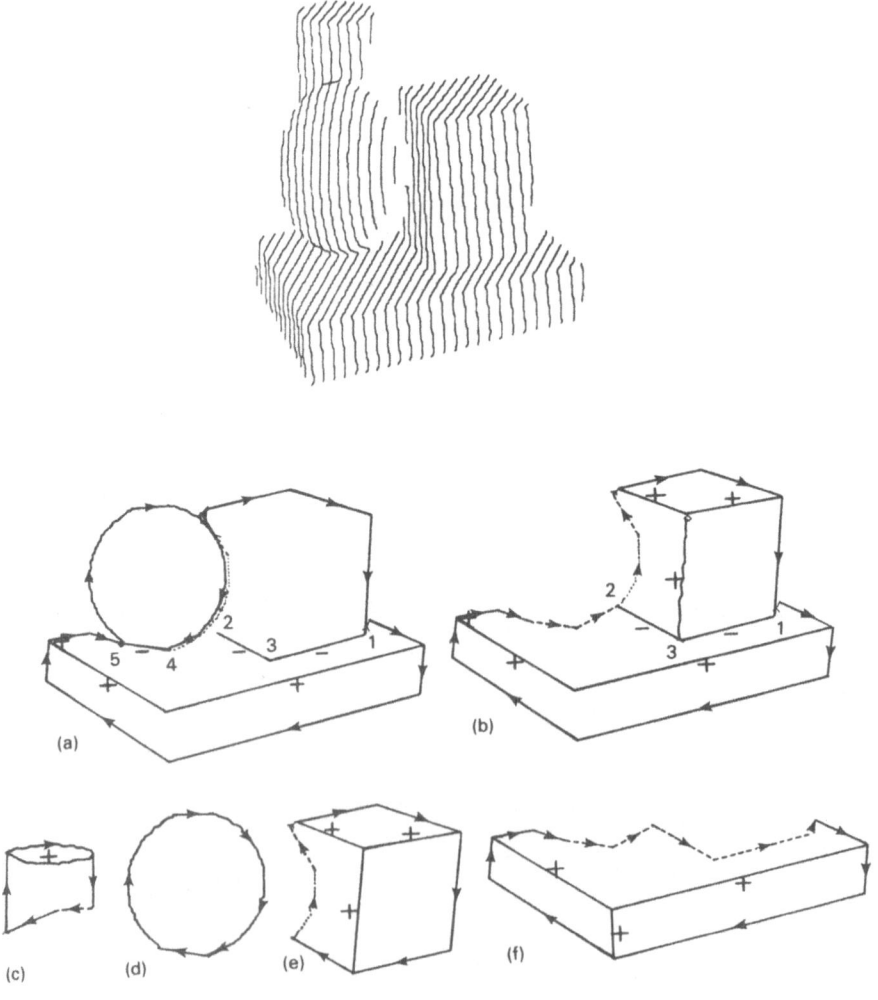

Fig. 8. Example of scene description process.

ple, used this kind of model for simple polyhedra. This straightforward method has some limitations for handling more complex objects both in making models and matching to a scene.

Nevatia [28] represented an object by an assembly of 2-dimensional generalized cones and a few types of joints connecting them. He tried to recognize objects by comparing the final symbolic descriptions with the models stored in the same way. Marr [29] proposed a hierarchy of models based on generalized cones. There are three hierarchies in the models: two of which are concerned with the part-whole relation and one of which represents the hierarchy of the precision. These hierarchies can guide recognition efficiently.

Generalized cones, however, have an ambiguity in that an infinite number of them can represent a single object. Most of the work imposed strong constraints on the cones. These constraints often restricted the flexibility of the models. Perhaps we need better methods, especially for modeling mechanical parts.

Geometric Modeling of Mechanical Parts — An ideal computer vision system may need 3-dimensional models and a powerful interface to guide representation. For example, if there are a few different descriptions of 2-dimensional projections of an object, the interface synthesizes them according to the 3-dimensional model and attached constraints. Or if some descriptions are made, the interface suggests the interpretation. For such a system, we must determine the structure of 3-dimensional models which we can easily create and from which we can retrieve useful information.

There have been similar demands in the fields of CAD and CAM. They need 3-dimensional models of machines or their parts which can be easily created by interaction and used for drawing, retrieval, graphic display, planning of production and data for numerically controlled tools. Many systems which manipulate geometry in ad hoc ways are already available to the industrial world, but these systems lack some properties such as complete representation or applicability to various purposes.

Recently more flexible systems have been developed (see Voelcker [30] for survey). Most of them specify shape through constructive solid geometry i.e., the definition of objects as compositions of primitive solids via the basic operators. In PADL-1.0 [30], for example, the primitives are a block and cylinder, and the operators are movement, intersection, union, difference and assembly. COMPAC [31] and GEOMAP [32] allow defining a shape whose contour is composed of straight lines and arcs, and creating a prism or conical body by using the shape as the base and a revolutional body with a cross section of the shape. Moreover, since they allow a rotation operator, they can combine primitives in an arbitrary fashion. These systems usually have facilities to define tolerances and other information for specifying materials, production methods and so forth. In an interactive system of the GEOMAP, one can design mechanical objects by combining simple commands and handwriting [33]. Fig. 9 shows an example of the system's output synthesized from a model of a pump thus created. Plans include the development of systems for handling free-form surfaces defined by spline functions.

References pp. 203-205

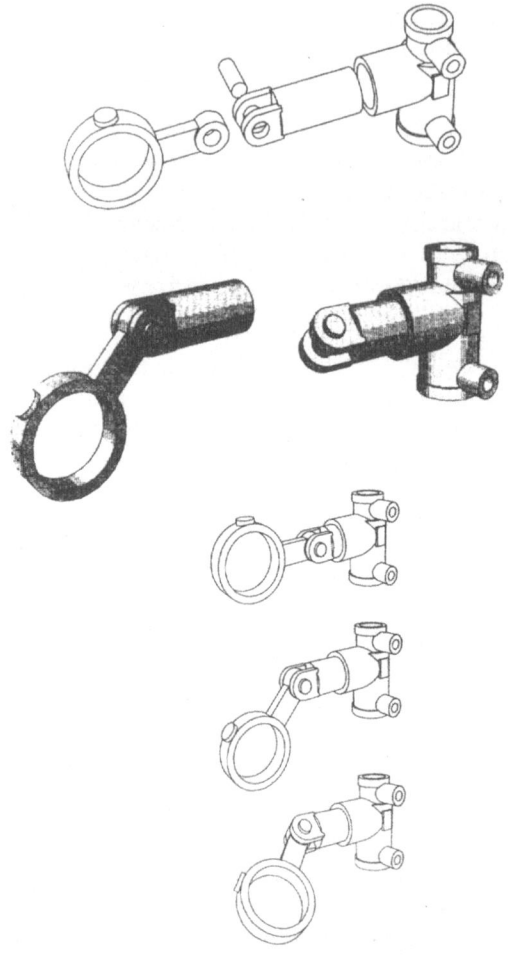

Fig. 9. Example of GEOMAP's output synthesized from a model of a pump.
(a) Drawing of parts.
(b) Shaded picture of subassembly.
(c) Movements.

Total Systems — There has been some effort to reduce the programming time for computer vision systems. Yachida [34] proposed a model-guided approach for recognition of mechanical parts. The system is shown example parts and taught important features of the image by an operator. The method allows the system to learn new objects easily. Bolles [35] developed verification vision which locates an object by applying local operators to locate the subparts in an image. The system is taught interactively an efficient sequence using the constraints of the subpart configurations and the confidences obtained for sample scenes. Thus it efficiently locates known objects at predicted places.

Teaching by showing may be a powerful approach to practical vision systems. For more complex objects, however, this approach may still require a great effort. Now I propose the use of geometric models described previously. In factories, many objects can be designed and produced by CAD and CAM. The unified geometric models can also be used for visual processing. Since the description of an object contains the information about its shape, tolerances and materials, the system can easily produce the predicted image given the specification of the lighting and the position. If the system has a collection of basic modules for feature extraction, it may be able to propose the first plan for locating the object. The plan might be robust because the system can make many simulated experiments considering the location change, tolerances and predicted noise. Examining the proposed plan, one may reject, reform or prepare for experiments based on it.

Another application is found in making the junction dictionary described before. We can even make an order made dictionary for a particular set of objects rather than just for trihedral objects.

CONCLUSION

One of the difficulties of computer vision is that a vast amount of computation is required for low level processing. Among three approaches to this problem taken in the PIPS project, one is introduced which employs image processing hardware composed of several special purpose modules. It is effective in practical applications such as IC chip manufacturing. Other examples of applications show improvements in some classes of processing by factors of 100, and that the total improvement depends on the weight of such kind of processing.

There still remain problems in range acquisition. Stereo vision is now within the reach of some applications by a flexible matching algorithm and the use of specialized hardware.

The other difficulty, that of writing computer vision programs, can be solved by employing unified geometric models of objects which can then be used for many tasks, such as CAD and CAM in a factory.

REFERENCES

1. Eijiri, M., Uno T., and Ikeda, S. A process for detecting defects in complicated patterns. *Computer Graphics and Image Processing* 2 (1973), 326-339.
2. Olsztyn, J. T., Rossol, L., Lewis, N. R., and Dewar, R. An application of computer vision to a simulated assembly task. Proc. 1st Int. Jt. Conf. on Pattern Recognition (1973), 505-513.
3. Kashioka, S., Ijiri, M., and Sakamoto, Y. A transistor wire-bonding system utilizing multiple local pattern matching techniques. *IEEE Trans. on Systems, Man and Cybernetics* SMC-6 (1976), 562-570.
4. Baird, M. L. An application of computer vision to automated IC chip manufacture. Proc. 3rd Int. Jt. Conf. on Pattern Recognition (1976), 3-7.

5. Matsushima, H., et al. Arrayed processor for image processing. TR-IE-78-11, 45-53, Inst. of Electronics and Communication Engineers of Japan, 1978. (in Japanese)
6. Mori, K., Kidode, M., Shinoda, M., and Asada, H. Design of local parallel pattern processor for image processing. *Proc. AFIPS* 47 (1978), 1025-1031.
7. Mori, K., et al. Research and development in image processing and pattern recognition system. PIPS Project Rep. 69-78, 1977. (in Japanese)
8. Levine, M. D. Computer determination of depth maps. *Computer Graphics and Image Processing* 2 (1973), 131-150.
9. Gennery, D. B. A stereo vision system for an autonomous vehicle. Proc. 5th Int. Jt. Conf. on Artificial Intelligence (1977), 576-582.
10. Julesz, B. *Foundations of Cyclopean Perception.* The U. of Chicago Press, Chicago, 1971.
11. Marr, D., and Poggio, T. Cooperative computation of stereo disparity. AI Memo 364, MIT AI Lab., 1976.
12. Sugie, N., and Suwa, M. A scheme for binocular depth perception suggested by neurophysiological evidence. *Biological Cybernetics* 26 (1977), 1-15.
13. Yasue, T., and Shirai, Y. Binocular stereoscopic vision for object recognition. Bul. Electrotechnical Lab. 37 (1973), 1101-1119. (in Japanese)
14. Shirai, Y., and Suwa, M. Recognition of polyhedrons with a range finder. Proc. 2nd. Jt. Conf. on Artificial Intelligence (1971), 80-87.
15. Agin, G. J., and Binford, T. O. Computer description of curved objects. Proc. 3rd Int. Jt. Conf. on Artificial Intelligence (1973), 629-640.
16. Oshima, M., and Takano, Y. Special hardware for the recognition system of three-dimensional objects. Bul. Electrotechnical Lab. 37 (1973), 493-501. (in Japanese)
17. Ishii, M., and Nagata, T. Feature extraction of three-dimensional objects using laser tracker. *Pattern Recognition* 8 (1976), 229-237.
18. Johnston, A. R. Infrared laser rangefinder. JPL Rep. NPO-13460, 1973.
19. Nitzan, D., Brain, A. E., and Duda, R. O. The measurement and use of registered reflectance and range data in scene analysis. *Proc. IEEE* 65 (1977), 206-220.
20. Popplestone, R. J., Brown, C. M., Ambler, A. P., and Clowford, C. F. Forming models of plane-and-cylinder faced bodies. Proc. 4th Int. Jt. Conf. on Artificial Intelligence (1975), 664-668.
21. Sugihara, K. Dictionary-guided scene analysis based on depth information. PIPS-R-No. 13, 48-122, Electrotechnical Lab., 1977.
22. Sugihara, K., and Shirai, Y. Range data understanding guided by a junction dictionary. Proc. 5th Int. Jt. Conf. on Artificial Intelligence (1977), 706.
23. Shirai, Y. A step toward context sensitive recognition of irregular objects *Computer Graphics and Image Processing* 2 (1973), 298-307.
24. Kyura, N., and Shirai, Y. Recognition of three-dimensional objects using region method. Bul. Electrotechnical Lab. 37 (1973), 996-1012. (in Japanese)
25. Oshima, M., and Shirai, Y. Representation of curved objects using three-dimensional information. Proc. 2nd USA-JAPAN Computer Conf. (1975), 108-112.
26. Roberts, L. G. Machine perception of three-dimensional solids. In *Optical and Electro-Optical Information Processing.* J. T. Tippett, et al (Eds.), the MIT Press, Cambridge, MA., 1965.
27. Falk, G. Interpretation of imperfect line data as a three-dimensional scene. *Artificial Intelligence* 3 (1972), 101-144.
28. Nevatia, R., and Binford, T. O. Description and recognition of curved objects. *Artificial Intelligence* 8, 1 (1977), 77-98.

29. Marr, D., and Nishihara, H. K. Representation and recognition of three-dimensional shapes. AI Memo 416, MIT AI Lab., 1977.
30. Voelker, H. B., and Requicha, A. A. G. Geometric modeling of mechanical parts and processes. *Computer* (1977), 48-57.
31. Spur, G. Status and further development of the geometric modeling system COMPAC. Proc. Geometric Modeling Project Meeting, Computer Aided Manufacturing International Inc., Arlington, TX, 1978.
32. Kimura, F., and Hosaka, M. Program package GEOMAP. Ibid.
33. Hosaka, M., and Kimura, F. An interactive geometrical design system with handwriting. Proc. IFIP (1977), 167-172.
34. Yachida, M., and Tsuji, S. A machine vision for complex industrial parts with learning capability. Proc. 4th Int. Jt. Conf. on Artificial Intelligence (1975), 819-826.
35. Bolles, R. C. Verification vision for programmable assembly. Proc. 5th Int. Jt. Conf. on Artificial Intelligence (1977), 569-575.

DISCUSSION

R. J. Popplestone *(University of Edinburgh)*

In our experiments we found it to be quite difficult to pick up the line in hardware if there was white paint on the surface.

Shirai

We project the beam from one direction and observe from another direction. If we don't see the line because of low contrast, then we know that. We can't measure range at that point.

Popplestone

One fails to get a model of the object?

Shirai

Every laser range finding system will have similar problems. If we cannot observe the reflected light, we cannot get the data. That is a common problem.

OPTICAL COMPUTING FOR IMAGE PROCESSING

A. D. GARA

General Motors Research Laboratories, Warren, Michigan

ABSTRACT

The application of optical computing to sensor-based robots will be described in terms of processing image information in real-time by the methods of spatial Fourier transform filtering. The basic requirements will be reviewed and advances in optical device technology which have led to real-time operation will be described. In particular, recent results on 2-dimensional correlation by optical matched filtering demonstrate the ability to determine the location and orientation of objects in low contrast scenes. These advances have resulted in a laboratory demonstration of a vision system for robot control.

IMAGE PROCESSING BY OPTICAL METHODS

Introduction — Optical computing is the ability to perform mathematical operations on an image by altering the spatial light distribution using optical or electro-optical components. The attractive feature of this type of image processing is the fully parallel nature of the operations, i.e. every point in the image is processed simultaneously. The operations generally are applied to a two-dimensional image and can be linear or nonlinear. The linear operations (such as the addition or subtraction of two images) [1], differentiation of an image intensity distribution [2], and convolution or correlation of two distributions [3] have received the most attention. Recently optical methods have been developed for nonlinear processing. Examples have been reported of logarithmic transformations for level slicing [4] and optical analog-to-digital conversion for optical planar logic operations [5].

The analysis of the above operations and the methods for implementing them are simplified by recognizing that, under certain conditions, a Fourier transform relationship exists between the spatial light distributions in the front and back focal planes of a

References pp. 234-235

lens [6]. These two planes thus play a unique role in optical processing and input images, filters, output images, and detectors are found in abundance in these planes.

While the potential for high-speed parallel optical processing has generated considerable interest, the number of applications have been limited by technological problems [7]. In this paper we will review briefly the concepts of optical processing to illuminate those problem areas and then describe how some of these have been overcome. We will then show examples of current work on real-time image edge enhancement and real-time optical correlation, both of which have potential application in robot vision systems.

Spatial Fourier Transformation of Image Distribution — The basic concepts of optical processing are contained in Fig. 1. A uniform intensity plane wave of monochromatic light (as from a laser) of wavelength λ illuminates the "input" in plane P_1. The input is usually in the form of a film transparency or photographic plate on which is recorded the scene information to be processed. This of course places a severe restriction on any "real-time" capability of optical processing and has been one of the major difficulties limiting its use. In the work to be described in this paper this limitation has been effectively removed. However for the present discussion we will consider the input to be in the form of a thin transparency. It is normally desirable that this transparency only affect the amplitude of the incident plane wave and not the phase. If this is the case let $g(x_1,y_1)$ represent the effect of the input transparency on an incident plane wave of unit amplitude (e.g. $g(x_1,y_1) = 1$ for points in P_1 that are totally transparent and $g(x_1,y_1) = 0$ for optically opaque regions). For a sufficiently

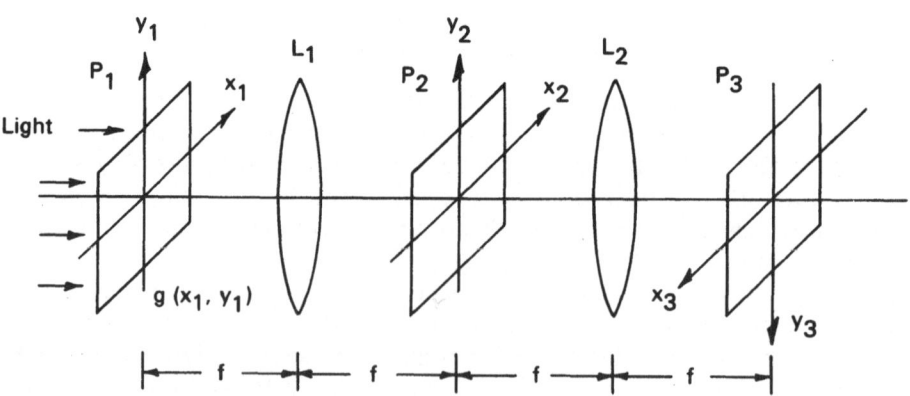

Fig. 1. The basic configuration of optical elements for image processing. The input is in plane P_1, and f is the focal length of the lenses, L.

OPTICAL COMPUTING FOR IMAGE PROCESSING

large diameter of L_1 the field in plane P_2 can be given (to within a multiplying constant) as [8],

$$G(x_2,y_2) = \int_{-\infty}^{\infty}\int_{-\infty}^{\infty} g(x_1,y_1) e^{-2\pi i \left(\frac{x_2 x_1}{\lambda f} + \frac{y_2 y_1}{\lambda f}\right)} dx_1\, dy_1 \tag{1}$$

where λ is the wavelength of the plane wave and f is the focal length of lens L_1. Discussions on existence conditions for the Fourier transform properties of a lens can be found in the above reference. In general one can infer tht infinite discontinuities do not exist in the real optical world since this would imply infinite optical energy. In addition the function $g(x_1, y_1)$ is zero beyond some finite x_1 and y_1.

The quantities $X_2/\lambda f$ and $Y_2/\lambda f$ are comparable to the frequency in Fourier transforms on temporal functions, hence these quantities can be identified as the *spatial frequencies:*

$$f_x \equiv \frac{x_2}{\lambda f}, \text{ and } f_y \equiv \frac{y_2}{\lambda f} \tag{2}$$

Thus, the lens L_1 displays in plane P_2 the decomposition of the object $g(x_1,y_1)$ into its spatial frequency spectrum. If the object has a periodic transmission character in one dimension, $g(x_1,y_1) = (\cos 2\pi f_0 x)$, which can be described as a sinusoidal grating the Fourier transform is a delta function at $f_x = f_0$ and $f_x = -f_0$. The finite extent of a real input and the finite aperture of the lens would yield a $[\sin(f_x \pm f_0)]/(f_x \pm f_0)$ behavior in P_2 rather than a delta function. If we follow the light through the second lens, L_2, we again find in plane P_3 a light distribution which is the spatial Fourier transform of the distribution in plane P_2. This will allow us, by applying Eq. (1), to recover the original image $g(x_1,y_1)$. However, since the Fourier transform taken twice on $g(x)$ is $g(-x)$, we show an inverted coordinate system in P_3 for convenience.

The availability of the spatial frequency spectrum of the two-dimensional image is the primary motivation for optical processing. The spatial spectrum can readily be altered by placing masks in plane P_2, and the altered spectrum transformed to produce an altered image in P_3.

References pp. 234-235

Filtering Operations on the Fourier Transform — The process of placing masks in the spatial frequency distribution is generally described as spatial filtering. The mask may only modulate the intensity distribution in the transform, or by introducing phase delays in the mask, the complex field distribution may be altered.

In general, if a filter with transmittance $T(f_x, f_y)$ is placed in plane P_2, the complex light distribution in plane P_3 is given by

$$U(x_3,y_3) = \int\!\!\!\int_{-\infty}^{\infty} (T(fx,fy)G(f_x,f_y)e^{-2\pi i(f_x x_3 + f_y y_3)} df_x df_y. \qquad (3)$$

The effect of the filter function on the image can be more readily seen using the convolution theorem of Fourier transforms. Equation (3) can be rewritten as

$$U(x_3,y_3) = \int\!\!\!\int_{-\infty}^{\infty} g(x_1,y_1)t(x_3-x_1,y_3-y_1)dx_1 dy_1 \qquad (4)$$

where $T(f_x,f_y)$ is the Fourier transform of $t(x_1,y_1)$. Thus the filtering operation in the spatial frequency plane is equivalent to a convolution of the impulse response of the filter with the original image. If one knows the convolution operation desired, the optical processor can be used to perform the operation provided that the corresponding filter can be physically realized. This is another one of the technological difficulties of optical processing which will be addressed later in the paper.

Two specific filtering operations which are important for robot vision applications are differentiation for edge enhancement and correlation for object identification and location.

The form of the filter function for differentiation can be seen from Eq. (1) since, apart from a constant multiplier we can write

$$\frac{\partial g(x_1,y_1)}{\partial x_1} = \int\!\!\!\int_{-\infty}^{\infty} f_x G(f_x,f_y)e^{2\pi i(x_1 f_x + y_1 f_y)} df_x df_y. \qquad (5)$$

Comparing this with Eq. (3) for the light distribution in the output plane of the processor, we can see that the filter for differentiation in the x-direction has the form $T \propto f_x$.

To perform a correlation operation, we note that, if $T = H^*(f_x,f_y)$, then

$$U(x_3,y_3) = \int\!\!\!\int_{-\infty}^{\infty} g(x_1, y_1)h^*(x_1 - x_3, y_1 - y_3)dx_1 dy_1, \qquad (6)$$

which is the cross-correlation of the input image $g(x_1,y_1)$ with the distribution $h(x_1,y_1)$. In particular, when $H^*(f_x,f_y) = G^*(f_x,f_y)$ the output plane contains the autocorrelation of the input image.

PRACTICAL ASPECTS OF OPTICAL IMAGE PROCESSING

Real-time Input — In the preceding section, we assumed a two-dimensional input function with a complex amplitude distribution $g(x_1,y_1)$ representing the image to be processed, where $[g(x_1,y_1)]^2$ is the intensity distribution in the image. For most applications, the actual scene will be three-dimensional, containing objects of various shapes and surface microstructure. If this geometry were illuminated with coherent light, the scattered light would not satisfy the simple two-dimensional distribution given above. Even if the scene were planar on a macroscopic scale, the surface microstructure would scatter light nearly uniformly in all directions giving rise to a rich distribution of spatial frequencies, even for constant intensity surfaces where $[g(x_1,y_1)]^2$ = const. Most objects of interest have this type of surface which diffusely scatters light and the surface scattering would be the dominant contribution to the spatial frequency spectrum. The microstructure of the surface of the object however is generally not of interest. For example, if the processing operation to be performed is correlation, the vision system wants to look for objects with a particular shape only, not a particular surface microstructure. To remove this difficulty, the input-images are usually in the format indicated in the previous section, i.e. planar transparency recordings of the scene to be processed.

For most applications however, "real-time" operation requires an input which is real-time, not a photographic transparency. The mechanics of film exposure and development severely limit the rate at which optical data may be acquired and processed. What is needed is a device with some of the characteristics of photographic emulsions (such as high spatial resolution, good image contrast, and optical flatness to avoid the diffuse scattering problem) but which can respond very quickly and be erasable at a repetition rate approaching television rate.

Such devices have been the subject of considerable research in recent years and have been the key to the progress that has been made in optical processing. The devices that have been developed can generally be classified into two categories depending on the manner in which the scene information is recorded on the device. In Fig. 2 we show an optically addressed and an electrically addressed configuration. In both cases, the purpose of the input device is to convert the scene information from an image derived from incoherent, diffusely scattered light to a coherent light image. For this reason the devices are often referred to as incoherent-to-coherent image converters or image transducers. (Some authors also use the term spatial light modulator.) The local intensity of the incoherent image in plane P_1 alters the transmission of the uniform coherent

References pp. 234-235

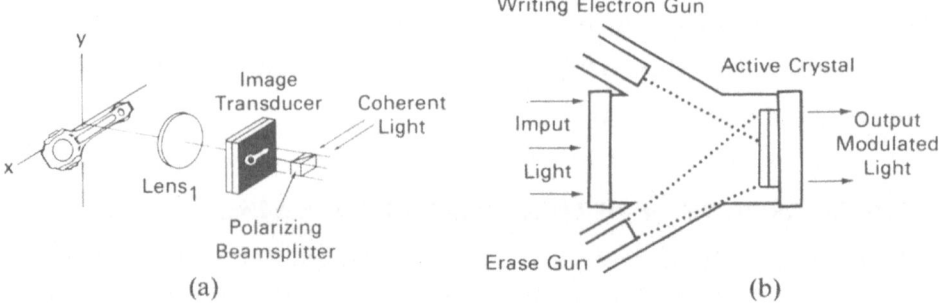

Fig. 2. (a) An optically addressed and (b) an electrically addressed image transducer. The writing electron beam may be driven by the video signals from a television camera.

illumination which passes through the active material. The coherent light emerging from the device will have a spatial modulation proportional to the local scene intensity.

A variety of electro-optic effects, materials exhibiting these effects, and device fabrication techniques have been explored. A summary of the most significant types of image transducers and their uses has been given by Thompson [7] and Casasent [9]. In this paper we will discuss only the device we have used, the liquid crystal light valve developed at the Hughes Research Laboratories [10]. The device (see Fig. 3) is used in the optically addressed reflection mode. The essential mechanism for the image conversion is an induced optical birefringence in the liquid crystal layer.

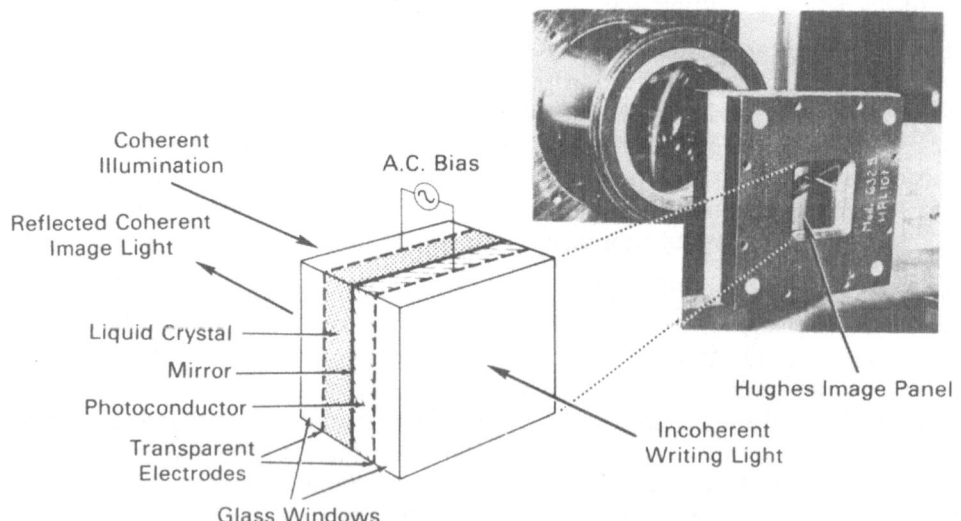

Fig. 3. The liquid crystal image transducer developed at the Hughes Research Laboratory. The liquid crystal is approximately 3 micrometers thick (Cross section not to scale).

The device first converts the input image intensity distribution (writing light) into a spatial distribution of photoelectron-hole pairs in the CdS photo-conductor, with an attendant spatial redistribution of the applied ac voltage across the liquid crystal layer. The voltage induces an optical birefringence in the liquid crystal approximately proportional to the local writing light intensity. A linearly polarized readout light beam passes through the liquid crystal, reflects from the dielectric mirror, traverses the liquid crystal a second time, and exits the device as elliptically polarized light. An analyzer, placed in the path of the emerging beam, passes only the component of polarization that is perpendicular to the original polarization direction of the readout beam. The spatial intensity distribution of the readout beam after the analyzer will be approximately proportional to the spatial intensity distribution of the input image. This entire process is reversible since the voltage distribution will disappear when the input light is removed. With no input light the readout light will remain linearly polarized perpendicular to the analyzer; hence no output light is passed.

The detailed features of this device can be found in the literature, and only a summary will be given here. The operating characteristics which are important for image processing are listed in Table 1. These are based on measurements made in our laboratory on a particular device. These values in some cases are not in agreement with other published data [11]. The differences are due mainly to the structural variations that are present from one particular device to the next.

TABLE 1

Characteristics of the Hughes
liquid crystal image transducer

Image Format	25 mm square
Resolution	25 micrometers
Contrast	100/1
Cycle Time	30 milliseconds (minimum)
Lifetime	>1000 hours
Sensitivity	1 microwatt/cm^2

As can be seen from Fig. 4, the device has a large useful dynamic range. The test data shown for white light illumination, however the cadmium-sulfide photoconductor has a peak spectral response in the blue green region [12].

A finite time is necessary for the device to respond to an input light pulse (Fig. 5). This response time is a function of the input light intensity as well as the parameters of the drive voltage, as shown in Fig. 6.

The rise time corresponds to the time to reach 90% of a steady-state output for turn-on, and the decay time indicates a return to 10% of the steady-state output for turn-off. (Turn-on and turn-off refer to input writing light.) We observed that with the addition of a weak bias illumination in the writing light, the response time improved dramatically.

References pp. 234-235

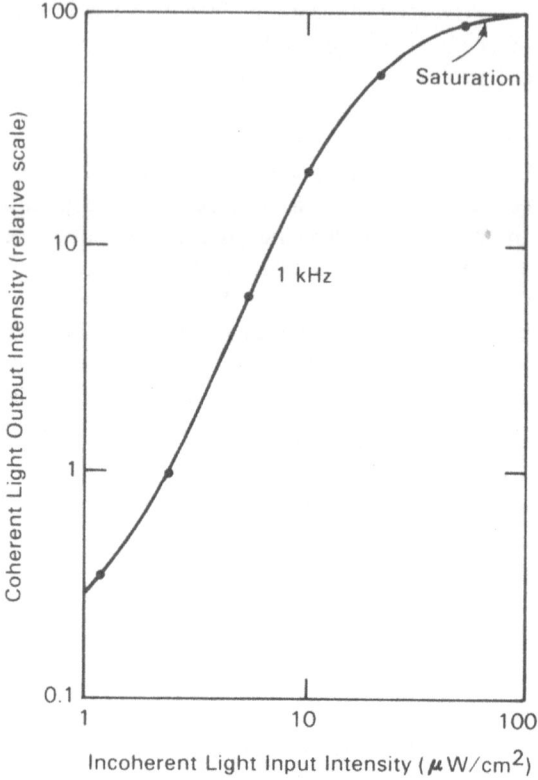

Fig. 4. The sensitivity of the Hughes device to white light input (tungsten filament).

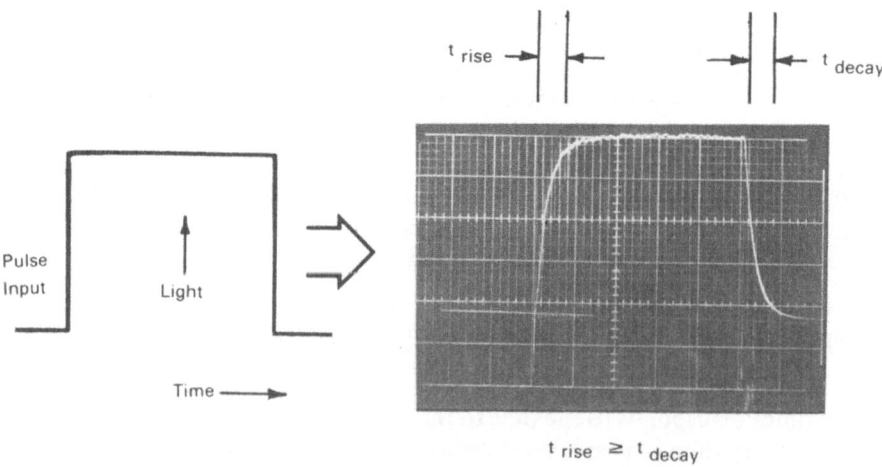

Fig. 5. Oscilloscope trace of Hughes transducer response to an input light pulse of ½ second duration at saturation irradiation level.

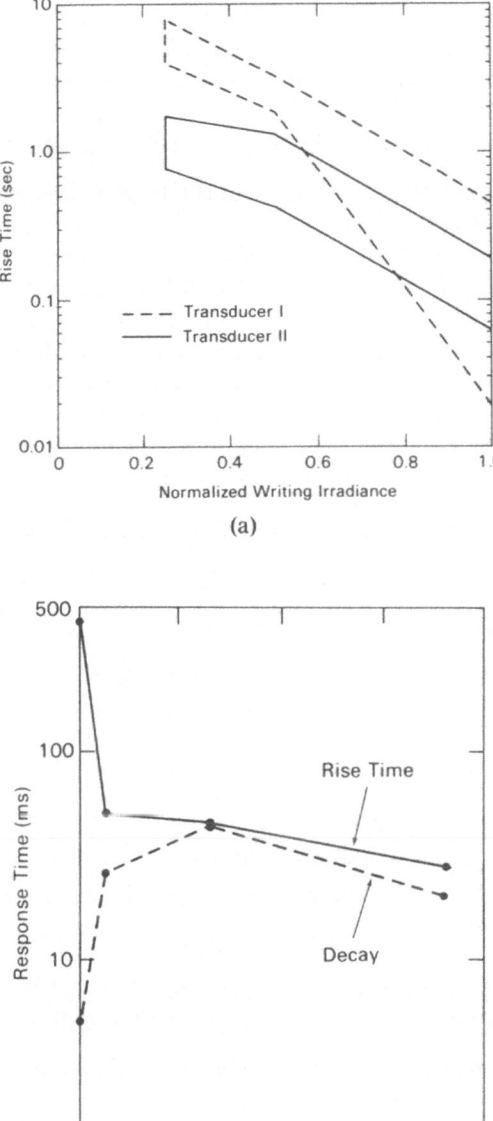

Fig. 6. (a) Rise time of two Hughes image transducers as function of normalized writing irradiance. The full and solid lines are the envelope of all measurements made on the two devices. Writing irradiance was normalized by the input irradiance which produced saturation. (b) Effect of a bias irradiance on rise time and decay time of Hughes image transducer II.

References pp. 234-235

The finite response time of the image transducer places a limit on the speed of object motion in the scene. Leading edges of an image (dark-to-light transitions) are degraded by the finite rise time and trailing edges (light-to-dark transitions) are blurred by the finite decay time. This loss of edge definition (or contrast) results in a reduction of the intensity of the higher frequency components of the spatial frequency spectrum.

Some of the limitation on performance of the Hughes liquid crystal transducer can be related to the cadmium sulfide photoconductor. These are cosmetic quality, input light sensitivity, and speed of response [12]. A silicon photoconductor transducer is presently under development at Hughes which will increase the input sensitivity by an order of magnitude. The time response of the silicon photoconductor will be faster than the liquid crystal, which will then place a limit of 25 milliseconds cycle time on the device, independent of input light intensity. In addition, the cosmetic quality will be improved since the silicon will not require mechanical polishing, which causes the pits and scratches that are common in the cadmium sulfide devices. The silicon based transducers are projected to be available within one to two years [13].

Filter Synthesis — The filter function, $T(f_x, f_y)$ in Eq. (3) for the most general case requires placing into the Fourier transform plane (P_2 of Fig. 1) a mask which will alter both the amplitude and phase of the light distribution in P_2. The amplitude modulator portion of the filter can be approximated by darkening a photographic emulsion or by controlled evaporation of some absorbing material on a glass substrate [14]. The phase modulation part however is very difficult to fabricate in even the simplest cases. Consider the one-dimensional differentiation filter $T(f_x) \propto f_x$. This requires an amplitude transmission profile which is zero at the origin in plane P_2 and increases linearly in both the positive and negative frequency directions. This change of sign between the two half-planes can be synthesized by a π phase difference in crossing the $f_x = 0$ point. i.e. $T(f_x) = e^{i\pi}|f_x|$ for $f_x < 0$ and $T(f_x) = |f_x|$ for $f_x > 0$. The π phase difference can be accomplished by evaporating a thin dielelectric film of proper optical thickness onto one half of the filter. Obviously one has a discontinuity at $f_x = 0$ and the quality of the differentiation operation will be influenced by the width of this phase change region. The amplitude portion of this filter is again not trivial since neither photographic emulsion nor metallic film evaporation can in practice give zero transmission at a point. Also, as the optical density of either absorber is increased the thickness increases result in a path length distortion. In practice the differentiation filter is approximate. We have investigated many approaches to this approximation.

A test signal and processor configuration used for a computer simulation and an experimental comparison of various edge enhancement filters in one dimension is shown in Fig. 7. The test signal s(x) is a one-dimensional rectangular pulse whose Fourier transform $S(f_x)$ is multiplied by the filter whose amplitude transmission is $T(f_x)$. The product is again Fourier transformed by the second lens to produce the output. The actual edge enhanced output for each filter is judged in terms of its approximation to a true derivative. A true band limited differentiation filter operated on the rectangular pulse input is shown in Fig. 8

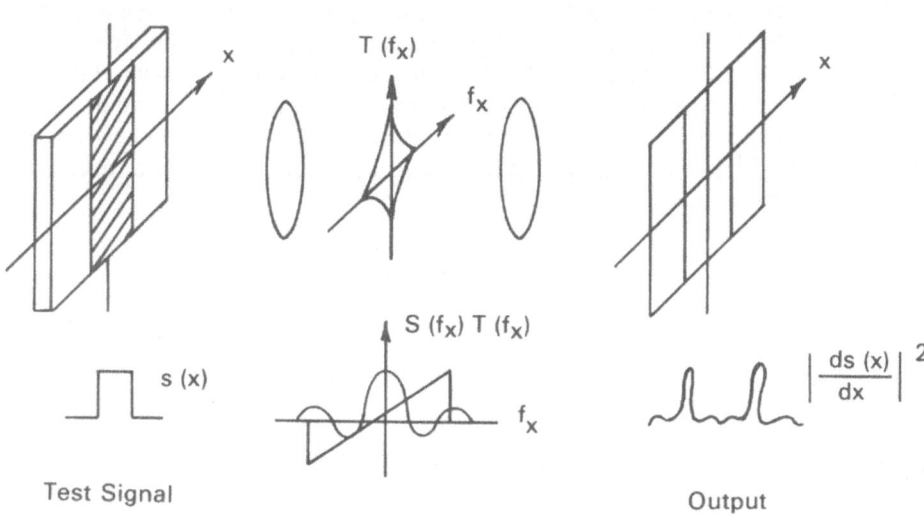

Fig. 7. Procedure for determining the effect of a one-dimensional differentiation filter, $T(f_x)$, on a slit input signal $s(x)$.

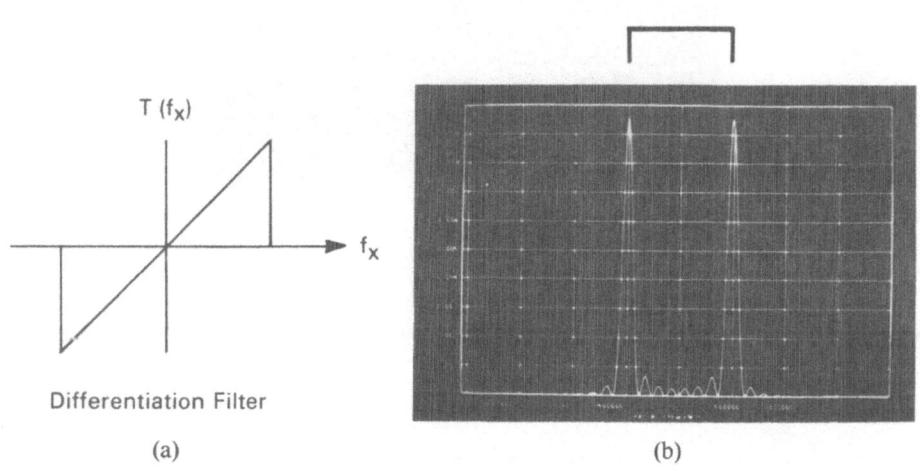

Fig. 8. (a) A band-limited true differentiation filter function, and (b) computer simulated effect on a slit input.

References pp. 234-235

The filter shown at the top of Fig. 9 is commonly called a high pass filter and is easily fabricated by placing an opaque strip on a flat glass plate. A high frequency cutoff of four times the low frequency stop was adequate for the computer simulation, shown in the center, to define the action of the filter. This filter has a double peaked edge with the null corresponding to the position of the input pulse, as shown. The bottom curve is the data obtained from an actual scan of an optical processor output using the above filter and input pulse. These data were obtained with a linear diode array detector. The input pulse width is 1.07 millimeters. The filter blocks from zero to 1 cycle/millimeter. The Fourier transform lenses have a focal length of 600 millimeters. The position of the peaks in the experimental traces correspond well to those predicted by the computer simulation. The relative height of the peaks is not accurately reproduced due to detector noise limitations.

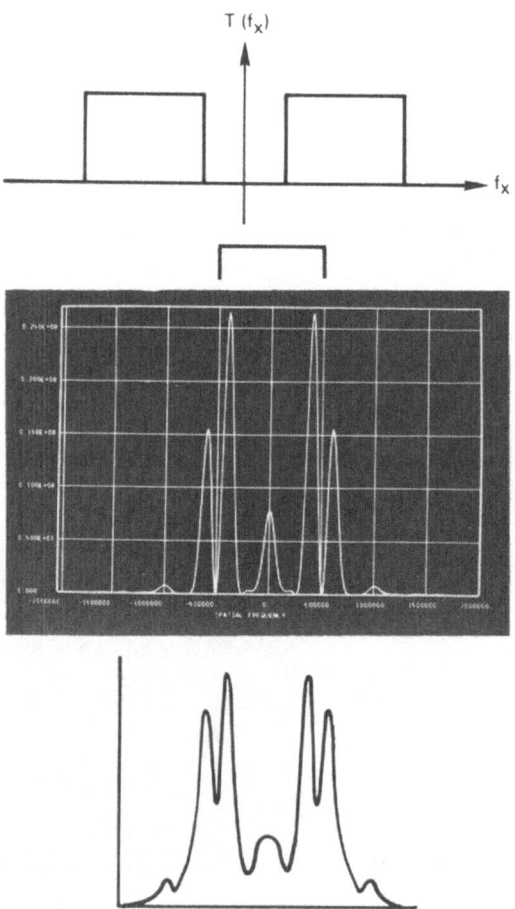

Fig. 9. A high pass filter (top) showing the computer simulated output (center) and the measurement output (bottom).

The response to one-half of a high pass filter (shown in Fig. 10) is similar to that from a high pass filter except for the double peak at an edge. This makes it less accurate in locating the edge (note that the peak intensity does not occur at the physical edge). However, it provides more total light in the edge compared to the side lobes. The input pulse is 1.07 millimeters, the same as for Fig. 9.

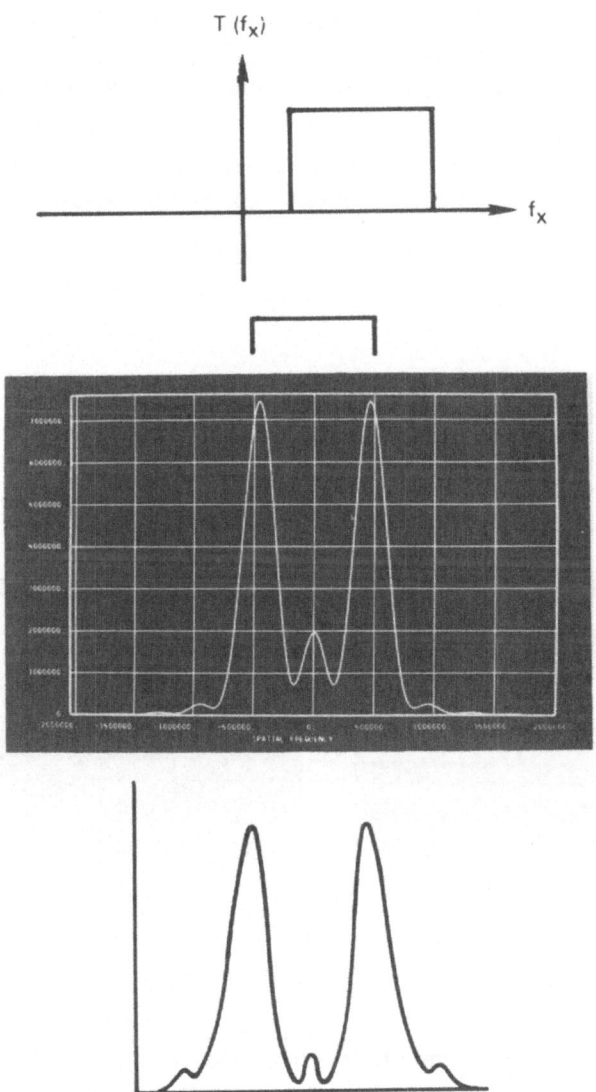

Fig. 10. A one-half high pass filter (top), the computer simulated output (center), and measured output (bottom).

References pp. 234-235

Two filters are shown in Fig. 11 for which we ran computer simulations but have no experimental curves. The first, on the left, has a transmission which increases linearly with the magnitude of the spatial frequency. This filter performs similar to the high pass filter in Fig. 9. In addition to the reduced side lobe at the center of the output, the peak edge signal is not sensitive to input pulse width. This filter can be fabricated by controlled exposure of photographic emulsion or by variable thickness thin film deposition. We do not have the facilities for the latter and the former is very difficult to control due to nonuniformities in photographic emulsion response and sensitivity to developing temperature and time. Neither technique can provide the continuous approach to zero transmission at zero frequency, hence a discontinuity will always exist in this region. For many applications the effect of the discontinuity can be minimized by a stop.

The half-plane linear amplitude filter on the right has the same effect as the half-plane high pass in eliminating the double peak at the expense of edge location accuracy.

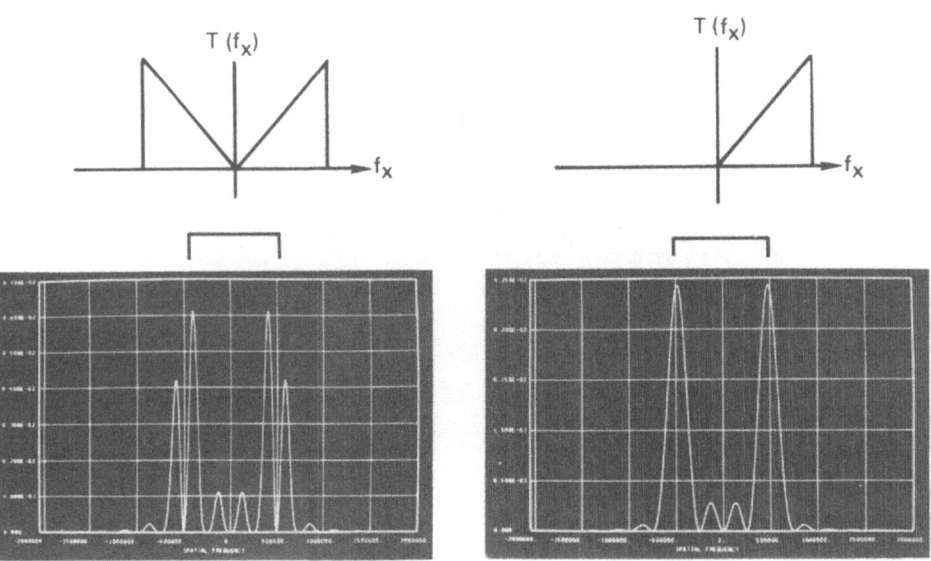

Fig. 11. A graded high pass (left) and a one-half graded high pass filter (right) showing the computer simulated output (below).

A phase filter with constant amplitude transmission is shown in Fig. 12. This filter should provide good edge location and has the additional advantage of maximum light output since no energy is being attenuated. However, the discontinuity at zero frequency is difficult to fabricate and the asymmetry in the experimental scan shown at the bottom is a result of that difficulty. This filter was made by depositing a thin film on one-half of a flat glass plate of sufficient film thickness to produce a π phase retardation in transmission. The width of the π step region was measured to be 250 micrometers. This produces a prism effect which makes one edge more intense than the other edge. When a 200 micrometer low frequency stop was placed on this discontinuity, this phase step effect is reduced. This also, of course, reduces the light efficiency of this type of filter.

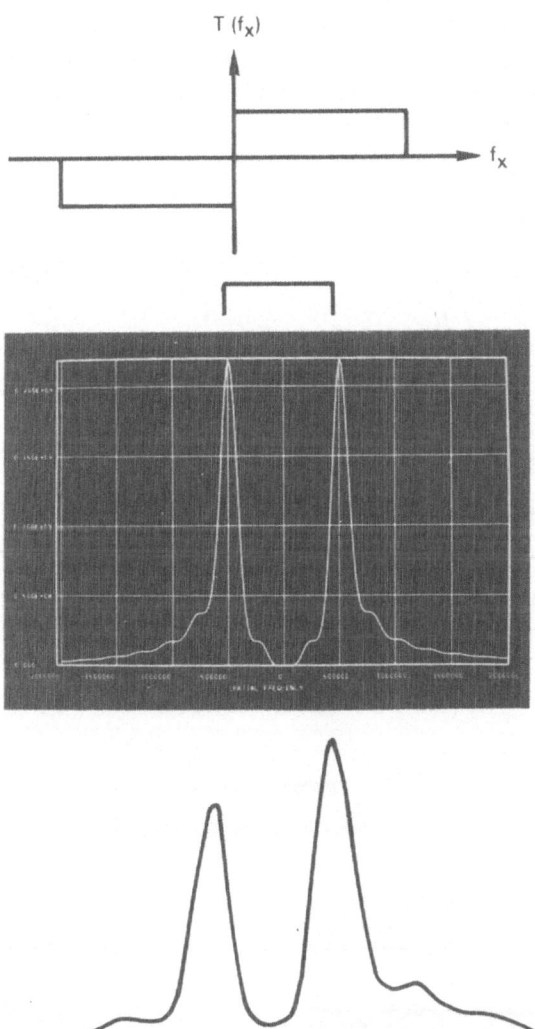

Fig. 12. A π phase-shift filter (top), the computer simulated output (center) and the measured output (bottom). The asymmetry in the measured output is a consequence of the finite width of the phase step.

References pp. 234-235

The differentiation filter characteristics can be approached using a technique described by Yao and Lee [15]. A holographic technique is used to make a filter which, over a limited region near zero frequency, approximates a linear differentiation filter (Fig. 13). A hologram is made in the Fourier plane using two point sources as the object, whose relative phases are displaced by π. The edge enhanced output will have an edge width given by Δ, the physical separation of the two points. We have made a computer simulation that shows that an optimum value for BW is about one-half the period of the sinusoid. Using a smaller portion of the sinusoid reduces the light level with no significant improvement in edge definition (i.e., side lobe reduction). The output signal from the processor using a series of one-dimensional holographic filters on the 1.07 millimeters horizontal input pulse is shown in Fig. 14. The dimensions listed correspond to Δ, the separation of the two out of phase point sources used to make the filter. This also is the width of the edge.

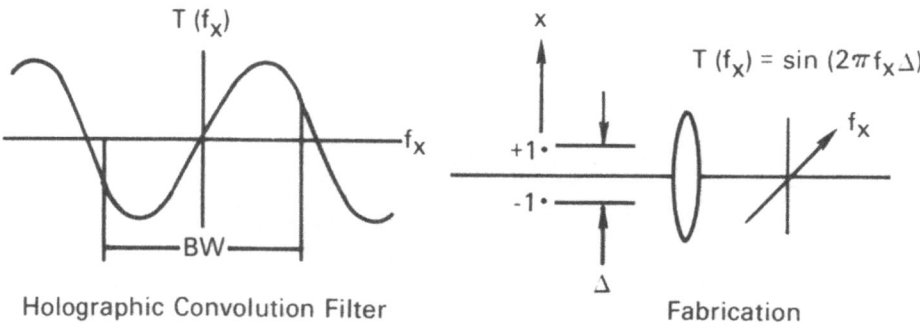

Fig. 13. Method for preparing a holographic filter which, over a limited region, gives a transmission function approximating a true differentiation filter.

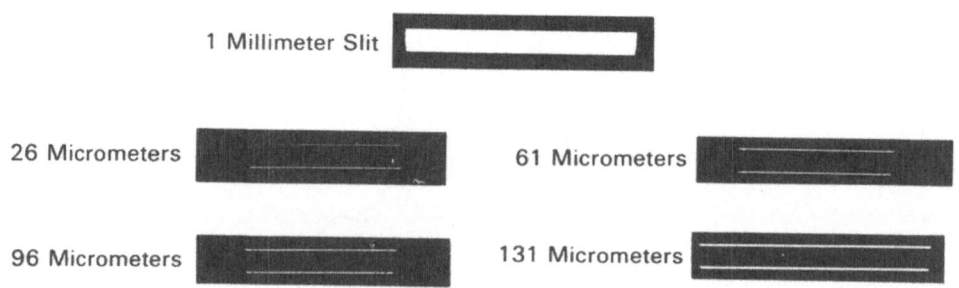

Fig. 14. The edge enhancement of a slit input using a holographic filter. The width of the edge line indicated is controlled by the separation Δ of the two point sources in Fig. 13.

OPTICAL COMPUTING FOR IMAGE PROCESSING

An Air Force resolution target was used for determining resolution limitations of each filter and edge signal dependence on input bar width. A graphic example of the resolution capabilities and edge signal strength as a function of input pulse width is shown in Fig. 15. The edge output on the left was made using a holographic filter with a point separation of 26 micrometers, corresponding to a sinusoid period of 40 lines/millimeter (the band limit of the transform lens). The output on the right was made using the half-plane high pass filter. Along with the obvious expected loss of resolution (only one-half of the available bandwidth is used) there is a variation of edge signal intensity with input pulse width as expected. Haagen [16] showed that the signal strength should vary inversely as the pulse width.

Holographic Filter   Half-Plane High-Pass

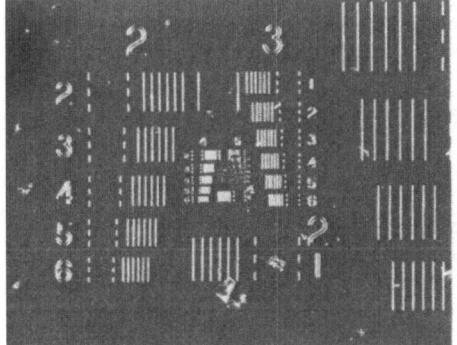 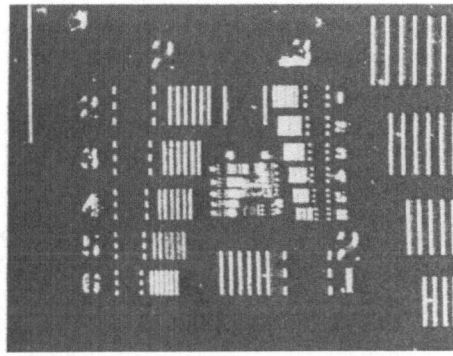

Fig. 15. Comparison of the edge enhancement results using (left) a holographic filter and (right) a half plane high pass filter on a resolution bar target.

In Fig. 16, we show a technique for combining two one-dimensional differentiation filters to obtain a two dimensional gradient filter. The optical processor is divided into two parallel channels by interrogating the input with two plane waves at different angles. Each wave goes to a different filter, one an x-derivative and the other a y-derivative. The two one-dimensional differentiated images are combined in the output plane. By making the two plane waves have orthogonal polarizations, the two outputs combine incoherently, thus giving a gradient, as shown. Coherent addition results in the component of the gradient along a 45° line.

EXAMPLES OF CURRENT STATUS OF REAL-TIME APPLICATIONS

Differentiation for Image Edge Enhancement — While the holographic filter gives the best approximation to a differentiation filter it may not be the best choice for a practical, real-time application. The filter requires precise positioning and complex optics are needed to multiplex two filters to achieve directional independent differentiation. In addition, the use of these filters requires a high degree of spatial coherence

References pp. 234-235

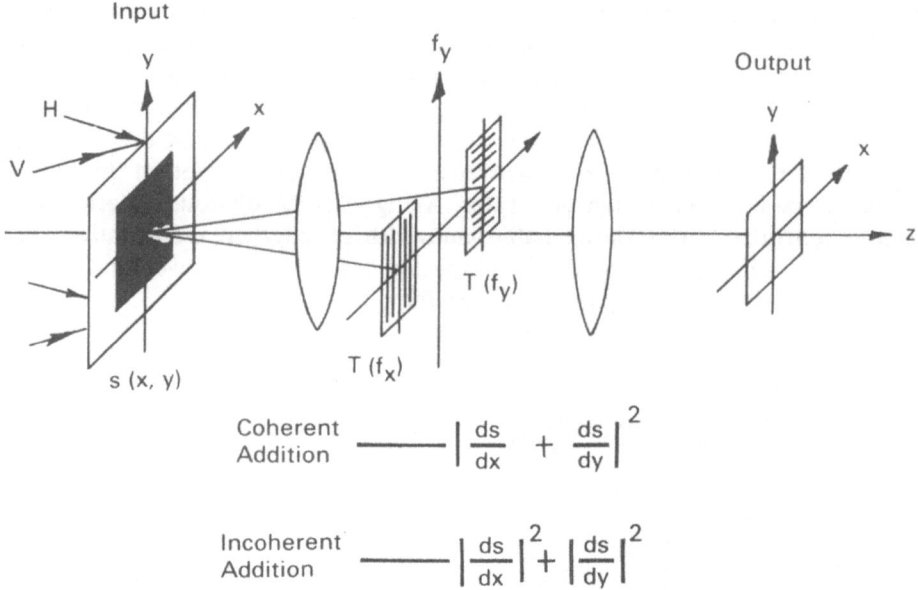

Fig. 16. A method for performing two-dimensional gradient operation. The x and y derivative channels are combined incoherently in the output plane.

in the laser illumination, hence optical noise suppression techniques which degrade spatial coherence could not be applied [17].

Lewis [18] has reported results of real-time edge enhancement with a two-dimensional version of a high-pass filter (Fig. 9) incorporating noise suppression. A diagram of the experimental apparatus is shown in Fig. 17. The filter is a circular opaque stop, 1 mm in diameter (corresponding to a low frequency cutoff of 7 cycles/mm spatial frequency in the input plane). To suppress coherent optical noise, the coherence of the readout beam was reduced using a spinning ground glass.

Edging results for an automotive connecting rod placed against a black foundry conveyor belt are shown in Fig. 18. The top pictures show a photograph of the object and an optically edge enhanced image as it appears on the television monitor display. The bottom right picture is a binary computer graphics display of the optically processed image. Black portions of this display correspond to regions of the edge image whose video level exceeds a fixed threshold level. The lower left picture is a computer graphics display of a digitally edge enhanced image. To produce this result, the input scene was digitized using the same television camera and approximately the same lighting conditions. This image was then enhanced with an eight point weighted gradient [19] filtering program. The edging program which was written in PL/1 and run on an IBM 370/168 computer required 4 seconds processing time.

This type of optical processing could be used to reduce the quantity of picture data for a digital picture processing system.

OPTICAL COMPUTING FOR IMAGE PROCESSING

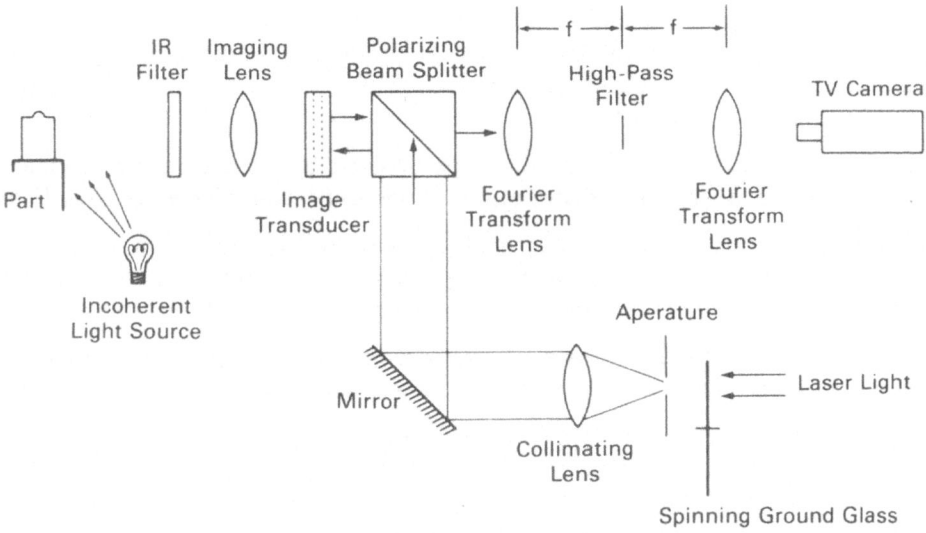

Fig. 17. Real-time high pass optical filtering system. The spatial coherence of the laser light is first degraded by the spinning ground glass to reduce the coherent optical noise in the edge-enhanced image detected by the TV camera.

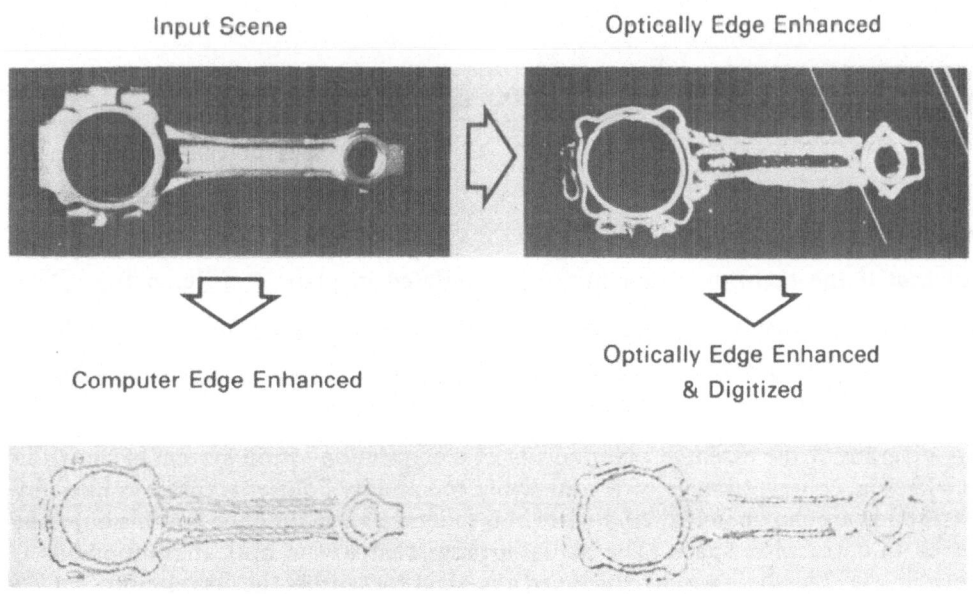

Fig. 18. Optical edge enhancement results for a connecting rod.

References pp. 234-235

Correlation for Identfying and Locating Parts — We saw previously that the output of the optical processor can display the correlation between the real-time input image and a stored model. The Fourier plane filter for this application has a transmission proportional to $G^*(f_x,f_y)$ the conjugate of the transform of the model. The problems associated with direct synthesis of this type of filter are severe, since in general $G^*(f_x,f_y)$ will be a complex distribution of intensity and phase. A significant simplification of this procedure was made by Vander Lugt [3], who proposed a holographic method for correlation filter synthesis. In this method, the model of interest is placed in the input plane and the transform $G(f_x,f_y)$, along with a second coherent beam (the reference beam, R) are recorded on photographic emulsion in plane P_2. For simplicity, the reference beam is a plane wave of unit amplitude, $R(f_x,f_y) = \exp(-2\pi i f_x \cos\theta)$, making an angle with the f_x axis.

Following the usual methods of holography the developed recording will have a transmission function containing the term $G^*(f_x,f_y)R(f_x,f_y)$ which contains the desired function. The transmission function contains other terms as well, but these will be physically separated in the output plane and not overlap the term of interest provided the angle θ is chosen large enough. Note that the correlation distribution in the output plane (P_3) is modified by the presence of the factor R in the filter. This has the effect of yielding a correlation distribution between a model $g(x_1,y_1)$ and the real-time image $h(x_1,y_1)$ given by

$$U(x_3,y_3) = \int\int_{-\infty}^{\infty} h(x_1,y_1) g^*(x_1-x_3+f\cos\theta, y_1-y_3) dx_1 dy_1,$$

i.e. the correlation function is centered at the coordinates $(f\cos\theta, 0)$ in plane P_3. Note also that if the real-time input image is translated in plane P_1, i,e, $h'(x_1,y_1) = h(x_1-a, y_1-b)$, then the correlation peak is shifted in P_3 by the same amount. Thus the system can identify if the model is in the scene by the strength of the correlation and the position of the model by the position of the correlation signal.

A diagram of the essential components of a correlation vision system is shown in Fig. 19. The coherent image of a connecting rod and its Fourier transform intensity distribution are shown in Fig. 20. Figure 21 is the correlation function for this object as viewed in correlation space. The actual arrangement of the optical components is shown in Fig. 22, where we have included a method for sensing the correlation spot. In addition we show an image rotator which is necessary for correlation detection when the objective can have an arbitrary orientation in the input plane. Since the Fourier

OPTICAL COMPUTING FOR IMAGE PROCESSING 227

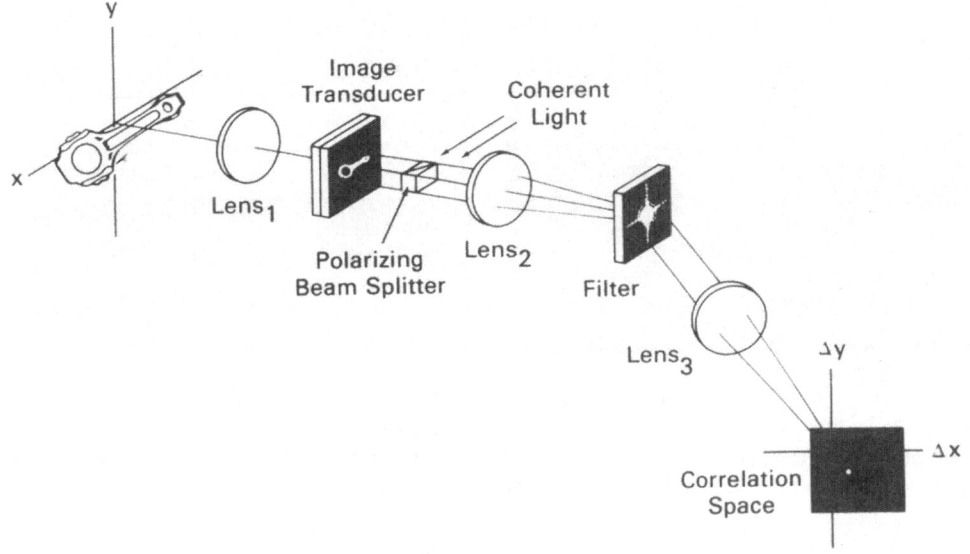

Fig. 19. Essential components of a real-time optical correlation vision system. The object on the left is illuminated with ordinary (non-laser) light.

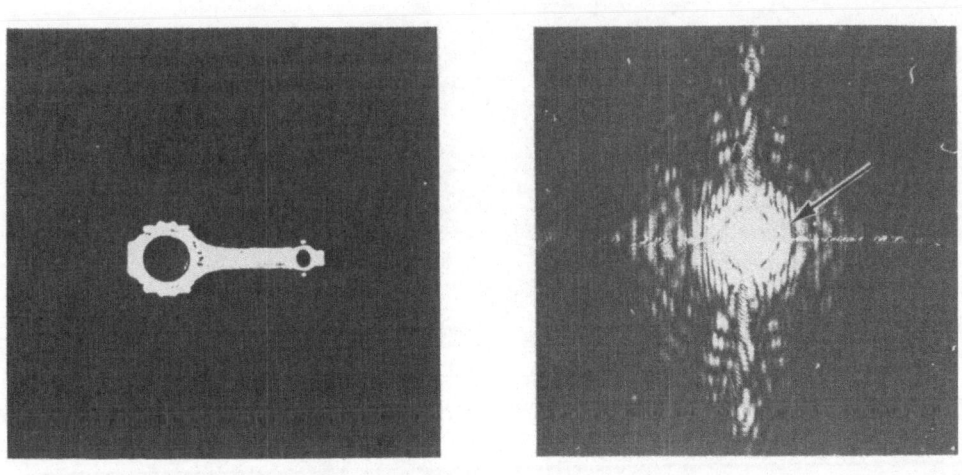

Fig. 20. The coherent image (left) from the image transducer and (right) the intensity distribution of its Fourier transform. The arrow indicates the position where the filter was optimized, corresponding to a spectral component at 3 cycles/millimeter.

References pp. 234-235

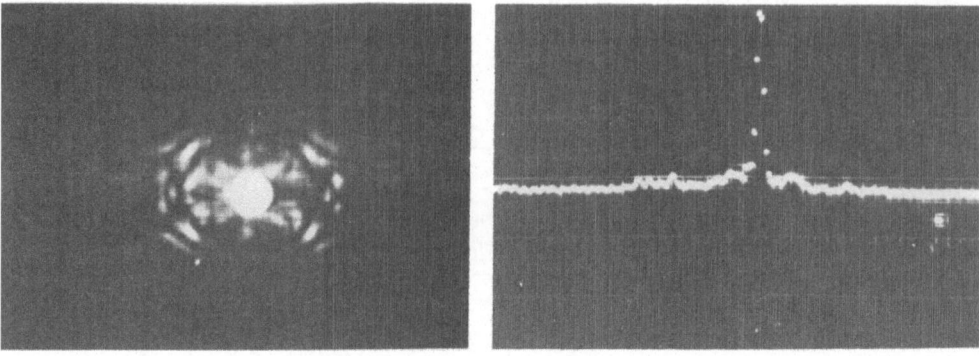

Fig. 21. The intensity distribution (left) in correlation space with the central peak overexposed to show the full distribution. The oscilloscope trace (right) shows the relative irradiance along a horizontal line through the correlation distribution.

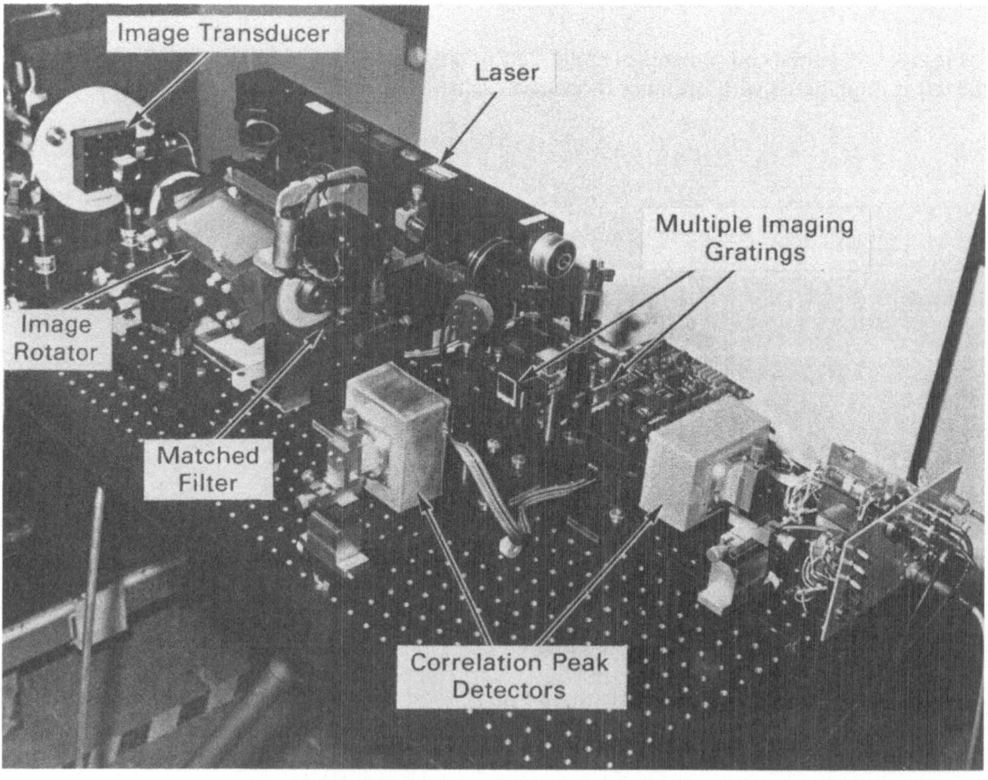

Fig. 22. The optical correlator used as a vision sensor for a robot.

transform is not invariant to rotation, it is necessary to either 1) rotate the scene itself, 2) rotate the coherent light distribution between the image transducer and the filter, or 3) rotate the filter. In all three methods, the correlation signal will peak as a function of rotation angle, as shown in Fig. 23.

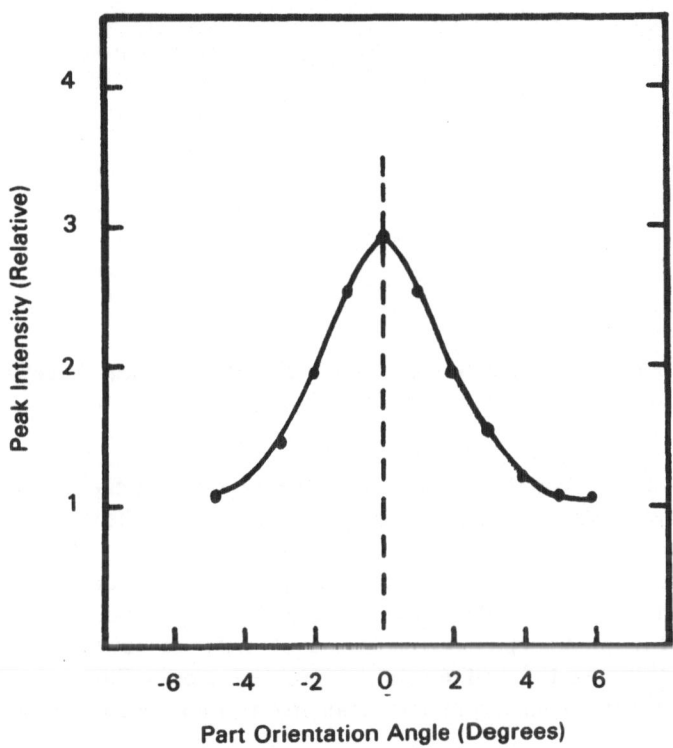

Fig. 23. The relative magnitude of the correlation peak near the angle match condition.

To search for the correlation peak, a scan of the correlation space is made for every increment in angle of the image rotator. An area camera is best suited to this purpose for maximum signal detectability since the correlation signal is a spatially isolated high intensity small region. However, to reduce the amount of data for each scan from N^2 (where N is the number of resolution cells in each dimension of correlation space) to 2N, we use a system of two orthogonal linear diode array detectors. A series of three gratings, each producing three equal intensity images of the correlation space, produces a line of correlation peaks normal to the axis of each linear detector. Thus, as shown in Fig. 24 the x-coordinate of the correlation signal can be obtained independent of the y-coordinate. A loss of signal and signal/noise accompany the reduction in data rate, however the correlation signal in Fig. 21 was obtained with this method, showing adequate signal is still available.

References pp. 234-235

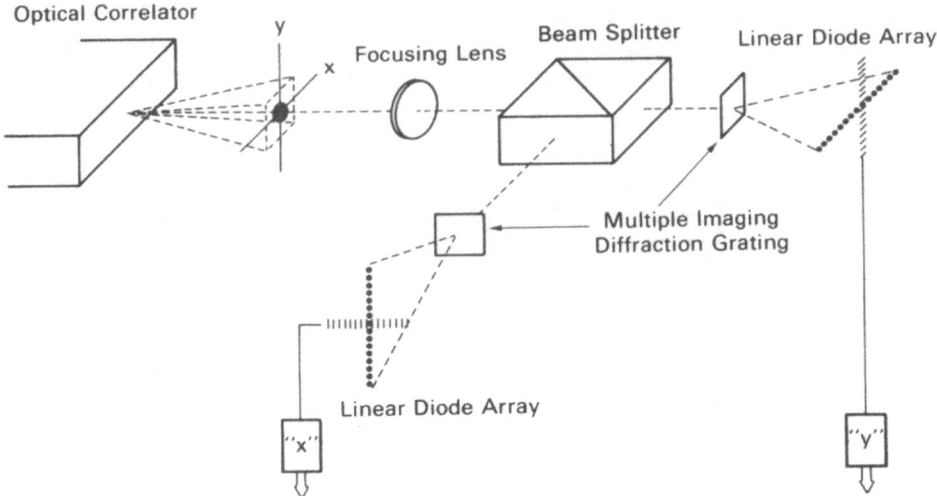

Fig. 24. A method for sensing the correlation peak using two linear diode array detectors. Multiple images of the correlation peak are created by the diffraction gratings.

The relationship between the coordinates of the correlation peak and coordinates of the model in object space is determined by the focal length of the lenses in Fig. 19, the angle 0 of the reference beam, and by the demagnification ratio of the image on the image transducer. The data in Fig. 25 taken for a connecting rod on a moving conveyor show that correlation space coordinates are linearly related to conveyor coordinates with an error of less than 1 part in 130 over a 0.6 meter object search distance. The scale factor (in this case 180 millimeters in object space per millimeter of correlation space) can be adjusted by changing the focal length of any one of the three lenses in Fig. 19.

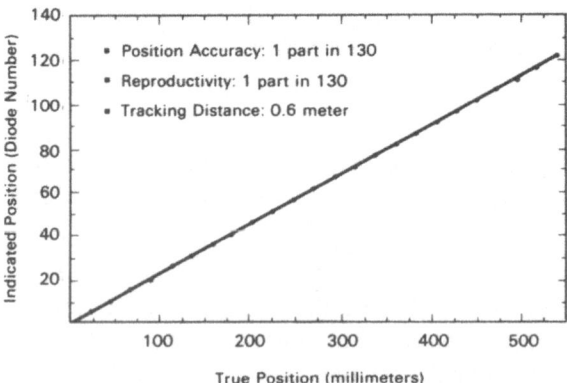

Fig. 25. Data showing the linearity between coordinates in correlation space (diode number) and object space (true position). The slope can be adjusted by the choice of lens focal lengths in Fig. 19.

OPTICAL COMPUTING FOR IMAGE PROCESSING

Fig. 26 shows the decrease in the peak value of the correlation signal as a function of conveyor speed. The signal loss is consistent with the rise and decay time of the image transducer under the conditions of weak bias illumination (Fig. 6). Bias illumination is provided by the weak reflectivity of the conveyor belt surface. The tracking accuracy of the system was tested at conveyor speeds up to 300 millimeters per second. No loss of accuracy could be detected at speeds below 250 millimeters per second and a tracking error of 1 part in 130 was detected at speeds above 250 millimeters per second.

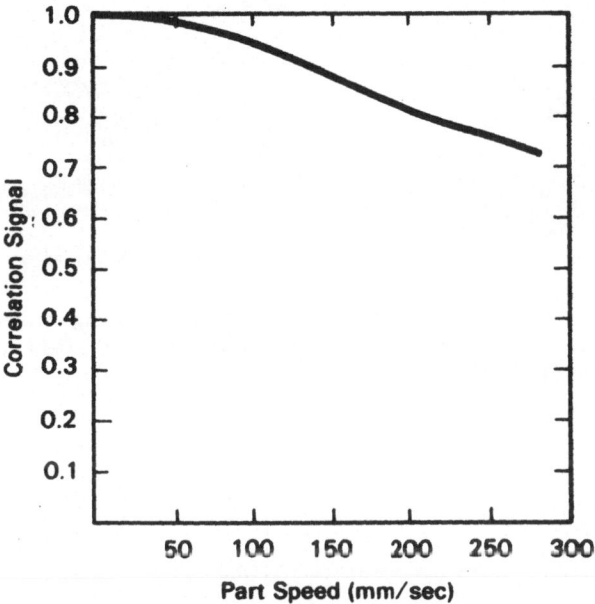

Fig. 26. The effect of object motion on the strength of the correlation peak. The decrease can be attributed to the finite response time of the liquid crystal image transducer (Fig. 6).

This optical system was tested as the real-time vision input to a Unimate robot (Model 2000). The tests were conducted to prove the feasibility of the method and to determine what problems needed to be solved for practical implementation. The test configuration is shown in Fig. 27. A 0.7 by 0.7 meter area of the conveyor belt is viewed by the optical system with the aid of a mirror attached to the optical platform. The search procedure and robot interface are diagrammed in Fig. 28. For this test a ready signal from the robot initiates the search procedure. When the image rotator, which is continuously rotating, passes the $\theta = 0°$ position, a start pulse initiates the scan of both linear detectors (each containing 256 photodiodes) sensing the light distribution in correlation space. The detectors scan once for every degree of rotation until the image has been rotated through 360°. A peak search determines the angular orientation giving the largest diode signal and the diode address of the peak signal. This information is used in the transformation to belt space for the location and orientation of the part on the belt.

References pp. 234-235

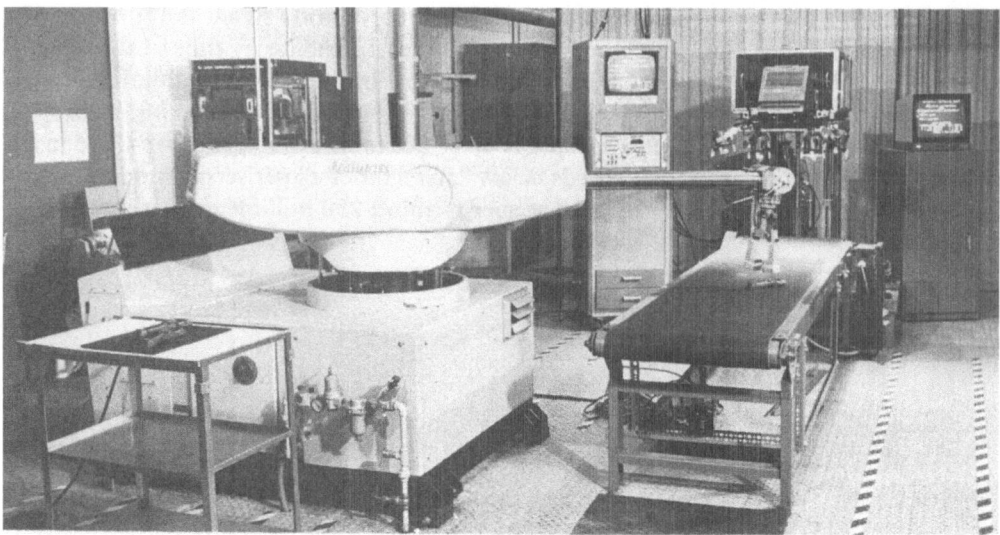

Fig. 27. The robot arm in position to grasp the part on the moving conveyor. The monitors indicate the coordinates of the part on the belt when the part was in the field of view of the vision system (lighted area on the belt).

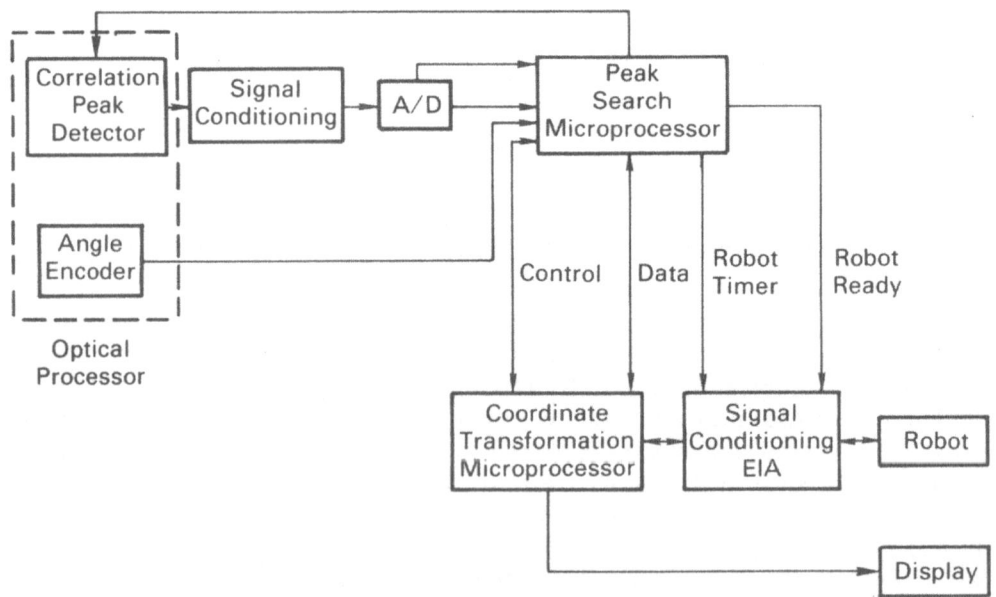

Fig. 28. The signals from the linear diode array detectors are processed in the peak search microprocessor to obtain the peak coordinates in correlation space. These data are transformed to belt space for the robot tracking and pick-up functions which are performed by the robot minicomputer.

If, after the 360° rotation, no peak is found above a fixed present threshold, the search continues. When a peak is obtained, the part coordinates are transferred to the robot minicomputer which tracks the belt position for pickup. The search may then be interrupted until further notice from the robot, or the search could be programmed to continue and stack part locations as they are found.

The time to complete a 360° search was set to be one second for the test. The limiting factor on speed is determined by the strength of the correlation signal from the linear detector. This signal strength depends on the output power of the laser, the light transmitting efficiency of the optical system (particularly the filter), and the sensitivity of the detector. A higher search rate is possible with the present system. The search rate does not limit the belt speed, set to be 60 millimeters/sec, as long as the part stays in the field of view during the 360° rotation.

The peak search was designed to find only one peak, therefore only one part could be found during the 360° rotation. If more than one part is in the field of view, the system finds the part giving the highest correlation signal. Under the best conditions, the correlation signal from the connecting rod was 20 times the background level observed with no part in the field of view. The fixed threshold was set to reject the smaller cross-correlation signal from other cast iron parts. Again under the best conditions, the vision system found all connecting rods (with a wide range of reflectivity) and rejected the other parts. However, drift of the optical alignment and the operating characteristics of the image transducer demanded frequent correction for optimum performance.

SUMMARY

It has long been recognized that optical methods have enormous potential for image processing. One of the limiting technological problems has been the inability to deal with real-time inputs. In this paper, we have shown two examples of real-time processing using a liquid crystal image transducer as the input light modulator. Optical edge enhancement may find use as a preprocessor to reduce the data load for digital image processing. Optical correlation has been demonstrated as a potential robot vision system.

The successful implementation of optical processing in an industrial environment will depend on many factors. The image transducers now available need to be more reliable and lower in cost. The continuing development of these devices, particularly for large screen display applications, may provide the characteristics needed. The complexity of the optical processor requires careful design and plant hardened construction. The availability and cost of complex optical systems will parallel the development of a large scale optical industry. Applications such as optical communications, optical printers and readers, and optical inspection may provide the basis for this development.

References pp. 234-235

ACKNOWLEDGMENTS

The analysis for the comparison of differentiation filters was provided by Robert L. Lewis. The robot vision tests were conducted as a joint program between General Motors Research Laboratories (GMR) and General Motors Manufacturing Development (GMMD). Michael Kaplit of GMR is responsible for the microprocessor system that interfaced the optical correlator with the robot. The robot control minicomputer was operated by Carl Rosetti and Don Mick of GMMD. Reed Farrar of GMMD contributed the design and construction of the image rotator and provided valuable assistance in the operation of the optical correlator.

REFERENCES

1. Lee, S.H. Mathematical operations by optical processing. *Pattern Recognition* 5 (1973), 21.
2. A variety of methods for performing differentiation have been reported. A representative survey would include:

   Lohman, A.W., and Paris, D.P. Computer generated spatial filters for coherent optical data processing. *Appl. Opt.* 7 (1968), 651.

   Eguchi, R.G., and Carlson, F.P. Linear vector operations in coherent optical data processing systems. *Appl. Opt.* 9 (1970), 687.

   Yao, S.K., and Lee, S.H. Spatial differentiation and integration by coherent optical correlation method. *J. Opt. Soc. Am.* 61 (1971), 474.
3. Vander Lugt, A.B. Signal detection by complex spatial filtering. *IEEE Trans. on Information Theory* IT-10 (1964), 139.
4. Kato, H., and Goodman, J.W. Non-linear transformations and logarithmic filtering. *Opt. Commun.* 8 (1973), 378.

   Dashiell, S.R., and Sawchuch, A.A. Optical synthesis of non-linear non-monotonic functions. *Opt. Commun.* 15 (1975), 66.
5. Strand, T.C. Non-monotonic non-linear image processing using half-tone techniques. *Opt. Commun.* 15 (1975), 60.

   Lee, S.H. Review of coherent optical processing. *Apply. Phys.* 10 (1976), 203.
6. An important paper establishing the concepts for the use of Fourier methods in optical processing is: O'Neill, E.L. Optical filtering in optics. *I.R.E. Trans. on Inf. Theory* 2 (1956), 56.
7. Thompson, B. J. Hybrid processing systems — an assessment. *Proc. IEEE* 65 (1977, 62.
8. Goodman, J. W. *Introduction to Fourier Optics.* McGraw-Hill, New York, 1968.
9. Casasent, D. Spatial light modulators. *Opt. Eng.* 17 (1978), 307-384.
10. Jacobson, A., et al. A real-time optical data processing device. *Inf. Disp.* 12 (1975), 22.
11. Bleha, W.P., et al. Application of the liquid crystal light value to real-time optical data processing. *Opt. Eng.* 17 (1978), 371.
12. Fraas, L.M., et al. Novel charge-storage-diode structure for use with light-activated displays. *J. Appl. Phys.* 47 (1976), 576.
13. Grinberg, J. private communication.
14. Vander Lugt, A. A review of optical data processing techniques. *Optica Acta* 15 (1968), 1.
15. Yao, S.K., and Lee, S.H. Spatial differentiation and integration by coherent optical correlation method. *Opt. Soc. Am.* 61 (1971), 474.

16. Haagen, W. Edge enhancement of photographic imagery. Contract AF 33 (615)-1267, Rpt. RF1801-26, The Ohio State U., 1968.
17. DeVelis, J.B., et al. Review of noise reduction techniques in coherent optical processing systems. *Proc. S.P.I.E.* 52 (1974), 55.
18. Lewis, R.W. Real-time coherent optical edge enhancement. *Appl. Opt.* 17 (1978), 161.
19. Duda, R. O., and Hart, P. E. *Pattern Classification and Scene Analysis.* John Wiley and Sons, New York, 1973.

DISCUSSION

C. Brown *(University of Rochester)*

This method seems to be relatively dependent on the autocorrelation of the part. How robust is it over a wide range of parts? If you have an autocorrelation which is not so sharp as it is on a conrod it looks like this method could fail.

Gara

Absolutely, we do depend on seeing the object on the belt.

Brown

What happens when you don't know the shape of the object? Or if the part is a circle or has fewer properties for finding orientation?

Gara

On a circular symmetric part, perhaps one wouldn't want its orientation. Usually one does want the orientation but the features which give the part orientation may be too small to give us a change in correlation as a function of orientation. One has to deal with that problem.

A. Rosenfeld *(University of Maryland)*

If you have two indentical parts in different positions, but in the same orientation you presumably get two peaks and your x, y bits are going to perhaps give it ambiguity as to which peak is which.

Gara

That's right. The way we have the system set up, we only look for the largest signal. If one could remove that object from the scene which gave the largest signal the next largest one would give the next signal.

Rosenfeld

If they were approximately the same you have two x values and two y values and four possible places to look.

Gara

We only pick the largest correlation signal. If the x signal is the largest for one rod it will also be the largest in the y channel for the same rod, so we will only pick one correlation spot.

R. R. Bajcsy *(University of Pennsylvania)*

One characteristic of Fourier transforms is that they are global transforms and the advantage of this to image processing is that you have the flexibility of having various sizes of windows and you can focus on various parts of the scene. One idea we have considered but done nothing about was that one could have an array of optical sensors of maybe 2×2 or 4×4 which would give you this model, and then compute several correlations instead of one correlation. Have you considered such an approach?

Gara

Yes, relative to some other problems. For example we have only looked at a part which had one stable position on the belt. The steering knuckle which Raj Reddy showed has many stable positions. In that case one would use a multiple input function or multiple channels in the processor. One could pick off portions of the input and send one portion to one channel and another portion to another channel.

An interesting development at the Hughes Research Labs is a CCD addressed image transducer rather than an optically addressed image transducer. With this, one has the capability to establish any image window one wants from a TV input.

B. H. McCormick *(University of Illinois - Chicago Circle)*

Has any thought been given to putting an identification on the casting. For example if I put a couple of circular areas in there, I wouldn't need to go through a rotation. I could obviously get the orientation.

Gara

That's right, then it would fit better into our patent which says look for two holes. I think I have heard of this being done in some cases for inventory purposes but not for pattern recognition. But I don't see why it couldn't be done on some parts. A problem is that it is an expensive task in some cases to do that.

D. Nitzan *(SRI International)*

How do you compare the two methods, your method and the CONSIGHT method in terms of performance and costs?

Gara

As you can imagine, we have done that. In terms of performance I think that the CONSIGHT method described by Steve Holland probably would have comparable coordinate positioning accuracy. Of course we could expand our input. Our search space was only 256 × 256 pixels in correlation space. Our input is capable of 1000 × 1000 pixels, so we could have expanded that and gotten higher accuracy. But for the application that he was addressing certainly someone in a foundry would look more favorable on the computer vision technique because it is a simpler system. We would be able to look at a stationary part. A part in a line of light needs to be moving, it cannot be stationary. I am not saying that this is a replacement for that system necessarily, or that this is competing with it necessarily, but it's another approach using another technology which may have application in other great areas.

I don't really think we can put a cost figure on this system. The image panel or transducer which I described is a research item developed at the Hughes Research Laboratory. They are in production now because that device has a potential commercial use as wide screen TV projector or a large screen graphics display flat panel. When they start rolling them off their production line then we can talk to you about cost figures — thousands of dollars eventually.

PROSPECTS FOR INDUSTRIAL VISION

J. M. TENENBAUM, H. G. BARROW and R. C. BOLLES

SRI International, Menlo Park, California

ABSTRACT

The proceedings of this symposium contain many impressive examples of what can be accomplished today in factories with computer vision. Compared with human vision, however, the capabilities of these specialized systems are still very rudimentary. In this paper, we will examine some important industrial vision requirements that are simple for humans to fulfill but well beyond the ability of any existing machine. We will try to understand what limitations of current systems make these tasks so hard for a machine and use this knowledge to outline the design of a general-purpose computer vision system, capable of high performance in a wide variety of industrial vision tasks. We will conclude by suggesting some promising research directions towards realizing such a system.

PERFORMANCE CHARACTERISTICS

Cost
Speed
Competence
Reliability
Flexibility
Trainability

The performance of a vision system can be evaluated along numerous dimensions. As researchers, we view competence as the most fundamental since it determines whether a task can be performed at all. If the basic competence exists, then it is up to the ingenuity of industrial engineers to bring factors such as speed, cost, and reliability within acceptable limits so that an application becomes cost-effective.

References p. 256

LIMITATIONS OF CURRENT INDUSTRIAL VISION SYSTEMS

        High Contrast
        No Shadows
        No Occlusion
        2-D Models
        Rigid Objects
        Standard Viewpoint

The competence of current industrial vision systems restricts their application, by and large, to situations where individual objects can be easily isolated in an image. Typically, objects are presented against a high contrast background (no occlusion) with lighting controlled to eliminate shadows, highlights, and other factors that would make scene segmentation difficult. Objects are recognized by extracting 2-D features which are matched against 2-D object prototypes. This limits the system's recognition to known objects observed from standard viewpoints. Many tasks naturally fit these constraints or can be readily engineered to do so, with structured light and other artifices. Examples include picking parts off a moving conveyor, bonding leads to semiconductor chips, and inspecting bottles for misaligned labels.

In many industrial vision tasks the required engineering can be unacceptably expensive, difficult, or time consuming.

TASKS BEYOND CURRENT LIMITATIONS

        Bin Picking
        Recognition of Parts on Overhead Conveyor
        Recognition of Nonrigid Objects
        Implicit Inspection
        Robot Vehicles

Bin picking, for example, is hard because parts in a jumble have low contrast and occlude each other, making it difficult to isolate individual parts in an image. Inspecting the components of a finished assembly is difficult for the same reason and less amenable to engineering simplification (e.g., first dumping the contents of the bin on a table). Swinging parts on an overhead conveyor are not neccessarily constrained to maintain a standard viewpoint. Nonrigid objects can assume a continuum of configurations and thus do not lend themselves to characterization in terms of fixed 2-D prototypes. Similarly, it is impractical to model in detail the appearance of all conceivable flaws (dents, scratches, blemishes and so forth) in an implicit inspection task. An archetypical example of a class of tasks that is inherently difficult to constrain is that involving robot vehicles in outdoor environments, such as construction site clearing and foresting.

Figs. 1-4 provide more specific examples of important industrial vision requirements that are easy for humans to fulfill yet significantly beyond the current competence of machine vision. In Fig. 1, the reader immediately recognizes the context, a bin-picking

PROSPECTS FOR INDUSTRIAL VISION

task, and is also cognizant of the layout of space; i.e., the area occupied by the bin and by the arm situated behind it. The representation of space facilitates complex spatial reasoning tasks such as planning approach trajectories for the arm that avoid collisions with the sides of the bin.

Within the bin (Fig. 2), the reader is aware of surfaces and the solid bodies they comprise. With such descriptions one can decide what part to grasp (based on shape and ease of access), how to pick it up (based on dimensions and inferred weight distribution), what the consequences of doing so will be (i.e., what fragment of the scene is likely to lifted along with the grasped body and what other parts are likely to be disturbed), and how the selected part can be mated with others on a pallet or in an assembly. The contemplated grasping operation can then be performed using 3-D visual servoing to keep the manipulator on its trajectory and orient it properly (Fig. 3).

Note that none of these functions requires any special assumptions about lighting or viewpoint or any prior familiarity with the specific parts or arm involved.

It is, of course, possible to recognize bodies by matching them with known object models. Matching can occur either before or after a part has been removed from a bin, independent of viewpoint and occlusion; complex, articulated objects like the arm are recognized as readily as simple rigid ones. Common imperfections ranging from surface marring (rust and scratches) to missing limbs are recognized as defects (as distinguished from artifacts like shadows or errors in matching) — even though defects are not explicitly represented in the model (Fig. 4).

Figs. 1-4. Bin picking: A "Tough Nut" for industrial vision system.

References p. 256

Why are these tasks that are so easy for humans to perform currently so difficult for computers?

COMPETENCE LIMITATIONS

> Representations
> Knowledge

The competence of any vision system is limited by the representations it uses to describe the world and the knowledge available for manipulating and transforming them [1 and 2].

REQUIRED LEVELS OF REPRESENTATION

> Images
> Pictorial Features
> Intrinsic Surface Characteristics
> 3-D Surfaces and Bodies
> 3-D Objects
> Space Map
> Symbolic Relationships

The ability to segment and describe the surfaces in a scene independent of lighting, shadows, and other viewing artifacts requires the ability to describe the color, distance, orientation and other intrinsic characteristics of the surface element visible at each point in the image. The ability to inspect surfaces requires, in addition, a generic model of flaws: what is a dent? a scratch? a discoloration? The ability to grasp and manipulate parts without expicitly recognizing them demands a 3-D volumetric representation of bodies occupying a region of space, together with a generic understanding of what constitutes a manipulatable body, and how various physical forms should be grasped. The ability to recognize objects independent of viewpoint demands the ability to extract view-invariant 3-D features from the body descriptions and match them with features of 3-D object models. The ability to reach for and manipulate objects in a scene requires a representation of the 3-D layout of space and descriptions of mechanical relations between objects such as "Supports," "Touches" and "Interlocks."

The importance of most of these levels of description has been recognized for many years in scene analysis research. With one key exception, they were all explicitly represented in the hand-eye systems developed at Stanford and MIT circa 1969/70 [3].

By contrast, current industrial vision systems use very impoverished representations, relying heavily on detailed models of particular objects to accomplish tasks.

REPRESENTATIONS IN CURRENT INDUSTRIAL VISION SYSTEMS

>   Images
>   Pictorial Features (Edges & Regions)
>   2-D Feature Attributes
>   Objects (Views)

Current systems often begin by thresholding the original grey-level image to obtain a binary array. Pictorial features (regions or edges) are extracted from the grey level or binary image and equated with surfaces and surface boundaries. Recognition is accomplished by matching 2-D attributes of these pseudo-surface features symbolically against 2-D models representing specific views of possible objects.

Given their impoverished representation, the gross inadequacies of current systems are hardly surprising.

LIMITATIONS OF CURRENT SYSTEMS

>   Weak Feature Extraction
>   No Invariance with Viewpoint
>   Restricted to Known Objects
>   No Descriptions of Surface Characteristics

Pictorial feature extraction is unreliable because region and edge-finding programs have no basis for distinguishing which image features correspond to significant scene events (i.e., surfaces and surface boundaries) and which do not (i.e., shadows, highlights, etc.). Equally distressing are cases where important surface boundaries are obscured by low contrast (see Fig. 5).

Significantly, human perception of surface boundaries does not appear to depend critically on contrast. In Fig. 6, for example, an intersection boundary is clearly perceived despite the absence of local contrast (Fig. 7); its presence is demanded by the integrity of the surfaces it joins.

Subjective contour illusions, like the so-called sun illusion (Fig. 8), appear to be an extreme example of this same phenomenon, where an edge is clearly perceived despite the complete absence of local evidence. A plausible explanation is that the edge corresponds to the boundary of an occluding disk-shaped surface, whose presence is implied by the abrupt line endings [4].

Even if an image could be reliably partitioned into surfaces based on regions and edges, the two dimensional shape features used for region description would still limit recognition to known objects observed from standard viewpoints; the ability to describe a new object so that it can be subsequently recognized from a different viewpoint requires three-dimensional description at the level of surfaces and volumes.

References p. 256

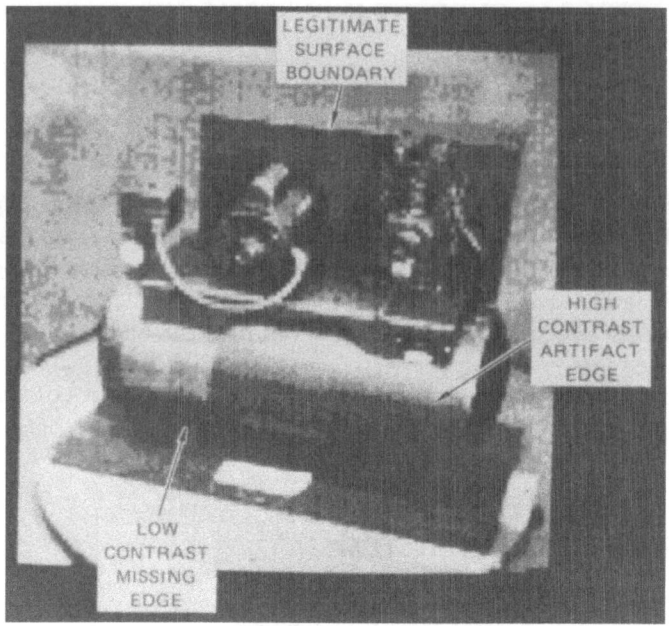

Fig. 5. Digitized image of compressor (5 bits at 120 × 120 resolution).

Current systems seldom have representations for interior characteristics of surfaces such as albedo, color, texture, and orientation. Such information is thrown away in thresholding to produce a binary image. Even in systems that don't threshold, uniformity of grey level is usually equated with uniformity of surface appearance in inspec-

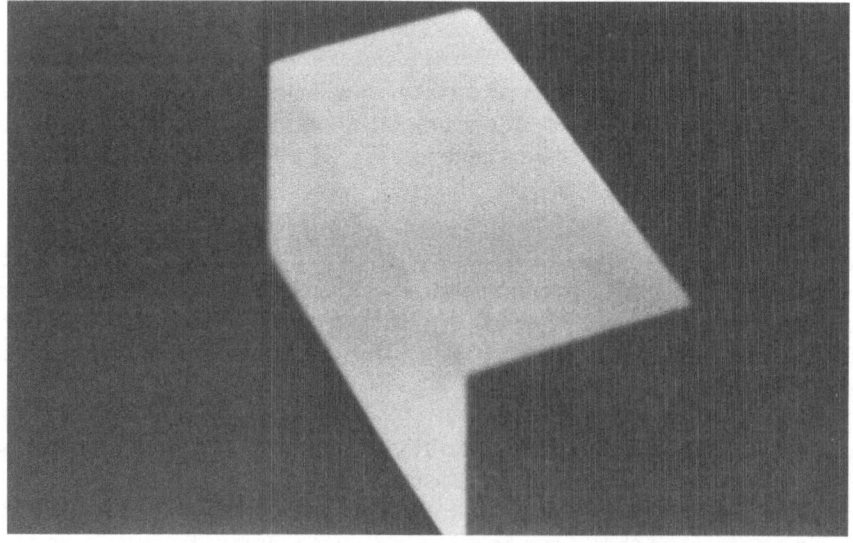

Fig. 6. Low contrast interior boundary.

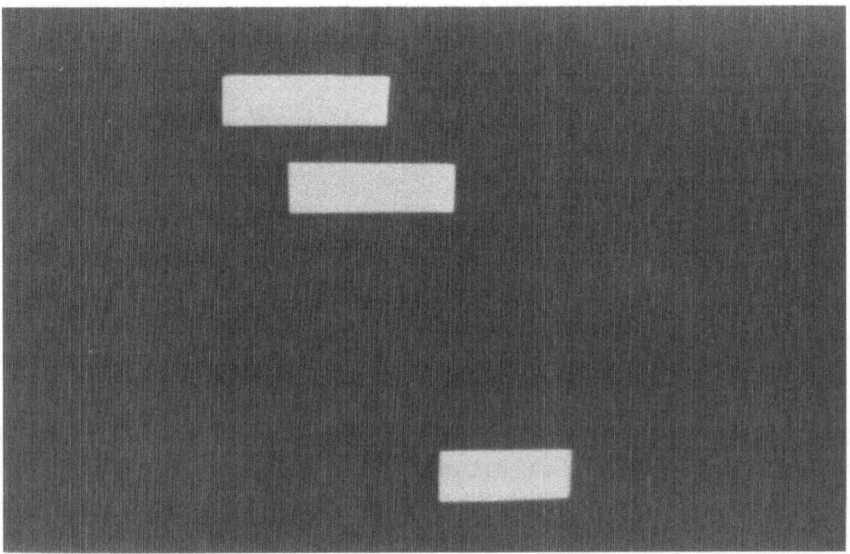

Fig. 7. The three homogeneous strips in Fig. 6 are typical cross sections of the surface boundary in Fig. 7, viewed through a mask. There is virtually no local evidence for the edge.

Fig. 8. A subjective contour.

tion tasks. The folly of this oversimplification might be exemplified by an example such as a painting robot which doggedly attempts to achieve a uniform finish by repeatedly painting over its own shadow.

To summarize, the limitations of current systems stem directly from their attempts to deal with 3-D and physical characteristics of scenes in terms of 2-D pictorial features

References p. 256

and 2-D object models. What is needed are many intermediate levels of representation to fill the enormous gap that exists between pictorial features and objects.

LEVELS OF DESCRIPTION

>Images
>Pictorial Features
>Intrinsic Surface Characteristics
>3-D Surfaces and Bodies
>3-D Objects
>Space Map
>Symbolic Relationships

We would like to single out one level of description—intrinsic surface characteristics—which we believe is fundamental to many basic inspection and manipulation tasks and a prerequisite to obtaining the other missing levels of description. Specifically, we believe it is important to recover the range, orientation, reflectance, hue, and other physical characteristics of the three-dimensional surface element visible at each point in the range. These characteristics are intrinsic properties of the surface and are independent of lighting and other viewing artifacts.

Humans seem able to infer such information throughout a wide range of viewing conditions, even when the scene does not contain familiar objects and when deprived of such powerful cues as stereopsis, color, and motion. No current scene analysis system incorporates this level of description; we believe this is a fundamental reason why their performance is so inferior to that of humans.

Intrinsic characteristics are conveniently represented by a set of arrays in registration with the original sensed-image array. Each array contains values for a particular characteristic of the surface element visible at the corresponding point in the sensed image. Each array also contains explicit indications of boundaries due to discontinuities in value or gradient. We call such arrays Intrinsic Images.

The primary intrinsic images are of surface reflectance, surface orientation, and incident illumination. Other characteristics, such as range, transparency, specularity, and so forth, might also be useful as intrinsic images, either in their own right or as intermediate results. The distance and orientation images together correspond to Marr's notion of a 2.5D sketch [2].

Fig. 9 gives an example of one possible set of intrinsic images corresponding to a monochrome image of a simple scene.

A concrete example of intrinsic images and their usefulness in computer vision can be seen in experiments by Nitzan, Brain, and Duda with a scanning laser range finder which directly measures the intrinsic properties of distance and apparent reflectance.

Because the range data is uncorrupted by reflectance variations and the reluctance data is unaffected by ambient lighting and shadows, it is easy to extract surfaces of

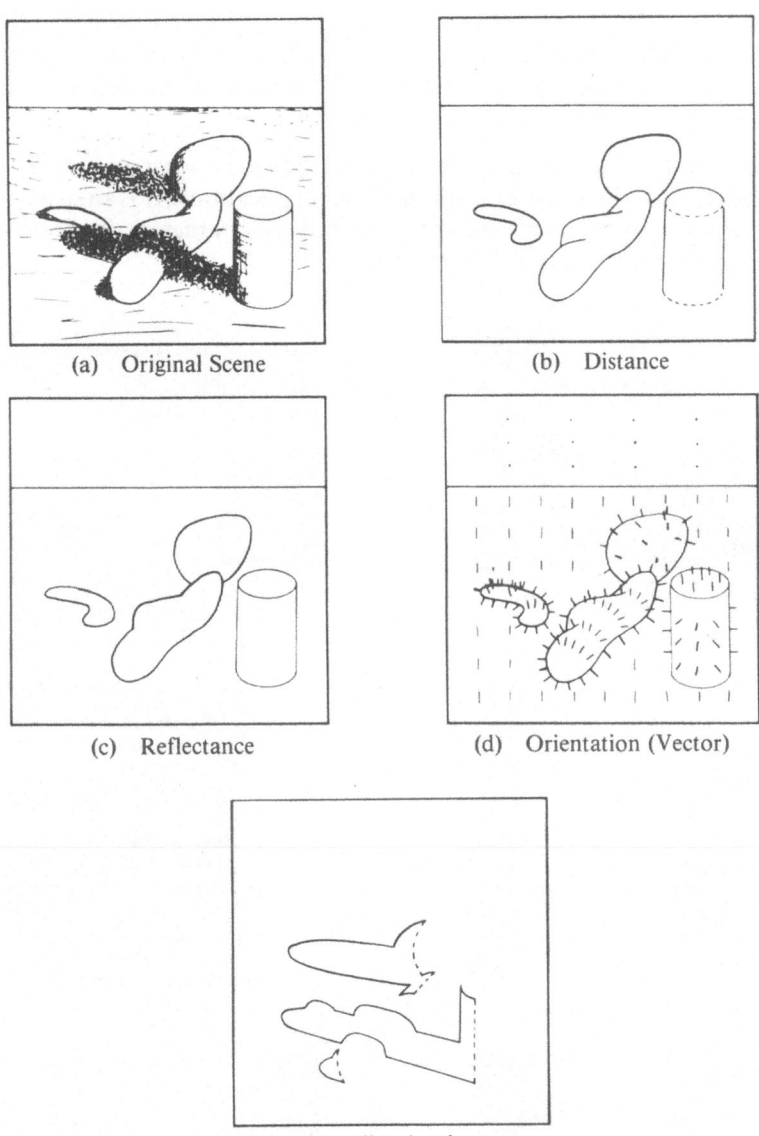

Fig. 9. A set of intrinsic images derived from a single monochrome intensity image. The images are depicted as line drawings, but, in fact, would contain values at every point. The solid lines in the intrinsic images represent discontinuities in the scene characteric; the dashed lines represent discontinuities in its derivative. In the input image, intensities correspond to the reflected light flux received from the visible points in the scene. The distance image gives the range along the line of sight from the center of projection to each visible point in the scene. The orientation image gives a vector representing the direction of the surface normal at each point. The reflectance image gives the albedo (the ratio of total reflected to total incident illumination) at each point.

References p. 256

uniform height (Fig. 10c) or reflectivity (Fig. 10e) and surface boundaries where range is discontinuous (Fig. 10d). Such tasks are very difficult to perform reliably in grey-level imagery; but with pure range and reflectance data, even simple-minded techniques such as thresholding and region-growing work well.

Laser range finders may eventually make good industrial sensors; but they are currently expensive, slow, and not appropriate in all applications. It is thus important to understand how it may be possible to recover intrinsic characteristics from ordinary grey-level imagery, as humans apparently do.

(a) A conventional photo of a scene

(b) Distance and reflectance images

(c) Extracted planar surfaces

(d) Discontinuities in range

(e) Thresholding reflectance

(f) Corrected view of cart top

Fig. 10. Experiments with a laser range finder.

RECOVERY OF INTRINSIC CHARACTERISTICS

The central problem in recovery is that the desired information is confounded in the sensory data. When an image is formed, whether by a camera or an eye, the light intensity at each point in the image is determined by the incident illumination, the surface reflectance, the surface orientation, and other intrinsic characteristics of the corresponding point in the scene. In the simple case of an ideally diffusing surface, for example, the image light intensity, L, is given by Lambert's Law:

$$L = I \times R \times \cos i$$

where I is intensity of incident illumination, R is reflectivity of the surface, and i is the angle of incidence of the illumination (see Fig. 11).

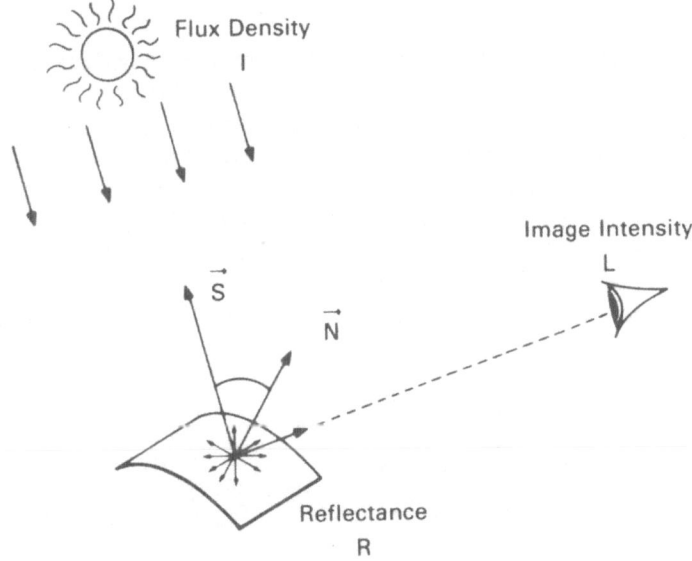

Fig. 11. An ideally diffusing surface.

Since the observed light intensity at a single point could result from any of an infinitude of combinations of illumination, reflectance and orientation, one might be tempted to conclude that unambiguous recovery is impossible. However, recent research at SRI, building on work by Marr and Horn at MIT, has led us to believe that the information can be unscrambled, at least in simple domains like that depicted in Fig. 9.

The secret of recovery necessarily lies in exploiting constraints derived from assumptions about the nature of the scene and the physics of the imaging process. Since surfaces are continuous in space, their characteristics (reflectance, orientation, range) are generally continuous across an image, except at surface boundaries. Incident illumination also usually varies smoothly over a scene, except at shadow boundaries. Elementary photometry tells us that where all intrinsic characteristics are continuous, image

References p. 256

brightness is continuous; conversely, where one or more intrinsic characteristics are discontinous, a brightness edge will usually result.

The pattern of brightness variation in an image can provide important clues to the local behavior of the intrinsic characteristics. A well-known example is the determination of surface shape from shading for surfaces of uniform reflectance under distant point-source illumination [5].

In [5] it was assumed that image intensity was continuous and that variations were due solely to variations in surface orientation. When a scene contains multiple objects, that may be differently colored or cast shadows, such assumptions are invalid. One is then faced with the problem of deciding what physical characteristic (or characteristics) are, in fact, responsible for an observed intensity variation and which characteristics are discontinuous across intensity edges.

The pattern of brightness variation on either side of an intensity edge can sometimes provide strong clues to the type of scene event responsible, (shadow or surface boundary) and thus to which intrinsic characteristics are actually discontinuous at that point. A simple example is the characteristic penumbrae and high contrast of shadow edges, indicating a discontinuity in illumination. The interpretation of brightness edges as scene events is also important because knowing the type of scene event sometimes allows explicit values to be determined for some of the intrinsic characteristics. For example, at an extremal occluding boundary, where an object curves smoothly away from the viewer, the surface orientation can be inferred exactly at every point along the

TABLE 1
The Nature of Edges

Region Intensities		Edge Type	Region Types	Intrinsic Edges Intrinsic Values			
LA	LB			D	N	R	I
Constant	Constant	Occluding sense unknown	A B shadowed	EDGE	EDGE	EDGE RA RB	IA IB
Constant	Varying	1 Shadow	A shadowed B illuminated		NB.S	RA RB	EDGE IA IB
		2 A occludes B	A shadowed B illuminated	EDGE DA DB	EDGE NA	EDGE RA	EDGE IA
Varying	Varying	Inconsistent with domain					
Constant	Tangency	B occludes A	A shadowed B illuminated	EDGE DA DB	EDGE NB	EDGE RA RB	EDGE IA IB
Varying	Tangency	B occludes A	A B illuminated	EDGE DA DB	EDGE NB	EDGE RB	EDGE IB IA
Tangency	Tangency	Not seen from general position					

LA and LB refer to variations of intensity along side A and B of an edge. Intensities are either constant, varying, or varying in accordance with the assumed orientations along an extremal boundary, the so-called tangency condition.

boundary. A test for extremal boundaries can be made by determining whether the observed brightness variation along the edge is consistent with the expected orientation.

We have compiled an exhaustive catalog of edge interpretations for the simple scene domain of Fig. 9, together with the constraints and values implied by each interpretation (Table 1).

The ability to classify edges in this way, at least for simple domains, suggests the following recovery paradigm.

STEPS IN RECOVERY

>Edge Detection
>Edge Classification
>Establishment of Boundary Conditions
>Propagation of Boundary Values into Regions
>   (Edge Refinement)

Edges are extracted in the intensity image and classified according to the catalog. The edge interpretations may provide values for one or more intrinsic characteristics at surface points along one or both sides of the edge. Values at interior surface points are determined by propagating in these boundary values, obeying continuity assumptions. Discontinuities may occur where values propagated in from opposite boundaries meet. This most often occurs when a physical edge does not result in a visible intensity discontinuity (as was the case in Fig. 6). In such events, missing edges can be hypothesized, establishing new boundary conditions, and the propagation process repeated, until a consistent set of boundaries and values is obtained.

The ability to refine the initial edge description as an integral part of the recovery process is essential to a practical theory. Extracting an ideal line drawing in a grey-level image is known to be very difficult, if not impossible. However, it is usually possible to extract a fairly good approximation containing a few gaps or extraneous lines. All that is required by the theory is that the initial line drawing be good enough so that any errors will show up as inconsistencies during the propagation process.

Another attractive feature of the model is its dependence on local computations which can be performed rapidly in parallel by low-cost LSI processor arrays. One possible implementation is sketched in Fig. 12. In essence, it consists of a stack of registered arrays representing the original intensity image (top) and the primary intrinsic arrays. Processing is initialized by detecting intensity edges in the original image, interpreting them according to a catalog of appearances, and then creating the appropriate edges in the intrinsic images (as implied by the descending arrows).

Parallel local operations (shown as circles) modify the values in each intrinsic image to make them consistant with intra-image continuity and limit constraints (for example, reflectance must be between 0 and 1). Simultaneously, a second set of processes (shown as vertical lines) operates to make the values at each point consistent with the

References p. 256

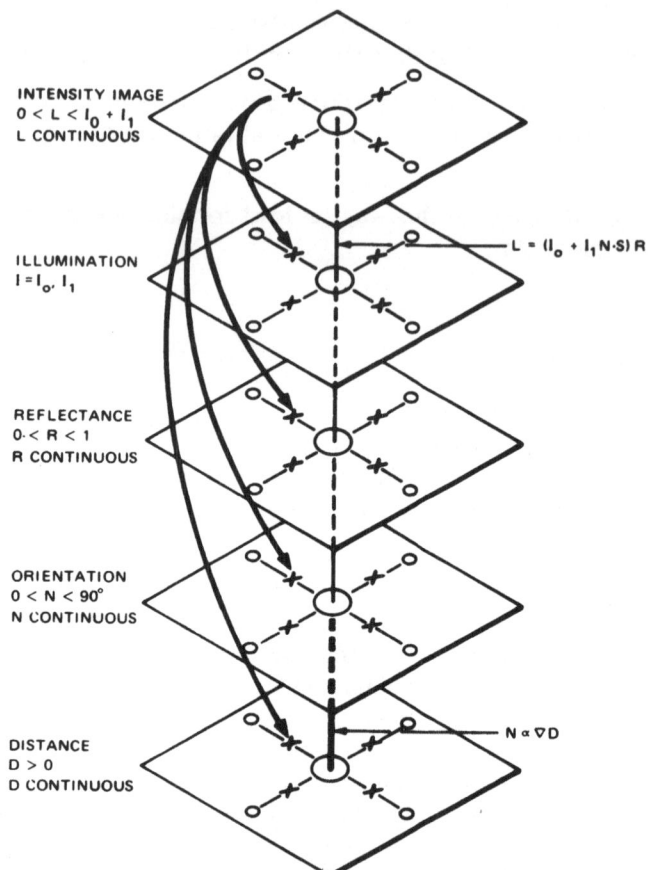

Fig. 12. A parallel computational model for recovering intrinsic images.

corresponding intensity value, as required by the inter-image photometric constraint. A third set of processes (shown as X's) operates to insert and delete edge elements, which inhibit continuity constraints locally. The constraint and edge modification processes operate continuously and interact to recover accurate intrinsic scene characteristics and to perfect the initial edge interpretation.

The action is envisaged to resemble an analog computer: as the value in one image increases, corresponding values in other images increase or decrease to maintain consistency with the observed intensity at that point. Within each image, values tend to propagate in from boundary conditions established along edges. This resembles relaxation processes used in physics for determining temperature or potential over a region from boundary conditions.

Our model can be regarded as a generalization of Horn's lightness model [6], Woodham's shading model [8] and Marr and Poggio's cooperative stereopsis model [7] for simultaneous recovery of geometric and photometric attributes. The model is significant, despite the simplicity of the domain, because it demonstrates for the first time the theoretical possibility of simultaneously recovering orientation, reflectance, and illumination from a single monochrome image, without recourse either to object prototypes or to primary depth cues, such as stereopsis, motion parallax, or texture gradient. We believe that these and many other well-known sources of information could be incorporated to improve overall performance, consequently extending the model's competence to deal with more complex scenes.

In particular, the current model can easily capitalize on simplifications that are often available in an industrial context. For example, it may be possible, by controlling lighting, to initialize values in the incident illumination array. Similarly, the reflectance array can be initialized when the albedo of a surface is known. The orientation array can be initialized using active technique such as grid coding [9] or photometric stereo [10], and range can be determined with a laser range finder.

EXPLOITING INTRINSIC CHARACTERISTICS

>Detection of Local Anomalies
>Interpretation of Anomalies as Scratches, Dents, Holes
>Segmentation into Surfaces and Bodies
>Extraction of Invariant 3-D Shape Features
>Object Recognition Based on 3-D or Generic Models
>Material Recognition Based on Reflectance
>Spatial Reasoning

Returning to the main theme of this paper, the key idea is that intrinsic images, no matter how recovered, provide direct access to 3-D and photometric properties of a scene that are not commonly available in computer vision systems. Such information facilitates all subsequent levels of description (eg., surfaces, volumes) and opens up exciting new areas for research.

At SRI, for example, we have an active project concerned with segmenting range and reflectance arrays from the range finder into surfaces and volumes. Surprisingly, finding surfaces is not quite as trivial as one might expect; range is often continuous across surface boundaries (eg., where floors meet walls) so that surface boundaries must be detected by discontinuities in local orientation. Since orientation is related to the derivative of range, it can be quite noisy (in practice a 1% range uncertainty can result in up to 30 degrees of orientation uncertainty). Thus orientation must be smoothed by spatial averaging. However, one must be careful not to average over the surface boundaries one is attempting to find. Hence, smoothing and boundary finding present a chicken and egg situation.

The above difficulties suggest that volumes, rather than surfaces, may be better candidates for initial scene partitioning because small perturbations in range have little effect on shape, symmetry axes, and other global characteristics of volumes.

Volumes are defined by surface elements that surround compact regions of space.

References p. 256

Often these elements cluster about a symmetry axis. Accordingly, we are thinking about using a 3-D symmetric axis transform to locate probable axes from the range finder data and representing the associated volumes with generalized cylinders [11]. Such a volume representation could then be matched to 3-D object models, building on the work of Nevatia and Binford [12] and of Marr and Nishihara [13].

The immediate possibility of recovering intrinsic images by mechanical or other means suggests many other interesting projects. These include detecting local anomalies and interpreting them generically as scratches, dents, discolorations and so forth, recognizing objects from generic descriptions (i.e., the type of description that allows recognition of all manner of nuts in Fig. 13), recognizing materials (wood, plastic, metal), and planning manipulation trajectories using the space map.

Fig. 13. Most readers will have little difficulty recognizing the nuts in this image: despite the variety of shapes, sizes and colors.

CONCLUSIONS

Although many industrial tasks fall within the competence constraints of current machine-vision systems, many others require a more general-purpose, high-performance vision capability approaching that of humans. In our current view, such a system would be based on an initial level of processing (Fig. 14) that transforms the information in the sensed image into an intermediate level of representation we call intrinsic images. Intrinsic images provide access to the physical and three-dimensional properties of a scene and thereby greatly facilitate all subsequent levels of perceptual processing.

PROSPECTS FOR INDUSTRIAL VISION

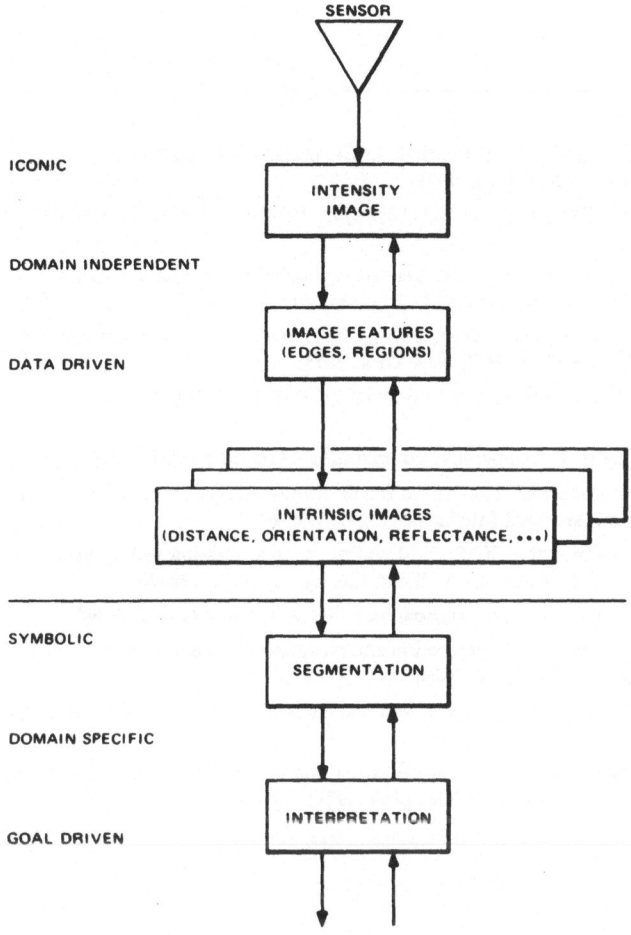

Fig. 14. Organization of a visual system.

In the near term, industrial vision systems can rely on active sensors that directly measure intrinsic characteristics. Therefore, researchers should now be seriously considering how to exploit this level of description. Ultimately, we hope to show the feasibility of recovery from ordinary grey-level video imagery, which would make many more applications practical.

Other key ideas in this paper include the need for multiple levels of representation that make explicit the information that must be recovered for each task, the use of 3-D models to achieve independence of viewpoint, and the use of generic models to allow manipulation, inspection, and description without explicit recognition. A system with these capabilities would exhibit much of the flexibility and robustness that makes human vision invaluable in factories today.

References p. 256

REFERENCES

1. Barrow, H.G. and Tenebaum, J.M. Representation and use of knowledge in vision. Tech. Note 108, Stanford Research Institute, Menlo Park, California (July 1975); an expanded version of a paper that originally appeared in *Sigart,* No. 52 (June 1975).
2. Marr, D. Representing visual information. In *Computer Vision Systems,* A. Hanson and E. Riseman (Eds.) Academic Press, New York, 1978.
3. Feldman, J. A. et al. The Stanford hand-eye project. Proc. Int. Jt. Conf. on Artificial Intelligence (May 1969), 521-526.
4. Barrow, H. G., and Tenenbaum, J. M. Recovering intrinsic scene characteristics from images. In *Computer Vision Systems,* A. Hanson and E. Riseman (Eds.) Academic Press, New York, 1978.
5. Horn, B. K. P. Obtaining shape from shading information. In *The Psychology of Computer Vision,* P. H. Winston (Ed.) McGraw-Hill, New York, 1975.
6. Horn, B. K. P. Determining lightness from an image. *Computer Graphics and Image Processing* 3 (1974), 277-299.
7. Marr, D., and Poggio, T. Cooperative computation of stereo disparity. *Science* 194 (1977), 283-287.
8. Woodham, R. J. A cooperative algorithm for determining surface orientation from a single view. Proc. 5th Int. Jt. Conf. on Artificial Intelligence (1977), 635-641.
9. Will, P. M., and Pennington, K. S. Grid coding: a preprocessing technique for robot and machine vision. Proc. 2nd Int. Jt. Conf. on Artificial Intelligence (1971), 66-70.
10. Metelli, Fabio. The perception of transparency. *Sci. Am.* 230,4 (1974), 90-98.
11. Agin, G. J., and Binford, T. O. Perception and recognition of curved objects. Proc. 3rd Int. Jt. Conf. on Artificial Intelligence (1973), 629-640.
12. Nevatia, R., and Binford, T. O. Description and recognition of curved objects. *Artificial Intelligence.* 8, 1 (1977), 77-98.
13. Marr, D., and Nishihara, H. K. Representation and recognition of the spatial organization of 3-D shapes. Proc. Roy. Soc. London B-200, 1140 (1978), 269-294.

DISCUSSION

J. Albus *(National Bureau of Standards)*

It occurs to me that the most important and interesting work has been with static images. In other words, take a photograph and process the daylights out of it. It seems to me that one of the most obvious things about the way humans operate is that their eyes are mounted in their heads which move around and if you want to see something, or if there is something ambiguous in the scene, you move your head or get up closer to it or look at it from the other side. Is anyone doing any work in the area where you move the camera around in some way? In other words, you make an initial guess as to what's in the scene, then you move the camera to view it from a better angle?

Tenenbaum

Yes. Certainly people use many different cues in order to extract the kind of information we are talking about. Computationally, range information obtained by moving the camera seems very valuable. Stereo is one obvious example and there is plenty of work that's been done. More recently, people have been getting interested in this approach, now that the technology is available for high processing speed capable of processing a continuous flow of imagery. There is work going on in Germany and England and at the University of Massachusetts that I am personally aware of.

There is no single cue that is going to solve all problems. What is very interesting is that although flow seems to be very important, it seems to break the world up in terms of which things hang together. Some recent experiments at Cornell University in which you take different range views came up with some very surprising results. Cornell is a significant place because that's where J. J. Gibson did all of his work on flow and in fact he is perhaps the only person who has written us a very critical letter about this work. He said that we missed the boat because we are looking at one image and that doesn't count at all. And of course I think he missed the boat because one would like to consider all cues, static cues as well as those of motion.

In the Cornell cue dominance work, surprisingly, it was found that stereopsis dominates motion. Binocular cues, like perspective and occlusion, tend to dominate when there is a conflict. So it's two different questions what people use and what's good for machines and certainly motion is an important thing.

If you are going to put an eye in a hand, and I think that is very valuable, one can look at objects from different points of view and resolve them. The model that we propose certainly seems capable of incorporating flow information. The parallel model is dynamic and there is no reason why we can't let the input image vary continuously and then allow temporal continuity constraints on how fast the boundaries of the various intrinsic attributes can change. They are very similar in form to the spacial continuity constraint that we are now using.

D. Nitzan *(SRI International)*

Marty, you have been talking about what we call non-contact sensors. There are others you did not mention. Sometimes we have both non-contact and contact sensors working together to obtain the information which is best for each application. I will give you an example. When I shave, I look in the mirror. But if I am still not sure how good it is, I use my fingers to be sure my face is really smooth. When we try to fit a peg in a hole, a camera could find the hole quite quickly. When I'm close to it I may resort to other types of contact sensors to get higher resolution and, if necessary, forget completely about sensors and resort to passive accommodation. What I'm trying to say is that there are a lot of techniques that have to be integrated but what is missing in all of this wishful thinking is the speed and cost under industrial conditions.

Tenenbaum

I did try to make the point that the model we were showing uses only local computations and once we can get the model worked out I see no difficulty of implementing the process in an LSI 11 and getting the cost down. Again I think it is premature to worry about cost now, but the general answer to your comment is that of course we would like to take any information that comes from any point and I think we have a mechanism for incorporating that information.

R. R. Bajcsy *(University of Pennsylvania)*

Along the lines that Dr. Albus mentioned there is another dimension that wasn't mentioned, and that is variable spacial resolution. I find that in work we are just finishing (the area of imagery taken from various altitudes) you can indeed save lots of time by recognizing on coarse resolution what is big and on finer resolution what is small. For example, you do not need to recognize big trees with the same resolution you need for a car or, perhaps, for some part of the car.

Tenenbaum

Variable resolution is undoubtably something that is very important. I really don't understand what to do with it. However, sometimes people comment that this relaxation form of solution we are using cannot possibly be right. Because relaxation converges very slowly, it's the worst kind of solution. There are ways of speeding up by going to a higher level of resolution and then passing the information back down. Certainly this may be one way of doing relaxation fast.

R. J. Popplestone *(University of Edinburgh)*

I might remark that it would be fairly difficult to build a machine that would, at a lower level of reasoning, be able to determine if a picture on a wall is actually a picture on the wall.

Tenenbaum

And why would that be difficult?

Popplestone

Well, it would have to be able to take different points of view on whether what you see is a picture or a scene.

Tenenbaum

One is certainly able to interpret a picture both ways. It is an ambiguity.

Popplestone

But there certainly is evidence in a projected picture on the wall that tells you that there is some sort of surface there. There is more information there than in the slide you are projecting.

B. Chern *(National Science Foundation)*

Since you mentioned that industrial applications motivate input to this, I'm a little curious, because I don't think people normally do what you are saying. Normally there is some sort of knowledge base which you have learned over a long period of time, and there is some feeling of what to expect and how your optical senses relate to that. You may, for example, reach out a hand and touch something to determine if indeed it is round or is sharp or is indented. Industrial scenes are going to be more structured because you generally have some idea of the kind of objects and the geometry of those objects. It would seem to me that what you want to be able to do is to use some of the information you have, whether it's from the computer aided design data base or other geometries of the work place or the parts, and then proceed to some sort of natural device where one determines that in this region it is really a cylindrical surface.

Tenenbaum

What you say is of course true. There are ways of incorporating knowledge about particular objects, even at this low level, into scene analysis. For example, if you know the actual size of the object you are looking at, and you recognize the object, then you can use that to calculate the distance to the object. But, perhaps more to the point, there has been a lot of attention paid to exploiting high level knowledge at the level of objects. We have done a lot of that work ourselves and we feel that there is no final answer to that problem. We are now, however, much more comfortable with this idea. The weakest link at this point is to provide some level of description on which those very same techniques can work, but work much more reliably than they can now on regions of uniform brightness, which you just assume correspond to surfaces, but which in fact may not if there is a shadow edge coming across it. So, that's an important consideration.

I just think, in agreement with Horn, the time is right to put some emphasis on low level processing. And, of course, the objective is to get a really general purpose system that doesn't need to be trained explicitly for many parts. This should work similar to the industrial worker who, when given a bin, is able to pick a part out and see if it's rotated, or whether it has a crack in it, with no training whatsoever. You were presumably able to do that from the bin pictures I showed you. So, in fact, while its possible to exploit higher level knowledge, it is not always necessary and not always desirable, especially if you are talking about flexibility and robustness.

SESSION IV
FUTURE ROBOT SYSTEMS

Session Chairman
J. A. FELDMAN

University of Rochester
Rochester, New York

STAND-ALONE VS. DISTRIBUTED ROBOTICS

J. F. ENGELBERGER

Unimation, Inc., Danbury, Connecticut

ABSTRACT

Research and development in robotics is proceeding down two paths which are at times complimentary and at times in conflict. Some look simply to emulate the human worker with a replacement machine; others look to redo the manufacturing process by distributing bits and pieces of intelligence and motor power to produce an unmanned manufacturing facility.

These concepts are combative but not completely polarized. The paper deals with both research directions and marshalls arguments on behalf of each. Whichever prevails, profound productivity gains will result.

INTRODUCTION

A decade ago we all had the chance to enjoy D. H. Clarke's movie "2001 A Space Odyssey". The closing passages were delicious in allowing each of us to plead privileged insight as to the deeper meaning. But, earlier on, there is less opportunity for wild-eyed speculation. We find humans in mortal combat with a robot. His name was HAL and we never saw him. He pervaded the space ship. The humans attacked his distributed intelligence and his communication channels. Like an octopus, he did not die with the loss of one tentacle.

Is this the inexorable direction for Robotics? Or could our future robot be a pottering artisan, self-contained machine doing pretty much as his human forbears had done? Certainly in the spectacularly successful movie "Star Wars", the two stand-alone robots, R_2D_2 and C3PO, struck an empathetic chord in American imagination that was never accorded to Hal.

References p. 270

And, when it comes to empathy, no robot has evoked it like the pathetic little robotized worker in Charlie Chaplin's "Modern Times". Fig. 1 is a publicity shot from that 1936 film in which Chaplin decried the dehumanization of man in the industrial work environment.

Fig. 1. Modern Times.

In 1936, human workers were plentiful, cheap and intimidated. Today, even high unemployment does not result in a cheap or intimidated worker. Fortunately, the technology at hand provides means for alternative solutions to robotizing a human worker.

One solution is to change the very nature of the workplace through computer aided design and computer aided manufacture, CAD/CAM, which is really Numerical Control reaching back to the drawing board and reaching forward to cutting tools and factory management. At the same time, the workplace becomes highly stylized through Group Technology classification of parts and siting of equipment. Orientation is preserved. The manufacturing system becomes so highly rationalized that the peculiar adaptability of the human being is no longer necessary, or only necessary for such menial tasks as "sweeping cuttings". So does Neil Z. Ruzig report of the McDonnell-Douglas parts fabrication plant in St. Louis [1]. (Please note that that is where the rationalization went out of the job—in the jumble of cuttings.)

The unmanned manufacturing plant is a virtual practicality in machined parts manufacturing. From East Germany through the USA to Japan there are isolated showplaces where batch manufacturing is carried on automatically under computer control. Rarely, if ever, has one of these systems been economically justified at the time of installation; but, the advance expense will inevitably serve these pioneers well in the future.

The situation is not so rosy, however, when it comes to operations other than machining. Foundry work, forge shops, and the most labor intensive activity of all, assembly, are not succumbing easily to automation except in very high volume manufacturing.

REQUIRED CAPABILITY OF INDUSTRIAL ROBOTS

What kind of robot will be best for manufacturing? Will it be an extension of DNC machining center technology with hardly any anthropomorphic associations, a HAL, or will it be a lovable stand-alone Chaplinesque robot, a C3PO, using appendages, sensory perception, and resident intelligence to carry out man's behest in the factory?

Probably, it will be a straddle. Even without robots, the computer is leaning heavily on the production process. Production control is rapidly being computerized and telecomputing feeds back the status throughout the factory floor. Inventory is managed, machine loading is decreed, and the workers get their daily job assignments from the computer. If only these unworthy ones would be more tractable and accept pay appropriate to the low skill levels demanded of them!

Let's continue the argument with reference to Fig. 2, owing to the National Bureau of Standards [2] and adopted by the Air Force as an integral part of its ICAM (Integrated Computer Aided Manufacturing) philosophy. Five levels of the hierarchy of control are shown and one presumes that the hierarchy continues on to the cosmic computer in the sky that makes the really big decisions.

This hierarchical system prevails in its counter parts, the DNC machining center, and the extrapolation to robotics is rational but perhaps a bit too seductive.

There will be operations that stem from a computerized data base that will be able to be defined with precision both as to product and as to the "world model" of the manufacturing arena. For the time being, these manufacturing situations will be very much in the minority. Even where it's all logical on paper, it will take uncharacteristic boldness on the part of a management to make the investment committment. Without unrestrained government support, it might not happen at all.

Let us reexamine Fig. 2. If we could build a stand-alone robot to replace humans in a large percentage of their more puerile factory roles, how high up the hierarchy must we go? Per the definitions, level four should be high enough. NBS says about the fourth level that "this level of control takes care of the complete operation of a robot in its associated work station"[2]. When we hire a human operator, we ask no more. The heritage of industrial engineering as espoused by Ford and glorified by Taylor [3]

References p. 270

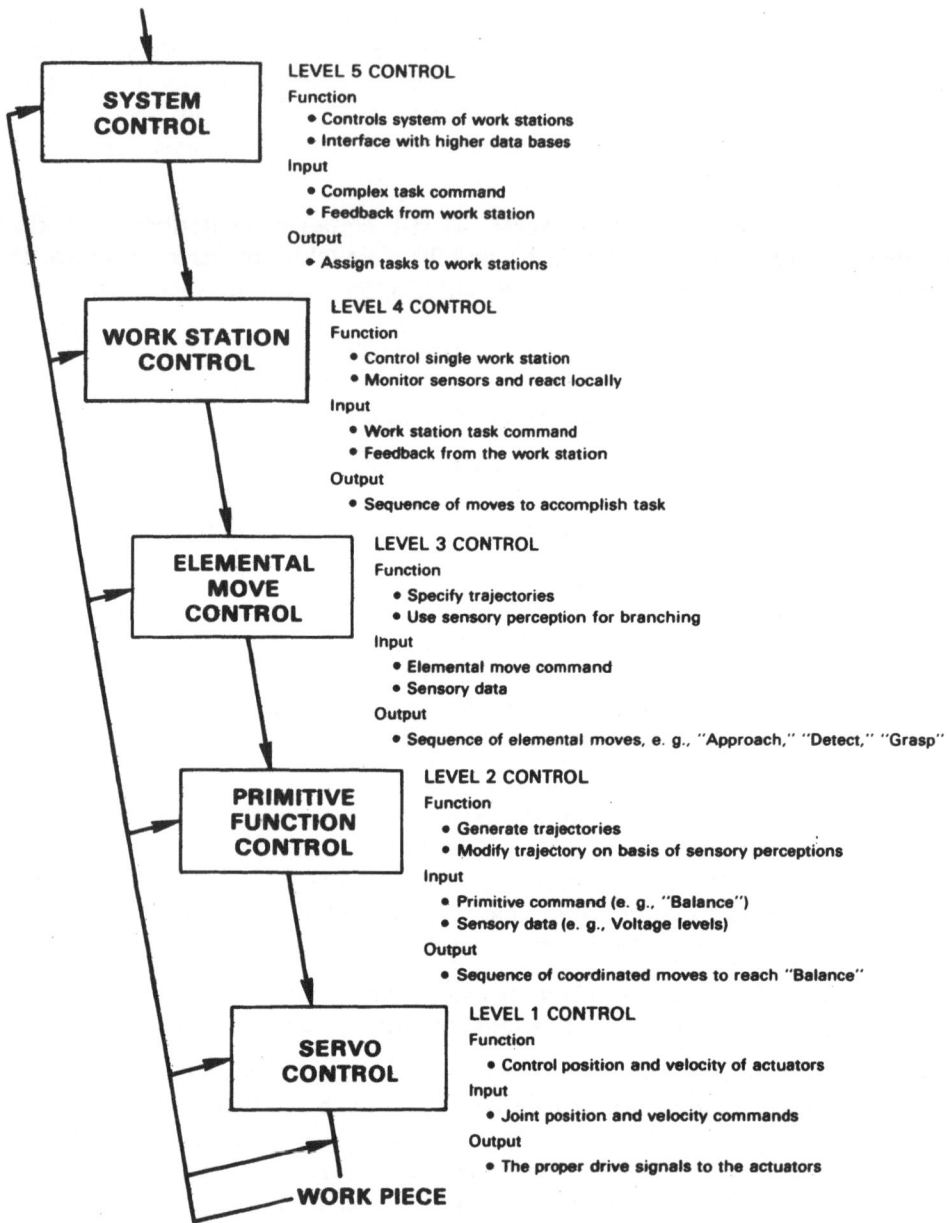

Fig. 2. Hierarchical control.

breaks the work down into simple tasks, easily learned, and imbued with no skill level bargaining clout. We don't give human operators job responsibilities as broad as the fifth level. This is part of the industrial engineering-supervisory role, a white color job.

It is enough in the stand-alone robot business to strive for a broadly useful blue collar worker. He needs no more than level four intelligence and manual dexterity.

And that is *one* tall order! Fig. 3 is a list of the attributes the stand-alone robot should have to function well at level four. Of these 1 through 10 are in hand; 10 through 20 are being sought diligently.

It should be noted that getting a robot to level four is where the action is. All of the higher levels are being developed anyway for manufacturing systems that have people at level four and lower.

The listing of what is needed is fairly specific in Fig. 3, certainly more specific than provided by the NBS report [2]. Missing in both, though, are qualitative observations. Just how good is the vision, how sensitive the tactile sense? For level two control, NBS sparingly defines a function as, "modifies trajectory on basis of sensory feedback". What a range of sensibility could be covered by that statement!

Did the vision module detect the presence of a single object on a back lighted surface or did it distinguish the uppermost part and its orientation in a random-scramble bin of parts? Both acts might be categorized as "modifying trajectory on the basis of sensory feedback". But what a world of difference!

1. Work space command with six infinitely-controllable articulations between the robot base and a hand extremity.
2. Fast, "hands on," instinctive programming.
3. Local and library memory of any size desired.
4. Random program selection by external stimuli.
5. Positioning repeatability to 0.3 mm.
6. Weight handling capability to 150 kilos.
7. Intermixed point to point and path following control.
8. Synchronization with moving targets.
9. Compatible computer interface.
10. High reliability (at least 400 hour MTBF).
11. Rudimentary vision: (a) orientation, (b) recognition data.
12. Tactile sensing: (a) orientation data, (b) physical interaction data, (c) recognition data.
13. Multiple appendage hand-hand coordination.
14. Computer directed appendage trajectories.
15. Mobility.
16. Minimized spatial intrusion.
17. Energy conserving musculature.
18. General purpose hands.
19. Man-robot voice communication.
20. Inherent safety (Asimov's law of Robotics).

Fig. 3. Robot attributes.

References p. 270

Did the fingers of the robot detect that there was indeed a part at the taught pickup point, or did they feel that a bolt was cross-threaded? Again, what a world of difference!

And these are the qualitative considerations down at level two, let alone level four, the suggested cut-off point for the stand-alone robot.

We can regress even further in the same vein. At level one, the control function is defined as, "controls position and velocity of individual actuator". Do we mean by that a bang-bang pneumatic actuator with mechanically adjustable stops and end position damping? Or do we mean the equivalent of the human elbow with infinite positioning capability, 5 Hertz frequency response, internal proprioceptive feedback, low space intrusion, and optimized energy consumption? One should refer to the work of P. Rabischong, et al. in "Is Man still the best Robot?" [4] to appreciate the standard set by a human operator "controlling position and velocity of an individual actuator".

There is so much needed to make a level four subhuman robot. Let's categorize its capability as that of a benign but motivated moron. That's enough for a level four robot; and, its required combination of sensory perception, musculature and intelligence *is* in the cards. As said already, that's where the robotics research and development action is.

If we develop a robot with level four capability, it will dictate the direction of manufacturing productivity improvement, particularly in all the labor intense activities. If we don't, the manufacturing process will seek other more disruptive methods that would be unrecognizable to a reincarnated Taylor [3].

An isolated example might be made of a project at the General Motors Technical Center. General Motors has mounted a major paint spray development program. It does include a manipulator arm, but the arm's significance is dwarfed by the distributed intelligence governing the system. Apparently, the view of the state of the robotics art and the urgency of a solution found against a stand-alone robot. The level four moronic robot that could emulate the humans now manning the paint spray booths did not appear to be in the cards. General Motors did not find imminent the stand-alone robot that could,

- observe the color of each oncoming car in turn
- select the appropriate paint
- hold doors, trunk lids, and hoods at convenient positions for painting
- adapt to varying line speeds while observing and applying proper paint coverage.

I forecast a HAL nightmare, when this highly sophisticated system goes beserk. Meanwhile roboticists have their work cut out to build robot sensibilities to a level four of such quality as to obviate the need for distributed robotics in the paint spray application.

ECONOMIC JUSTIFICATION OF ROBOTS

The robot attribute compendium of Fig. 3 is missing one crucial quality, economic justification. There's no place for robots that don't justify themselves economically over available, willing, human counterparts. This principle seems to be universal whether examined in the USA, Western Europe, Japan, the Comecon countries and even the third world. The measures differ but the principle is the same. With time, the pressure mounts for robotics. A well documented example can be made by comparing the cost of human labor versus robotic labor in the U.S. automotive industry. Fig. 4 shows the trend since the advent of the first industrial robot, the Unimate. In the U.S. automotive industry today, the economics are inexorable. (The inertia of the industry is a pretty powerful deterring force, however.)

But economics and time will prevail even if robots hobble along with present deaf, dumb and blind capabilities. Moreover, if skills are developed to the NBS level four at only the moronic level, watch out ICAM! The factory peasant will become an anachronism, but the workplace will remain familiar.

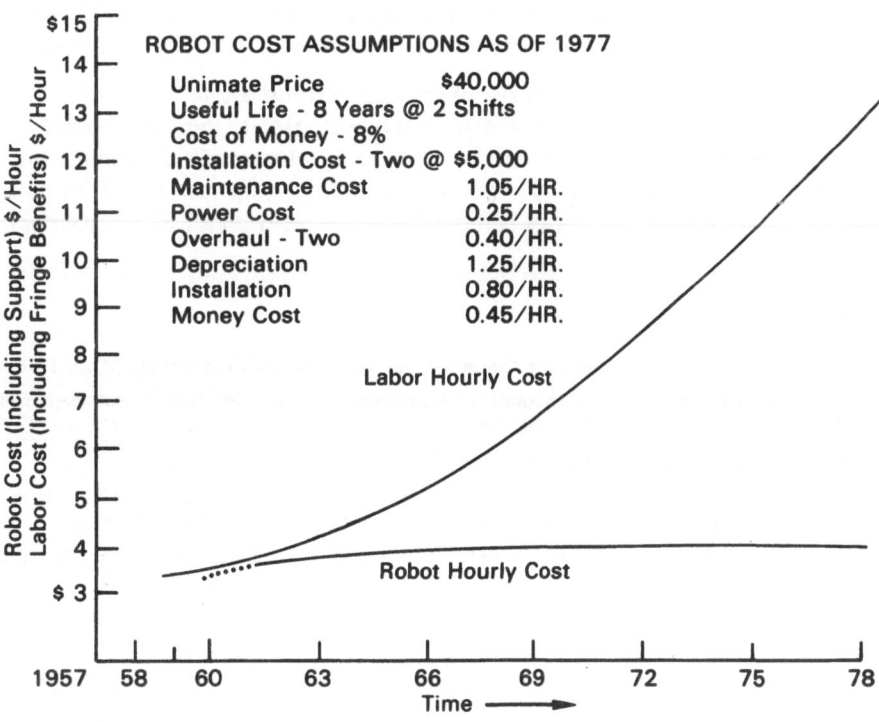

Fig. 4. History of labor cost and history of Unimate robot cost in the automotive industry.

References p. 270

Given our moronic, level four robot, just what can one predict for penetration of these robots into the manufacturing process? In "A Robotics Prognostication" [5] this author opined as follows:

"What can we expect of market growth? It seems that the pacing factors may be capital formation and sociological acceptance.

Let us say that the blue collar work force of just Western Europe and the United States who do work amenable to robotization is 80,000,000 people. Suppose just the most hazardous or stultifying 5% of these jobs are considered pressing candidates for robot workers. Finally, assume that an average robot price is $30,000. That pegs the gross market at 120 billion dollars.

This market is not going to be filled spot on. Rather, we can look to a 30% compound growth rate from a 1977 base of about 40 million dollars. The stricture of capital availability and government employment policies may result in a mature annual market of 3 billion dollars. This allows 50 years for robots to take over 5% of the blue collar work in Europe and the United States. The labor displacement impact would be a negligible percentage of normal attrition.

Of course the foregoing doesn't take into account the growth of robotics in Japan or in Eastern Europe or even in the third world, where the irrational fascination with high technology will probably create some kind of grotesque market."

Finally, this prognostication deals only with "stand-alone" robots. Unquestionably, the "distributed robot" will also find broad, albeit later, acceptance. Thus, one can safely predict that the engineers who would develop artificial humans will succeed and the engineers who seek to strip the manufacturing process of human-like activity, will also succeed. Robotics is indeed a fertile field on all of its frontiers!

REFERENCES:

1. Ruzic, N. P. The automated factory - a dream coming true? *Control Engineering*, (April 1978), 58.
2. Barbera, A. J. An architecture for a robot hierarchical control system. NBC Special Publication 500-23, U.S. Department of Commerce, National Bureau of Standards, Washington, D.C. (December 1977).
3. Taylor, F. W. A piece-rate system. *Trans. ASME,* 16 (1895), 856-903.
4. Rabischong, P., Peruchon, E., Pech, J. Is man still the best robot? Proceedings of the 7th International Symposium on Industrial Robots, Japan Industrial Robot Association, Tokyo, (1977) 58.
5. Engelberger, J. F. 1977 Joint Automatic Control Conference, IEEE Society on Control Systems, San Francisc, CA, (June, 1977).

DISCUSSION

B. H. McCormick *(University of Illinois - Chicago Circle)*

How much of your cost is maintenance or indirectly manpower?

Engelberger

It hasn't been broken out, but the cost of maintenance out of the $1.05 is about $0.60.

McCormick

I guess what I'm saying is, to keep it invariant over time, what is the effective labor cost of running a robot per hour?

Engelberger

Well, if the laborer who is running the robot is earning as much as the $14.00/hour man is, and it takes $0.60/hour for maintenance, then for every robot hour the fraction of the labor cost is .60/14.00. The cost of labor is very small.

D. F. Whitney *(Charles Stark Draper Laboratory)*

Have you allowed for inflation in the hourly rate of the people who are doing the maintenance and the overhaul?

Engelberger

I have, throughout the period. Back at the beginning our maintenance and overhaul costs were considerably higher because we didn't have such a good product. And you know that maintenance costs have gone down, otherwise the curve wouldn't be quite as level. Remember, on the cost of the machine, another thing has happened. As some of us have said here, it won't be too long before we can give away the electronics, you'll just have to buy the mechanics. And that's a big swing from when we started our development. Then the cost of the machine was 3/4 electronics and 1/4 mechanical.

Whitney

In other words, this figure is partly historical and partly projective?

Engelberger

No, everything is history here.

Whitney

Except the 8% cost of money. That isn't history.

Engelberger

At the moment in time, when I compared this in 1977, it was 8%. Now I can change the cost of money to 10% today and you can see what would happen. The $0.45/hour for money costs would go to $0.50. So it's not that big an impact.

P. H. Winston *(Massachusetts Institute of Technology)*

Are you using any Unimates in manufacturing Unimates?

Engelberger

In a secondary sense. We buy our castings for the Unimate from a guy who uses Unimates to make castings.

Winston

The reason I asked the question is that I'm intrigued by your numbers because I thought I would copy them down and bring them out whenever the journalists come around. They seem like very reassuring numbers from a social point of view. Then it occurred to me to wonder if over the period of the next 100 years, when robots start bootstrapping on themselves, if those numbers are still so reassuring. I wonder if you have thought about that?

Engelberger

Well, let me try some other numbers on you in 100 years for reassurance. From 1870 to 1970 the farm population in the United States went from 47% down to 4%. In that 100 years, as you undoubtably know, there have been disruptions. But if I look at it now, there are a lot of people going back to the farm rather than work in a factory. I wouldn't want to trade off which is a better job. The thing in this argument that I give the press always is that productivity is fundamentally good. It is always good. The only issue is what do you do with this blessing. If I give you more productivity, and there is more material wealth, then you have a problem dividing it up. We never do that very well. It's not a fair world, but you certainly can decide that I'll have 20 hour work weeks or 10 hour work weeks. Or you can decide to put 3% of the GNP to cleaning the air and the water. Who is going to distribute that cost? Divide those two problems up. The assumption that unemployment and the robot have a connection is very very tenuous.

J. Albus *(National Bureau of Standards)*

What year did you pick the $30,000 for the cost of robots?

Engelberger

I chose today, and made a volume extrapolation. Were I to design the machine today, it would be cheaper than today's machines, because of economies of scale.

Albus

You're saying that today a $30,000 robot, if it were designed today, would perform the same tasks?

Engelberger

Yes, in fact that list of goodies would be in there and it would still be a $30,000 machine.

J. A. Feldman *(University of Rochester)*

I had a comment and a question. When you started off and talked about distributed robots and stand-alone robots, I wasn't sure which one was which and I guessed wrong. That is, often, when I think about distributed computations, I think about all these separate stations, which one might distribute throughout a plant, which gives you a problem of coordinating their actions. The alternative would be something in which you had one single processor which, in the computer business, would be called a stand-alone thing. So the technical question is, from your point of view, is this problem of integrating the control of lots of these Unimates or PUMAs important? Is it something that one can just skip over?

Engelberger

Let's look at the tremendous work IBM is doing in telecomputing. They don't need robots at the end of the line. They are organizing the entire factory workplace and assigning jobs and getting the data all back to a central control location. That's being done independent of what this device is out there. Now the human being, I consider a stand-alone robot. It takes its guidance from the supervisory system, but it's a much different thing. The ICAM slides today don't mention the word robot at all. It's integrated computer-aided manufacturing. Somewhere down in that maze I think you'll find that there are arms. Maybe. So they are quite different approaches.

It's almost as if integrated computer-aided manufacturing is like making a drawing for NC. You put a dot in the corner and measure everything from the dot, so you can make tapes. We're going to put a stake in the ground and we're going to measure everything in that factory from that stake as NC. This approach requires a database, which is beautiful certainly in the airframe industry, and with that database we can make all these things in this factory and never have any people or little fixtures. They are quite different philosophies.

ROBOT ASSEMBLY RESEARCH AND ITS FUTURE APPLICATIONS

J. L. NEVINS and D. E. WHITNEY

Charles Stark Draper Laboratory, Cambridge, Massachusetts

ABSTRACT

Assembly research comprises the definitive description of how parts interact during assembly (called Part Mating Science), and the collection of part mating processes into systems that assemble products. Products can be assembled by four techniques: manual labor, special purpose machines, programmable systems (a new type of interest in mid volume production), or hybrids of the above. Programmable systems promise to exhibit some of the adaptability of manual labor plus the repeatibility of special machines, a good combination where parts variation and market uncertainties are present.

Part mating analysis first concentrated on rigid parts of simple shape. Geometry and friction were studied to yield criteria for successful insertions. This yielded the unique Remote Center Compliance, a device containing only springs which performs close clearance insertions. Four industrial firms are exploring how this idea can aid their assemblies. A fifth is applying part mating science and experimental techniques to their product design. Applications include robot assembly, special machine workheads, and manually operated insertion stations. Part mating research at Draper has been extended to non-rigid parts.

The prime focus of programmable systems is not simply to make them but to systematically determine the various options and to explore them in enough richness that practical systems can be constructed. This knowledge does not exist. Thus, industries setting out to build these systems do so in a climate of great unpredictability.

Research needs to be carried out in the following areas:

- economic justification, emphasizing reprogrammability

References pp. 320-321

- engineering requirements and specifications on assembly stations and sensors (visual and non-visual)
- scheduling and operation — particularly in the light of partial system failure
- supervisory and subsystem software
- construction of assembly test beds designed to gather data on system behavior under normal and failure modes
- research into programmable feeding and inspection systems.

INTRODUCTION

Assembly research has two principal thrusts—the definitive description of how parts interact as they are assembled together, called part mating sciences, and the ensemble of these part-mating processes into a system that assembles products. Systems that assemble products are of four principal types: 1) manually; 2) by special machines (sometimes called transfer machines) when the annual volume justifies them; 3) by a new class called programmable systems which are of particular interest to middle volume manufacturing; and 4) finally all three techniques can be used to form a hybrid system. The latter kind may be transitory to fully programmable systems when they emerge, but in any event they are of particular interest to the smaller manufacturer who seeks to minimize capital risk in an uncertain marketplace.

The principal achievement at Draper Laboratory has been the development of an integrated scientific knowledge base—both for categorizing assembly tasks and for determining new machine configurations necessary to economically perform industrial assembly. These developments were recently incorporated into a demonstration of programmable automatic assembly of an automotive alternator. The work has been divided into four broad areas: 1) analysis of insertion tasks; 2) task statistics; 3) economic analysis and assembly system configuration choices; and 4) programmable automatic assembly demonstration.

The next three sections—Part Mating Science, Programmable Systems, and results from an Experimental Programmable System Test Bed—briefly summarize the first phase of this work. The remainder of the paper describes our present research directions and the stimulation for them.

PART MATING SCIENCE

Round peg-hole insertion tasks with chamfered corners are found, by far, most frequently in the assembly of metal products with machined or cast parts. This task has been extensively analyzed, and a complete statement of the requirements on relative errors is now available. The important design variables have been identified and for many parts it has been determined that conditions for successful assembly can be met more easily than the clearances between the parts would make it at first appear.

These analyses have been amply verified experimentally and it has been shown that the conditions can be obtained from the blue prints for the parts (Fig. 1). The analysis and experiments also form a valuable pattern which can be followed in the analysis of other tasks.

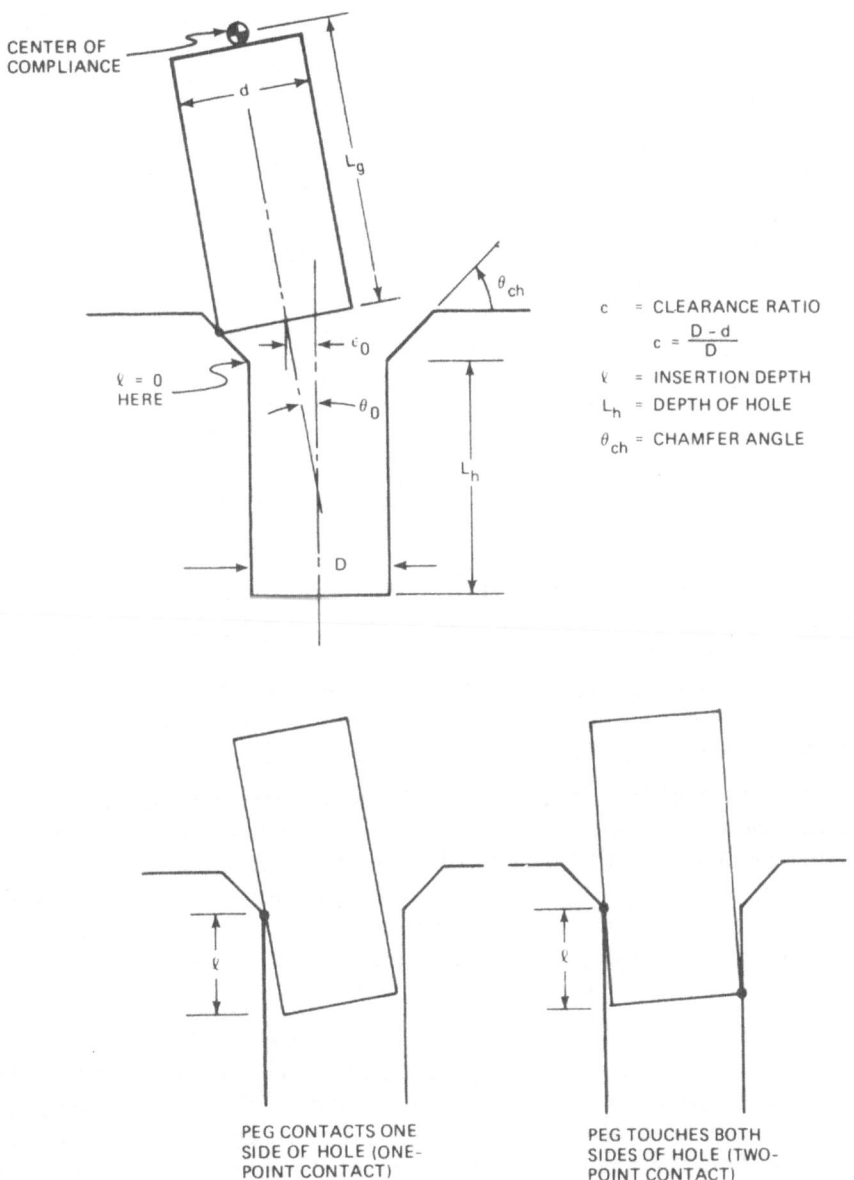

Fig. 1a. Basic geometric relationships.

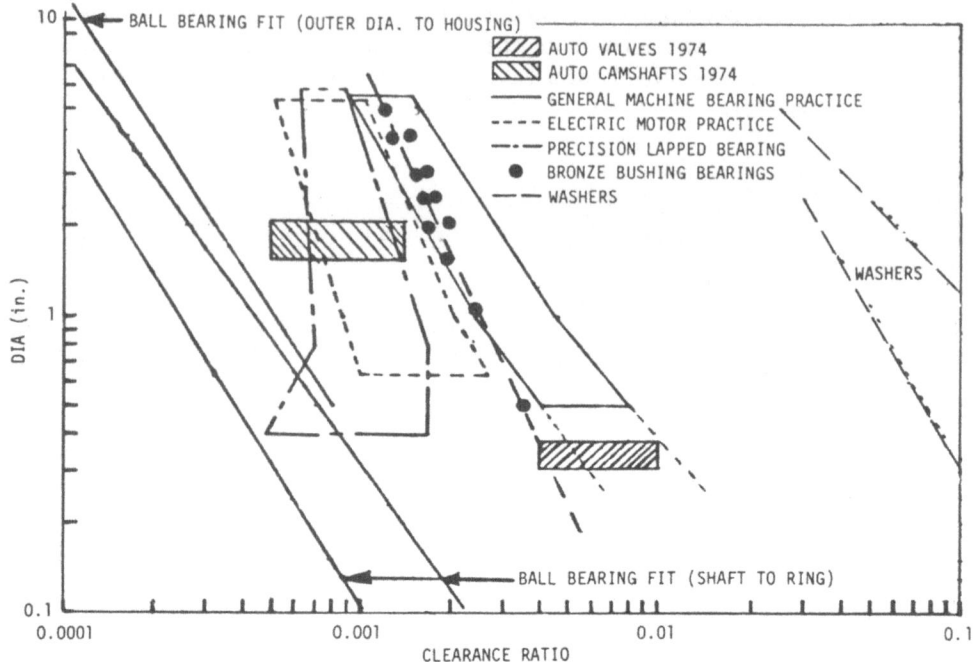

Fig. 1b. Industrial practice — diameter of parts vs. clearance ratio.

Cases where the peg touches one side of the hole and both sides of the hole have been considered. Two types of "jamming" during assembly have been identified, and the conditions for preventing them specified in terms of geometry, friction coefficient, and arrangement of the applied forces. The geometry of angular errors in screw thread mating has also been analyzed and shown to have looser error tolerances than typical peg-hole.

To test these theories a number of carefully constructed experiments have been carried out. One of the first experiments (Fig. 1C) involved the use of a pair of Unimate 5000 manipulators to test the capability of these kinds of manipulators to perform the assembly of simple, rigid, machined pieces without adaptability. What was shown was that assembly was possible but the probability of success could not be stated because success depended on unspecified compliances unique to each machine that were at least time and temperature dependent.

A second group of experiments was carried out to test friction and jamming theories (Fig. 1D). These early experiments were conducted on a second generation pedestal force sensor but the sensitivity of the unit to moment information was inadequate. This same sensor was used successfully to perform the first experiments for very large (Error $\leq$ radius of hole) where estimation techniques might be useful. To compensate for the lack of sensitivity in the sensor the test pieces were made 10 times oversized (Fig. 1E).

Fig. 1c. Programmability experiment. Two Unimate 5000 industrial manipulators were used to assemble pieces of a small gasoline engine where the nominal clearances of the parts were 0.001-0.002 inches (0.025 - 0.05mm). The positioning error of the manipulators was 0.020-0.028 inches (0.525 - 0.72 mm) and about 0.5° in angle (neglecting backlash). The experiment showed that this kind of system would work for pieces with slightly lubricated mating surfaces, chamfers, bevels and rotationally compliant grippers but the probability of successful assembly could not be stated. (Ref. 2nd Report, Exploratory Research in Industrial Modular Assembly, for the period 1 February 1974-30 November 1974, CSDL report no. R-850)

References pp. 320-321

Fig. 1d. First friction and jamming tests. Experimental system shown was used to verify coefficient of friction for various materials and to test jamming theories. Although system used 2nd generation force sensor system (pedestal configuration) its moment sensitivity was almost two orders of magnitude less sensitive than force readings.
(Ref. 3rd Report, Exploratory Research in Industrial Modular Assembly, for the period 1 December 1974-31 August 1975, CSDL report no. R-921)

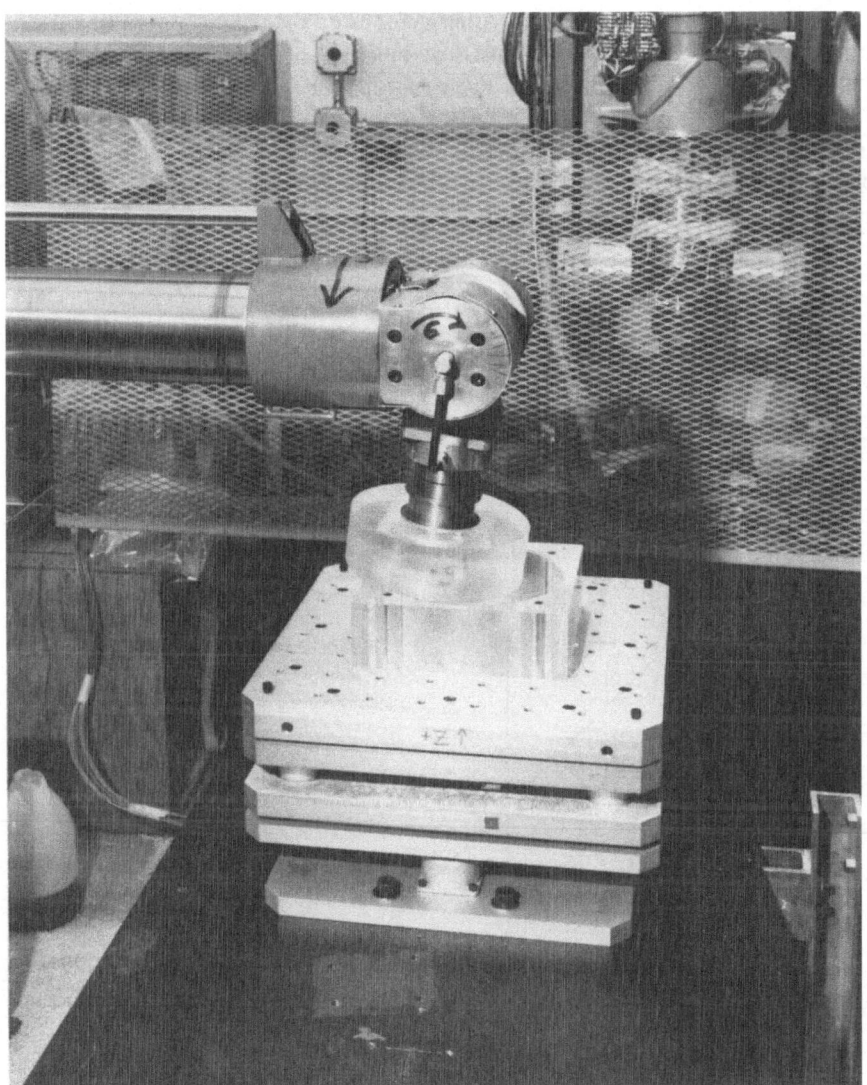

Fig. 1e. Tilt strategy estimator. Use of eximator for part mating errors approximately equal to the radius of hole. Pedestal force sensor measured tip-over moment from which estimator calculated the azimuth of tilt and, from this, the direction to the center of the hole from the center of the peg. The peg was then driven sideways until it struck the opposite side of the hole. The tilt was then removed and the peg inserted into the hole. The sequence of moves constitutes an information strategy without search.

The inaccuracies in the U-5000 manipulator and the lack of moment sensitivity in the force sensor required that the pieces be 10 times oversize.

(Ref. 3rd Report, Exploratory Research in Industrial Modular Assembly, for the period 1 December 1974-31 August 1975, CSDL report no. R-921)

References pp. 320-321

The first definitive part mating experiments required the development of a more sensitive force sensor array and a wrist with only one carefully specified compliance. This device was attached to a milling machine base and specific relative errors were imposed on insertion tasks using special test pegs and holes. All data were gathered and processed on-line by a computer. The results verified the conditions for one of the two predicted types of jamming and showed that unambiguous force and moment data could be obtained (Fig. 2).

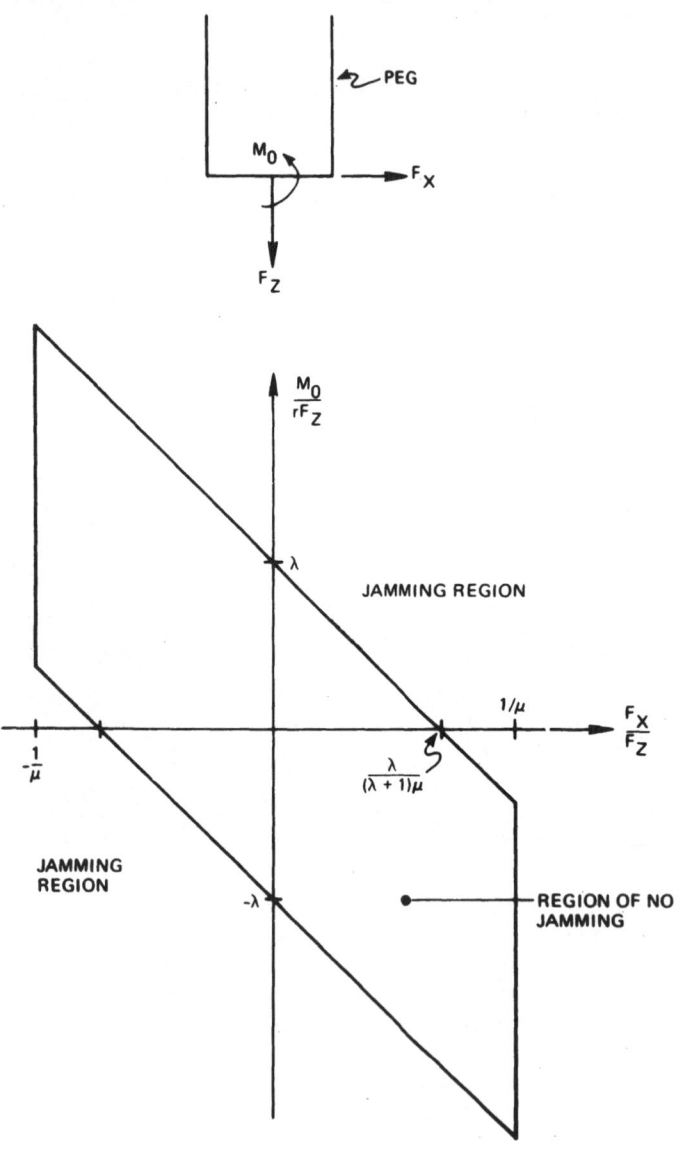

Fig. 2a.   Force moment relationships to avoid jamming.

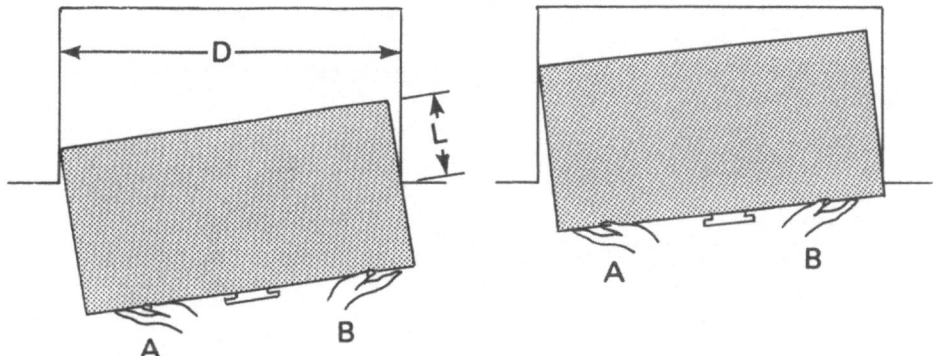

Fig. 2b. Difference between wedging and jamming was clarified during the development of compliant-gripper mechanisms. When, for example, a bureau drawer becomes wedged (left), it is literally locked. Any further application of force will deform the drawer or the bureau or both. Theory shows that wedging arises when the drawer is inserted as such an off angle that the ratio of L/D is less than the coefficient of friction (μ) when two-point contact first occurs. The only remedy is to pull the drawer out and start again. If, however, the ratio L/D is larger than μ at the time of initial two-point contact (right), wedging cannot result, although further movement can be impeded by jamming. The remedy is to break the two-point contact by pushing at A, thereby changing the direction of both the applied force and the applied moment. Compliance devices on preceding page apply forces in accord with this theory.

References pp. 320-321

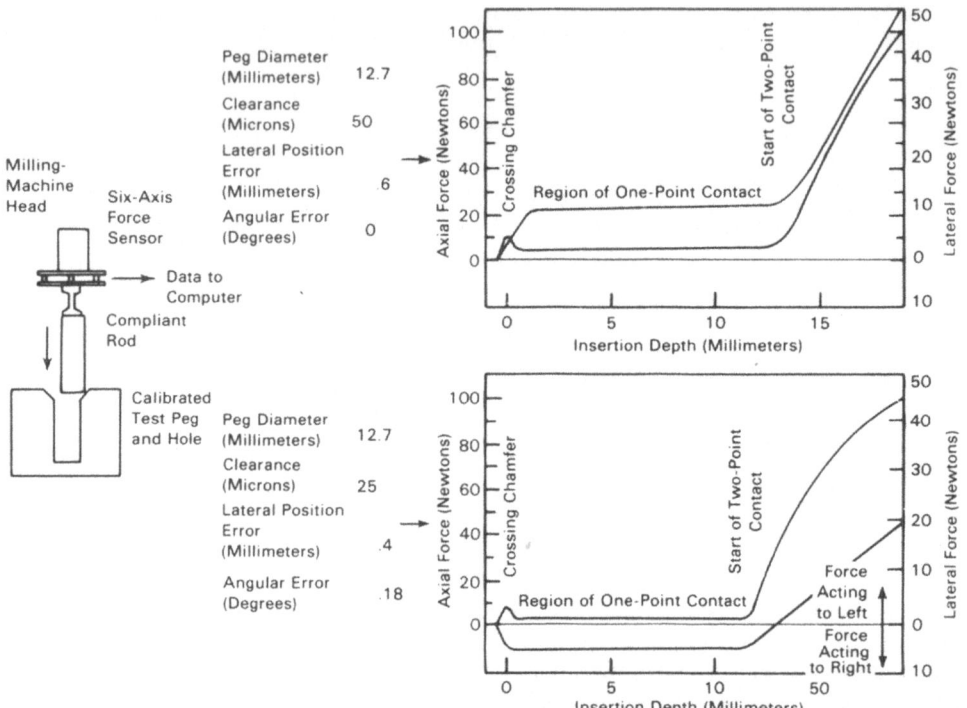

Fig. 2c. 1st definitive part mating experiment, depicted in this illustration disclose the forces that are developed when a steel peg is inserted into a hole under carefully calibrated conditions of clearance and misalignment. The forces and moments are detected and sent to a computer by a six-axis force sensor attached to the head of a milling machine. A simple compliant rod provides a limited amount of compliance between the force sensor and the test peg. In this pair of experiments the axial forces increase sharply by about the same amount after two-point contact takes place. The behavior of the lateral force in the second experiment (first pushing to the right and then to the left) is characteristic of combined lateral and angular error. The results of the part-mating experiments conformed to theoretical predictions.

Fig. 2d. 6-Axis force sensor array specifically developed for parts mating experiments.

The combination of geometric and friction analyses, plus the stimulus of the above data, gave rise to the invention of a totally new device for accomplishing assembly — the Remote Center Compliance (RCC) [1] (Fig. 3). This device accomplishes mechanically what active force feedback does with sensors and servos. It allows chamfered peg-hole insertions with 0.0003 clearance ratio with 1 mm (0.04″) or larger initial position errors and a few degrees of angular misalignment. The error tolerance of this passive compliance device is much larger than was originally thought possible for passive devices. Further, the device works for both negative as well as positive clearances.

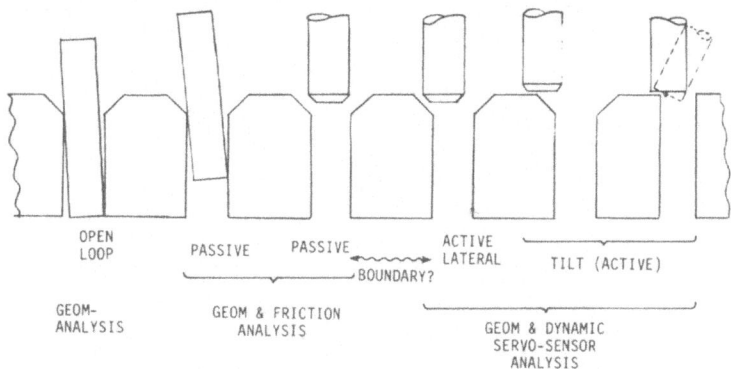

Fig. 3a. Error regions and strategy options.

References pp. 320-321

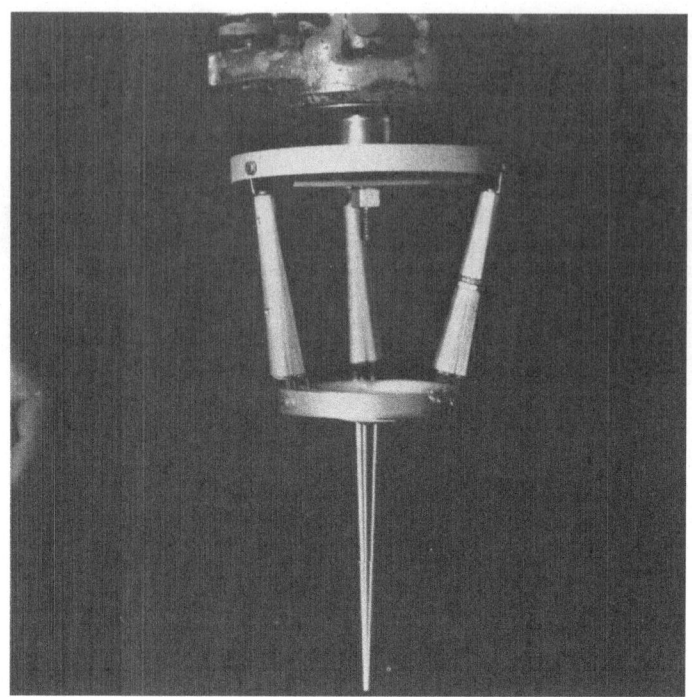

Fig. 3b. Concept Remote Center Compliance (RCC).

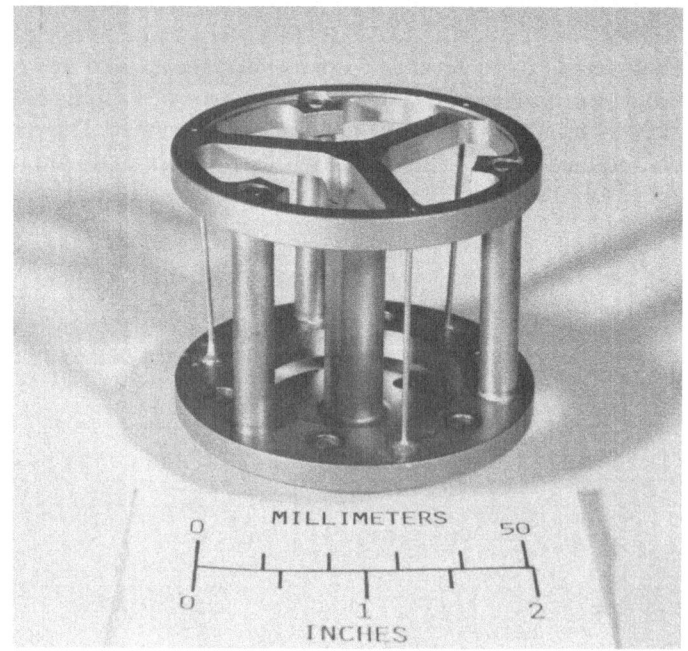

Fig. 3c. 1st practical RCC.

Fig. 3d. Model 4A Remote Compliance Center. The 4A RCC has been designed as a rugged and relatively foolproof unit with overall protection in all directions. It was done to provide a unit for evaluation by industry in the industrial environment. Several models of the 4A have been placed in industry. Its nominal specifications are:

Mass (weight)	1.36 kg	(3 lbs)
Focal length from base	20 cm	(8 in)
Lateral stiffness	100 N/cm	(55 lb/in)
Torsional stiffness	0.1 nM/mrad	(0.9 in-lb/mr)

PROGRAMMABLE SYSTEMS

To identify and quantify the kinds of assembly tasks required to assemble products, a study was made of ten products and subassemblies to answer two questions: what assembly tasks occur, and what gross directions relative to the assembly do the parts approach from? Eight of these products were of case or machined metal, one of molded plastic, and one of a variety of plastic, sheet metal stamplings, and wires. The latter product was the exception to the findings from the others: 70 percent of the parts arrive from one direction and 35 percent of all tasks are single peg-hole insertions. Adding a second direction antiparallel to the first picks up another 20 percent of the parts. Screw insertions represent 25 percent of all tasks, and ten other tasks make up the rest. This justifies the choice of tasks to analyze, although more extensive surveys

References pp. 320-321

could alter the results. It can be concluded that the first nine products form a group with quantifiable characteristics. In particular, because most parts arrive from one direction, it should be possible to assemble them with devices which have much fewer than six degrees-of-freedom. This has important economic and system configuration ramifications.

A second survey, shown earlier in Fig. 1B, examined industrial design practice with respect to sizes and clearances between parts. Here it was found that particular types of parts, made by certain manufacturing techniques, reliably fall into predictable clearance ratio ranges from 0.001 to 0.01. This confirms the choice of assembly task difficulty to analyze. It categorizes parts by their geometric properties and allows the analytical tools developed to be applied to general manufactured parts.

One of the principal problems facing programmable assembly researchers is that there are many technological options to consider for these very complex systems. We have chosen a fairly straightforward, but very powerful, economic modeling technique to help systematize these issues. Economic models have been made of programmable assembly machines and systems of such machines. Simple assumptions have been made concerning machine component costs and cycle times. Comparison models of special purpose assembly machines and manual assembly have been developed from similar assumptions so that relative costs can be obtained (Fig. 4). The economic model for programmable assembly identified the price-time product as a guideline for the design of assemblers to be used in programmable assembly systems. The price-time product is defined as the cost of the assembler (including installation in the system) multiplied by the average single part assembly time for that assembler. Thus, the price-time product measures the effects of the cost of the assembler and its speed. Assemblers capable of comparable motions which have equal price-time products, will, when combined into programmable assembly systems, result in the same cost of assembly for that product. For example, if assembler B costs $100,000 and assembler A cost $50,000, but assembler B can assemble one part, on the average, in 2 seconds while assembler A takes 4 seconds, then assemblers A or B, when used in the correct quantity in programmable assembly systems, will result in the same cost of assembly for that product since both have price-time products of 200,000 $-seconds.

Also the economic model can be used to obtain an upper bound on the price-time product for which an assembler cannot be economically competitive with either manual assembly or fixed automation assembly. For an assembler to be significantly economic, with a relatively wide range of production volumes, its price-time product value should be well below this bound. For assumptions based on typical values of the cost of system components, labor rate, and cost of capital used, the upper bound can be estimated at 293,000 $-second [2, 3].

The structure of this model has been examined, and conclusions have been drawn concerning the sensitivity of unit assembly costs to various factors. The sensitivity analysis is equally important for identifying research issues or directions that need to be examined. Without an analytical model, these vital sensitivity analyses cannot be performed.

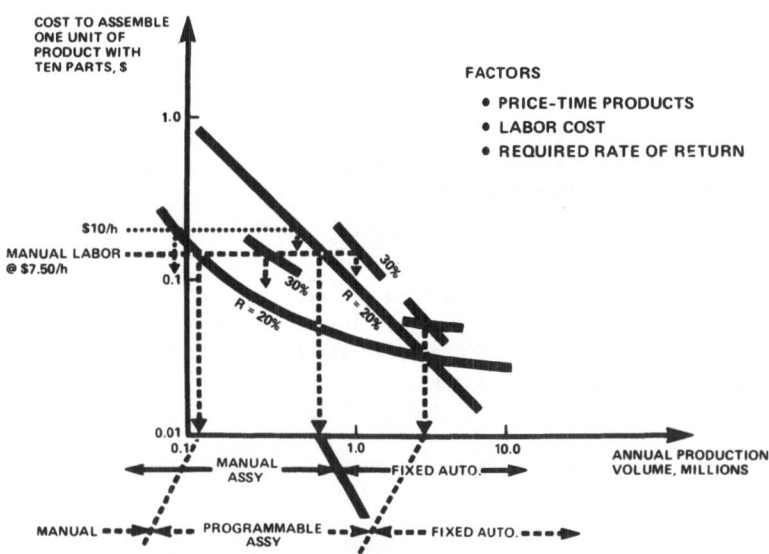

Fig. 4a. Economic relationships for manual, fixed automation and programmable assembly — 1st order.

Manual assembly cost:

MCPU = MATP * LABCST * NPART

where
MCPU = manual assembly cost per unit
MATP = manual assembly time per part, sec
LABCST = cost rate of labor, $/sec
NPART = number of parts in the product

Fixed automation (transfer machine) assembly cost:

$$TCPU = \frac{TSCST}{PAYPER * VOL}$$

where
TCPU = transfer machine assembly cost per unit
TSCST = transfer machine total cost
PAYPER = payback period in years

Next,
TSCST = NPART * TMCPP

where
TMCPP = transfer machine cost per part

Therefore

$$TCPU = \frac{NPART * TMCPP}{PAYPER * VOL}$$

This model is based on the payback period method rather than discounted cash flow for simplicity. Any particular case can be worked out using the more accurate method.

Fig. 4b. Economic models for manual and fixed automation.

References pp. 320-321

Programmable system cost:

$$PSCST = NSTA * STAP + NPART * TOLPP$$

where
- PSCST = programmable system cost
- NSTA = number of assembly stations
- STAP = single station price
- *TOLPP = tooling price per part

The number of stations required is

$$NSTA = \frac{VOL * NPART * PARTTIME}{NSPY}$$

where
- PARTTIME = assembly time per part, sec
- NSPY = 1.152 × 10^7 sec/yr for an uptime fraction of 0.8, a 250 day year and a 2 shift, 16 hour day.

The assembly system cost is then

$$PSCST = \frac{STAP * VOL * NPART * PARTTIME}{NSPY} + NPART * TOLPP$$

and, using the same payback period model as in eq(II-27), the assembly cost per unit is

$$PCPU = \frac{NPART}{PAYPER}\left[\frac{STAP * PARTTIME}{NSPY} + \frac{TOLPP}{VOL}\right] \longleftarrow$$

*Includes basic feeding mechanism (bowl feeders, hoppers, magazines, etc), feed tracks and chutes, and placement or escapement devices or conveying mechanisms that link the parts together.

Fig. 4c. Economic model for programmable assembly.

ASSUMPTIONS

NUMBER OF PARTS	=	10
PROGRAMMABLE STATION PRICE	=	$30,000
TOOLING PER PART	=	$7500
TRANSFER MACHINE COST PER PART	=	$30,000
PART STATION TIME	=	3 s
PAYBACK PERIOD	=	2 years to 4 years
LABOR COST	=	$7.50/hr to $10.00/hr
MANUAL ASSEMBLY STATION TIME	=	7 s
NUMBER OF SECONDS PER YEAR	=	1.152 × 10^7

Fig. 4d. Cost assumptions.

ROBOT ASSEMBLY RESEARCH

To adequately study the area of programmable systems, it was found necessary to construct a programmable system test bed. This test bed consists of a single electric arm originally constructed by the Bendix A&M Division (Dayton, Ohio), as a prototype arm for industrial assembly. Supporting the arm are sensors, tooling, parts feeding, minicomputer systems and a specially developed software system. To test the knowledge developed to date an automobile alternator with 17 parts was assembled on the system (Fig. 5). At present it can assemble the alternator in 2 minutes and 42

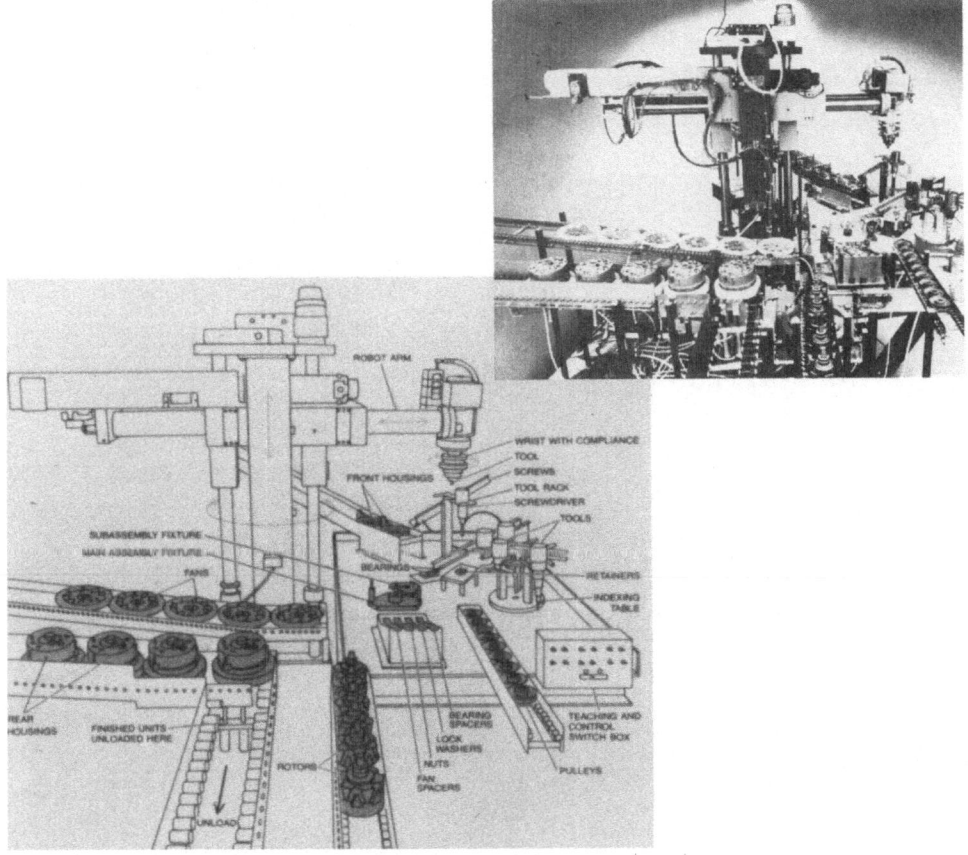

Fig. 5 Layout of the Robot Assembly Station places tools and parts within easy reach of the robot arm, which has four degrees-of-freedom. The assembly task requires six different kinds of tools, held on a table that "indexes", or turns, to supply the proper tool for each operation. The alternator's 17 parts are fed by gravity from 12 feeders. (The 17 parts include 3 screws which have only one feeder, and 3 long bolts, fed together with the rear housing). The assembly is performed on two different fixtures, one for the main assembly, the other for subassembly. The robot is operated by a computer that drives the four points to designated stopping at designated speeds. The points, speeds and tool operations are programmed with the aid of a control box and a simple keyboard language. The language names and sequences the points and tool operation. A major feature of the robot is a wrist-and-gripper mechanism that responds compliantly so that parts can be inserted into close fitting openings without jamming.

References pp. 320-321

seconds using 6 tools and 8 tool changes [4]. In principle it can assemble products about one foot cube in size requiring insertion motions vertically down, or (with appropriate tools) horizontally, or vertically down with some spin about the vertical axis. Electric motors, some types of gear boxes, electric entrance boxes and terminal board arrays are examples. This test has yielded much knowledge of a technical and economic nature concerning programmable assembly. Some of the results are shown in Figs. 6 and 7, and in the next section.

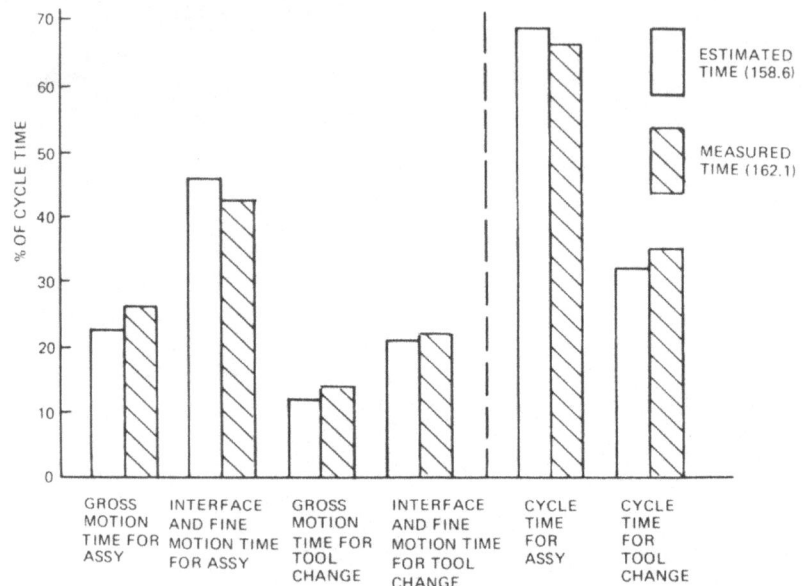

Fig. 6a. Assembly cycle time analysis from programmable system experiment.

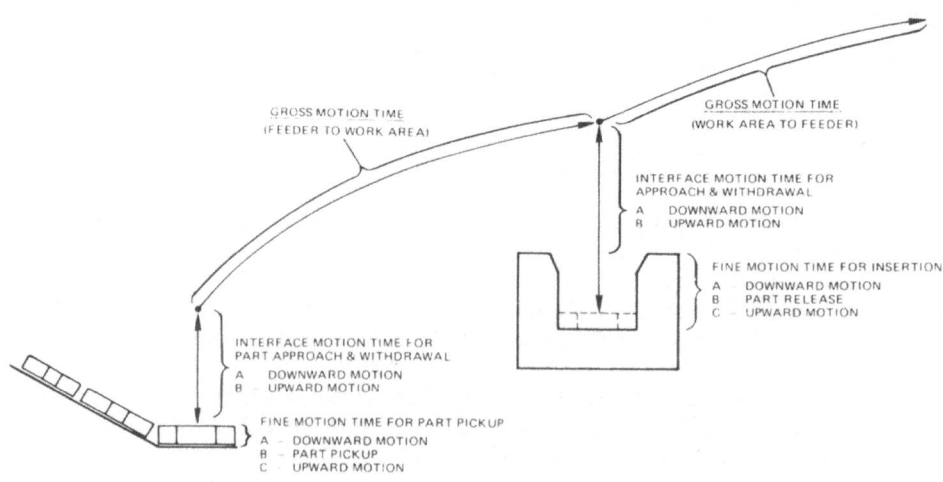

Fig. 6b. Arm trajectory definitions.

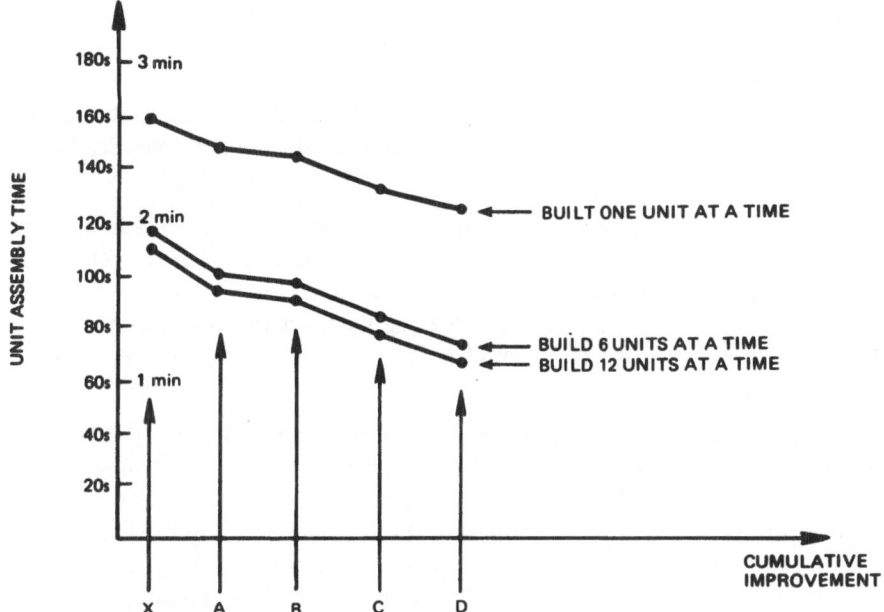

X: PRESENT LAB SYSTEM
A: DRIVE 3 SCREWS AT ONCE WITH A 3-HEAD SCREWDRIVER
B: DESIGN CHANGE TO ELIMINATE BEARING RETAINER PLATE, PLUS A
C: PICK UP ROTOR AND TIGHTEN ROTOR NUT WITH ONE TOOL INSTEAD OF TWO, PLUS A AND B
D: ELIMINATE SUBASSEMBLY STATION AND BUILD ENTIRE ASSEMBLY ON ONE FIXTURE, PLUS A, B AND C

Fig. 7. Effect on time to build one alternator of various improvements in tooling and product design.

RESULTS FROM EXPERIMENTAL PROGRAMMABLE TEST BED

The automobile alternator assembly experiment produced several significant results and conclusions.

Adaptable, Programmable Assembly of a Real Industrial Product was Accomplished — As of this writing, approximately 75 alternators have been assembled (many by disassembling the same sets of parts). Adaptability was demonstrated by the system's ability to absorb position error of 1.25mm (0.05 in.) or more, due in part to variation, uncertainty of part location in feeders, softness of feeder mounting, and lack of precise assembly fixture alignment. Programmability was demonstrated by the ability to program different assembly sequences using substitute tools when other tools were broken, to improve the program's reliability by teaching it "tricks", and to shorten the cycle time by a combination of new tools and more efficient teaching. The basic software system is sufficient to support programs and assembly sequences for a wide variety of products containing similar assembly tasks.

Performance of the System was Thoroughly Documented — A detailed time

References pp. 320-321

study was made. Working with documented time/motion studies we found that the system had the potential for reducing the present 2:42 assembly time to 65 seconds if many alternators were made at once, with tool change time spread over all. This short cycle time also required some tooling improvements and a design change. No change in software or arm performance was assumed. Further reductions are possible. See Figs. 5, 6 and 7.

Concept of Engineered Compliance was Shown to be Sufficient to Perform a Variety of Difficult Single-Direction Tasks Vertically — The impact of the RCC on system design style, design and debugging time and ultimate performance was profound. Rather than estimate a time need for debugging insertion tasks without the RCC, we prefer to state that some tasks could not have been reliably and repeatedly performed at all without it.

The Software and Teaching Techniques Appear Adequate — The four-axis arm design makes visualization of the geometric relationships by shop floor personnel easy. It thus appears feasible to divide the programming as we have done: an industrial engineer lays out the feeders and tools and writes the flow sequence in the high-level language; a shop floor teacher (or in the future a CAD/CAM data file) inputs the arm position data corresponding to this program. The advantages of this technique are its hands-on immediacy, its simplicity, ease of debugging, and relatively simple software. Potential weaknesses could emerge if extensive reteaching were to be needed in a great hurry.

PRESENT RESEARCH DIRECTIONS

The present research activities are stimulated by three things, namely: 1) to capitalize upon and to extend the present knowledge base and devices, 2) to use the output of present experiments to identify research issues to be explored, and 3) to pursue the generic problems in this area. For the latter, a number of lists of issues have been compiled for discussion.

Extending the Knowledge Base

1. Part Mating — This work falls into seven principal areas: 1) part mating experiments, 2) analysis of force fits, 3) high speed insertions, 4) optimal chamfers, 5) force feedback experiments, 6) RCC developments, and 7) force sensor development.

a. Part Mating Experiments — The part mating test bed (Fig. 8) has provided the basic tool for experimentally verifying part mating theory. During the past year the force histories of nearly 10,000 part-matings or separations have been recorded and hundreds of computer plots have been generated—Fig. 9 shows a typical data plot from the manual RCC insertion station shown in Fig. 10. In addition, one industrial firm has duplicated the part mating test bed for their own studies. Control of this system and methods of automating test data acquisition and processing are continually being improved. A range of force sensors and RCCs to support new part mating tasks are being developed.

Fig. 8. Draper Laboratory parts mating experimental test stand. The spindle and table of the vertical mill may be fitted with a variety of tooling, measuring or control hardware. Quantative data on mating forces and moments can be displayed, read into a digital data file, or plotted. As shown, the spindle is fitted with an RCC and 6 degree-of-freedom aluminum force sensor. The nominal maximum forces and moments of the aluminum force sensor are:

Axially
400 N (100 lbs)   Force
20 Nm (200 in lbs) Moment

Laterally
200N (50 lbs)   Force
10 Nm (100 in lbs) Moment
Resolution of one part in 2,000 of the maximum axial force is possible.

References pp. 320-321

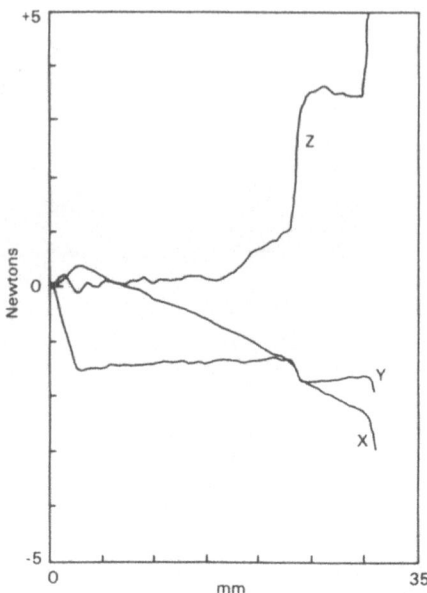

Fig. 9. Typical data from a manual insertion.

Fig. 10a. Fixture for aiding rapid, reproducible manual instrument bearing assembly. Instrument ball bearings are a light interference fit at operating temperatures and are assembled into a heated housing to provide assembly clearance. The assembly must be done quickly and at light force levels. It is common in unaided assembly that a ball bearing will jam and have to be driven home. A manual assembly stand assisted by the RCC to avoid jamming, and a six degree-of-freedom force sensor to document the assembly force history, is expected to yield significant improvements in the process. A brake is included in the design; activated by the force sensor, it can retard the drive and limit the seating forces exerted on the bearing. Any one of the six forces and moments measured may be selected for analog display (on the meter to the left of the assembly stand). Each or all of the measured quantities may be stored on a computer file or recorded for permanent record.

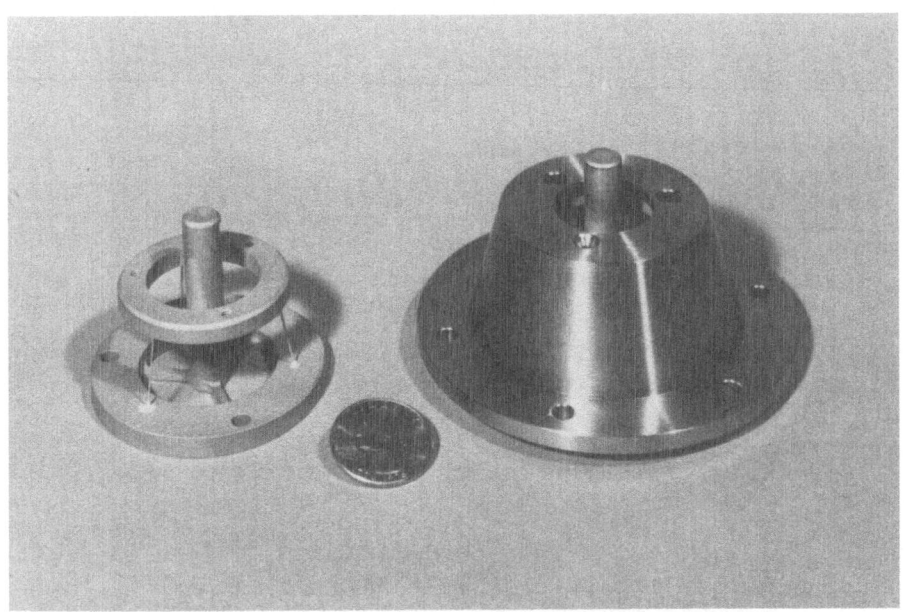

Fig. 10b. Small RCC. A small RCC designed for the manual instrument bearing installation stand. Its manual specifications are:

Focal Length from Base	12.5 cm	(5 in.)
Lateral Stiffness	185 N/cm	(100 lb/in.)
Torsional Stiffness	0.12 Nm/mrd	(1.2 in lb/mrad)

b. Theoretical Analysis of Force Fits — With the realization that the RCC was capable of solving both negative as well as positive clearance chamfered fits, and the development of the first application system (Fig. 11), a theoretical investigation of negative clearances was started. Its objectives are: 1) to find the insertion force for quasi-static force fits given the dimensions and materials; 2) to determine the parameters necessary for selection of an appropriate hammer for force fits; and 3) to calculate the forces developed during hammered force fits. Analytical solutions have been developed to describe the insertion force for a peg into a hole for any pair of materials which obey Hooke's Law and the minimum energy for further movement of a hammered force fit or a press fit have been calculated. Future activities will include modeling the behavior of pneumatic hammers, developing a model to account for effects of static preloads and fixture design on hammered force fits, and extensive testing to verify analytical models.

c. High Speed Insertions — The investigation of high speed insertions has caused a reexamination of the RCC (discussed below). One of the key questions to be answered is: During a high speed impact, is the center of percussion of the suspended mass more important than the center of compliance? To date, equations have been developed that can predict the motions of the mass suspended in the RCC after impact. The lateral and angular velocities of the suspended mass must be known along with the coefficient of restitution for the materials involved. An equation for an effective mass was also

References pp. 320-321

Fig. 11. Assembly fixture incorporating RCC. An assembly fixture incorporating the RCC for the purpose of centering and erecting a small brass ferrule to be driven into an interference hole in an aluminum casting. The purpose is to avoid possible problems with jamming and subsequent broaching by using the unique properties of the RCC.

developed for more precise determinations of the coefficient of restitution. These equations were developed by representing the suspended mass as a bipolar equivalent mass system. An experiment is planned to verify the validity of this action.

The amplitude of the bounces can be reduced by increasing the stiffness (reducing the compliance) of the RCC. However, if the stiffness is sufficiently large it is possible that the sliding motion during the insertion can be restricted. The three regimes that occur during the sliding part of an insertion are being analyzed so that the maximum stiffness that will allow insertion can be determined. The three regimes are sliding on the chamfer, one point contact sliding, and two point contact sliding.

Before the system can be modeled, the equations of motion for the various regimes must be developed. In addition to the three previously mentioned regimes, the free vibration situation must be considered. Differential equations have been developed for all but the two point contact situation. During an insertion several changes in the boundary conditions take place. As the insertion passes from one regime to another, changes in the direction and magnitude of the various forces acting on the suspended mass occur. Geometric restrictions also change with depth due to the chamfer. These effects must be accounted for in the model.

Several subtle nonlinearities may also be added to the model. While the migration of the center of compliance, due to deformation of the RCC, has been shown to be insignificant, a change in the magnitude of the latter stiffness due to deformation may be great enough to require modeling.

It is hoped that by varying the inertia and compliance elements of the RCC and suspended mass, a more rapid insertion can be realized. It may also be necessary to add damping to the system to reduce free vibration.

d. Optimal Chamfers — As part of a large part mating program or an industrial sponsor, some results on optimal chamfers have been obtained, of which some can be presented here. The problem is to find the optimal slope for a straight chamfer. Optimal is defined as yielding the smallest work to achieve insertion:

$$\text{Min} \int_{Z_0}^{Z_1} F_I dz$$

where F_I is the Z axis (or insertion direction) force required for insertion, Z_0 is where the part first strikes the chamfer (considered constant), and Z_1 is the bottom of the chamfer. If we assume that the part to be inserted is long compared to its diameter, and it is supported by a rotational compliance at the end far from the inserted portion, then F_1 is approximately:

$$F_I = \frac{K(Z_0 - Z)(1 + Z'\mu)}{Z' - \mu}$$

where Z is the travel distance of the part since striking the chamfer at Z_0, K is the effective lateral support stiffness, Z' is the chamfer slope (to be chosen optimally), and u is the coefficient of friction. Carrying out the minimization results in:

$$Z' = \mu + \sqrt{\mu^2 + 1}$$

Some examples are:

μ	Z'	Chamfer Slope in Degrees
0	1	45°
0.5	1.618	58°
1.0	2.414	67.5°
2.0	4.236	76.7°

References pp. 320-321

e. Force Feedback Experiments — Three questions are currently being looked at:

- What is the role, if any, of force feedback in teaching?
- How does force information aid in monitoring an assembly operation?
- What is the use of force information and estimation theory with a servo controlled wrist for assembly problems with large effort?

The latter problem is currently being looked at as a Ph.D. Thesis [5]. It is of interest because it offers a closed loop technique for assembling parts, albeit with a fairly large overhead cost.

The test matrix for examining the role of force information in teaching includes a rigid arm, an arm with an RCC, and the presence/absence of force information. Preliminary results from the first experiments indicates that there are two phases in teaching. The first phase is geometric and quasi-static. The arm is moved slowly through points which the operator estimates to be correct, or which are provided by off-line teaching software. Small errors in these points can be observed visually by noting the deflection of the RCC. These deflections seem easier to interpret than force readings on dials, although better force display would help. The second phase is dynamic. The arm moves at production speed and, due to controller errors, does not pass exactly through the statically taught points. Some touchup is needed, and the motions are too rapid to permit visual observation. Here a force record would appear to be better. One can envision an automatic force correcting loop which will adjust the taught points over several assembly cycles.

The role of force monitoring is quite a complex one. Most early researchers used only the binary force information to detect the presence or absence of a part, the torque level for setting a nut, etc. The combination of a six axis force sensor and an RCC provides a mechanism for measuring the complete force trajectory of a mating process. Already a device (Fig. 10) has been constructed for aiding the manual assembly of expensive, close clearance precision parts. This device offers full six axis force monitoring/recording, and force level set points for alarms coupled with brakes for the manual inserter if thresholds are exceeded.

The study in process seeks to examine the issue of assembly as a reproducible process that is monitored and controlled through force measurements.

f. RCC Developments — At present the stimulation for RCC developments comes from external industrial applications and internally from the desire to solve the chamferless insertion problem. New solutions to the latter issue are currently being explored on the laboratory level.

The success of the RCC has led to many inquiries about possible applications. One challenging application was a press fit to be performed at high speed. It was necessary to recheck the validity of the assumptions behind RCC design since it was originally intended to do clearance fits at low speeds.

The ideally behaving planar RCC consists, in part, of a four bar linkage with pinned ends. In its common realization the pinned ends are replaced by elastic wires. The question is whether the effects of this replacement on the kinematic behavior of the RCC are significant under any load condition, especially the large axial loads associated with force fits.

It was decided to ignore the portion of the linkage allowing translation, on the basis that its behavior is acceptable in either embodiment, to establish an elastic multi-element beam-column element computer model of the wires and manipulate the model with various loads, noting deflections and inferring the center of rotation for each set of loads.

The conclusions were that in the geometry range of existing RCC loads, the use of elastic links does not have a significant effect on the deformation geometry. The rotational compliance is very stiff to either a later or an axial force (at least when force levels are small compared with buckling critical forces), and while the center of rotation deviates greatly from its ideal location at the focus, under load of such forces the associated deformations are negligible. Significant deformations occur, of course, in response to the applied moment, but the center of rotation is at or near the focus. Under combined axial and lateral forces and an applied moment, the deformation response to the moment swamps that of the forces and the center, and while not at the focus, is very, very, close.

The RCC under axial loads was also studied since the spring constants are augmented by the presence of that load. These augmentations to the spring constants are:

$$K_X \sim F_V/H_L, \text{ and } K_\theta \sim F_V L \left(\frac{2L}{H} - 1\right) \qquad (6\text{-}1)$$

The second equation is a simplification of:

$$\frac{M}{F_V} = \left[\left(\frac{D}{2}\right)^2 \left(\frac{1}{H} - \frac{1}{L}\right) + L \left(\frac{2L}{H} - 1\right) \right] \theta \qquad (6\text{-}2)$$

The variables are defined in Fig. 12.

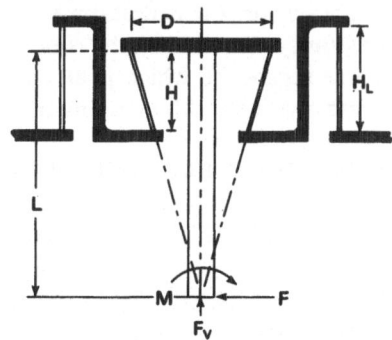

Fig. 12. Schematic of RCC for stiffness analysis.

The first RCC units sold to customers were designed to be "bullet-proof". While no failures were anticipated, any necessary repairs would have been very difficult. A new design (Fig. 13), built to the requirements of an industrial user, is now being tested. It offers the same function at much lower weight and cost (7075 aluminum vs. 17-4PH stainless steel). While it is less armored, it offers ease and economy of repair. The rate of the creation of new designs is also accelerating.

g. Force Sensor Developments — This work has focused on improving the performance and reliability of the present sensor and the construction and testing of a pedestal force sensor.

i. Sensor Improvements — The development of a more reliable sensor package is primarily an engineering problem. Particular aspects of the problem include gage heating, operating temperatures, and excitation; materials, material stress limits, and material specific errors or uncertainties; gage and bridge configurations, compensation, bridge configuration—specific nonlinearities and bridge sensitivity; surface preparation, adhesives, and gage backing; amplifier choice, size, location and grounding; sensor geometry and specialization for different magnitudes and mixes of loads. The goals are to be able to produce conservatively configured sensors for long life stable laboratory/industrial use and to be able to quickly configure and design sensors for specific requirements in a largely routine way.

ii. Pedestal Force Sensor — During the initial alternator assembly experiments a six axis force sensor was mounted on the wrist of the manipulator arm. This sensor provides information about insertion forces of various parts (notably the insertion of the bearing into the alternator front housing). It was then determined that mounting the force sensor on a pedestal under the workpiece would supply additional information during assembly. For example, while a wrist-mounted force sensor would indicate the angular displacement of a peg entering a hole, a pedestal force sensor would more readily indicate the lateral displacement of the peg relative to the hole. This information is of use not only in force feedback assembly, but also to speedup the teaching process by allowing the arm to detect errors in positioning.

Several considerations influenced the design of the pedestal force sensor. Rather than fabricate a new type of force sensor, the pedestal was designed to accept the six-axis force sensor which was previously mounted on the robot arm. The pedestal serves as a mounting point for the force sensor, to which is attached a plate serving as the assembly jig to monitor the forces involved with inserting the bearing in the alternator front housing, and fastening the bearing retainer (Fig. 14). One of the first experiments with this unit is to determine the value of force feedback during both teaching and automatic operation.

iii. Ultrasensitive Sensor — The current sensor design has a dynamic range of 2000 and a threshold of about 0.5 N (2 oz). Work has begun on a sensor with similar dynamic range and a threshold 10 to 20 times smaller. This sensor will support delicate part mating studies for an industrial customer.

Fig. 13. Model 4B remote center compliance. Designed to provide a lighter weight (1 lb. or .45 kg mass) unit capable of easier repair and lower cost. The focal length and elastic parameters are nominally the same as the Model 4A.

References pp. 320-321

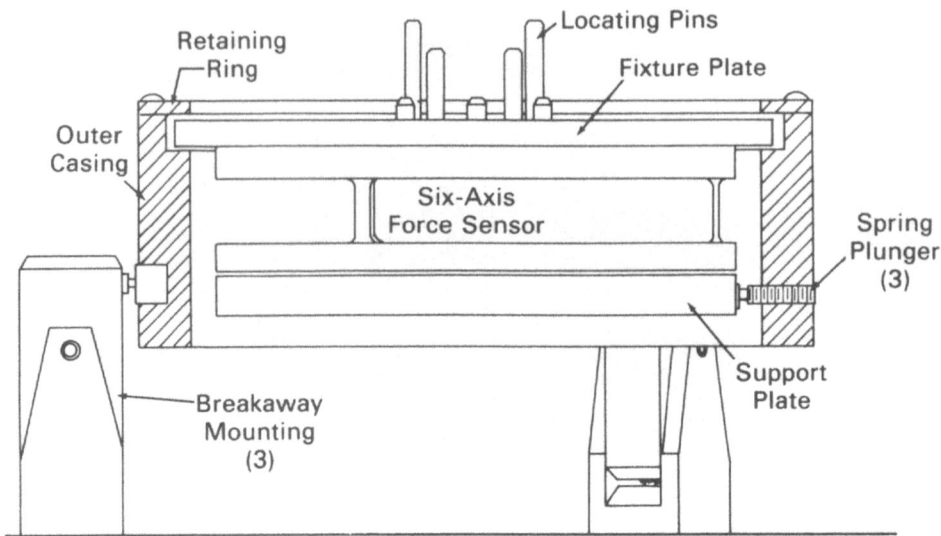

Fig. 14. Pedestal force sensor. Unit is protected from static overload force by a break-away mount. The break-away mounting is adjustable for static force as torque level. Once broken away, forces are transmitted to the outer casing directly by the fixture plate.

iv. Systematized Sensor Design Procedure — The above efforts have stimulated creation of complete design equations and criteria for high performance sensors having wide dynamic range, low threshold, and good linearity.

2. *Programmable Assembly Systems* — *a. Background* — The programmable assembly system (PAS) problem can be separated into parts similar to the flexible manufacturing system (FMS) problem: assembly requirements, system design, and system operation. PAS's are defined here as general arrangements of robot (or other) assembly stations and conveyors. In principle, work can proceed from station to station in any sequence (Conventional assembly lines are a special case of this type of arrangement).

There exists no data base even remotely comparable to machinability tables that describes assembly in a quantitative way. Part mating research is aimed at this problem, but there is a long way to go. Aside from individual insertions, data is needed on batch sizes, product size, number of parts, time required for operations and so on, to allow rational system sizing to proceed.

Designing PAS's involves a second set of unknowns. FMS designers currently can choose from standard catalogs of machining centers (although that may not be the best way to design systems). PAS designers have little to choose from right now and may be able to influence what becomes available. A basic question is: should the stations be more or less identical, or should they, like FMS stations, be specialized and different? Quite different design techniques, system designs, and system operating issues result from each choice.

Operating a PAS, especially with a resident mix of different products or a changing work pattern, presents a third set of difficulties. One can break operation down into two regimes: steady state and transient. The former, if it ever occurs, represents steady or repeating behavior while the latter is fluctuating or non repeating. Current research results indicate that good steady behavior (all stations about equally loaded and busy) can be achieved if the incoming work stream has equal work for all stations and the conveyor system has sufficient speed and carrying capacity. But this state may be achieved only quite slowly and may result in large in-process inventory. This in turn may make model mix and batch switching cumbersome (Fig. 15).

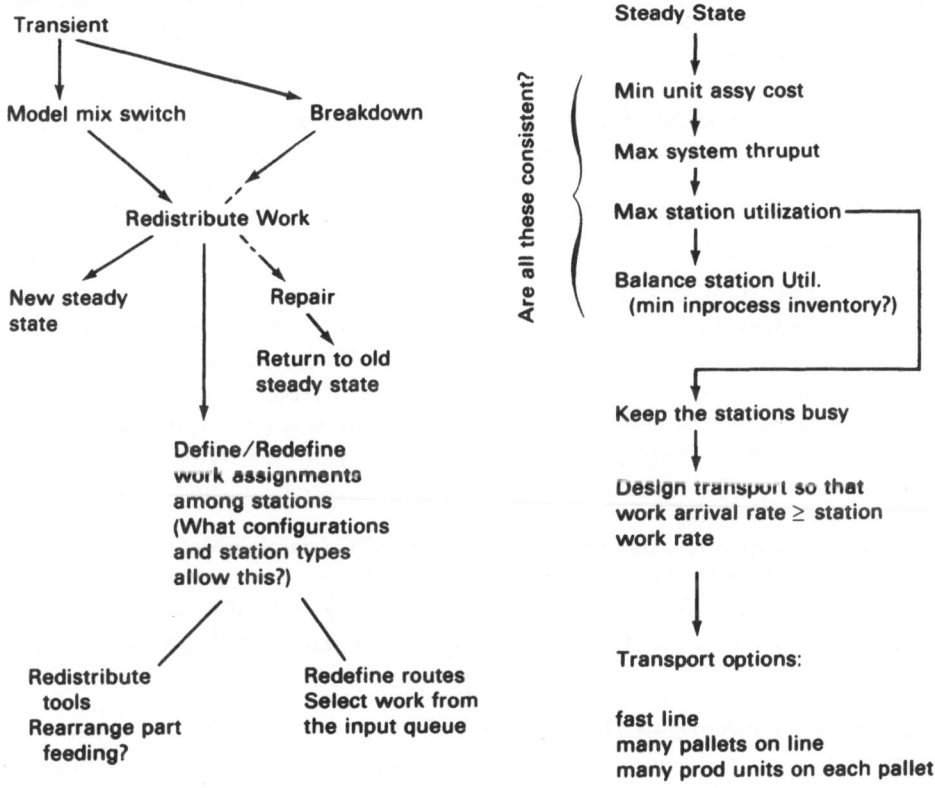

Questions: Where does min inprocess inventory fit?
Which configurations are easier to fix and which suffer least loss of assy cycles?

Fig. 15. Logic diagram for programmable system research.

b. Work Plan — The approach to this work currently being taken is threefold. First it is planned to take apart and reassemble a number of things. This is done for two reasons: as an aid in defining requirements and to classify and analyze the tasks or unit operations. The second direction is analysis and design of systems that can assemble several things. This approach requires both analysis and a multistation test bed for performing experiments. The third direction is to analyze the operational behavior of these systems. These are discussed briefly below.

i. Requirements Analysis — Our previous work on assembly requirements consisted of a survey of ten product items which were assembled and disassembled [6]. Many similarities emerged from these seemingly different items, and we found that several characteristics of them were important to the design of assembly systems. For example: pertinent to individual parts: size, weight, clearances, chamfers, support, dimensional stability, insertion distance, insertion type, arrival direction axis, and kinematics. Our work in part mating theory contributed to our ability to evaluate some of these characteristics.

This work needs to be expanded in two directions. First, an attempt must be made to extend this kind of survey to entire product units, to determine the ranges of variables like batch size, model mix, the real differences between models, the flow rate, and the need, if any, for multi-product mix in one assembly systems. This work, by necessity, will heavily involve industry.

Second, deeper examination of the parts and assembly tasks in products is needed. Prior work identified insertions of round pegs the most frequent task and found that many products have one dominate insertion direction. While this pattern may not hold up as the sample is broadened, we hope that some identifiable classes of products and tasks emerge. It may be that defining tasks in the human way (insert, twist, etc.) is too loose and will never converge to a finite set. A new way of identifying assembly tasks and new designators for tasks may be necessary.

It is hoped that the methods of group technology will be applicable. In the long term one can visualize a code to describe a product so that the overall requirements for an assembly can be read from the code. Product families capable of being assembled by the same system could be identified by similarity of the codes, as is presently done in metal cutting.

ii. System Design Issues — The laboratory single station programmable system test bed was first tested by assembling a 17 part automobile alternator described earlier. Since that was accomplished much data on this kind of assembly was gathered. To further broaden this work a second system consisting of two arms is now in the planning stage. In support of this work a pair of prototype assembly robots (Fig. 16) have been purchased from the Bendix A&M Div., Dayton, Ohio.

The principal issues to be explored with this kind of system are multi-arm interaction, system architecture questions, parts feeding, inspection, sensors and their interaction, and software. It is expected that the work will also explore workstation dependency on the presenting tasks in the products to be assembled and on overall production requirements. This information will come from the survey data discussed above. A major option for station design is the degree of specialization (as to tasks or

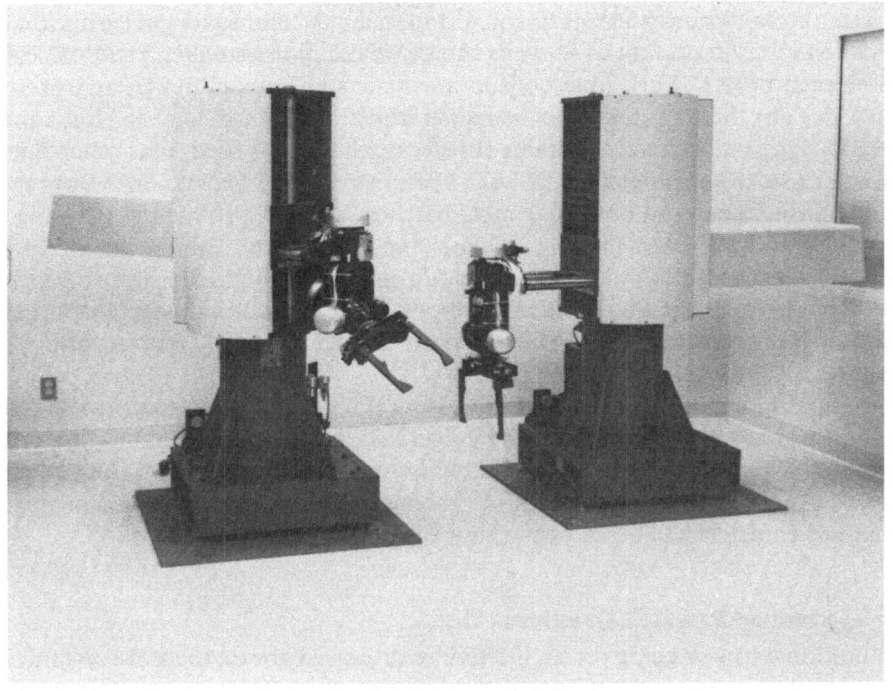

Fig. 16. Two arm programmable system test stand.

task types it can perform, number and/or direction of the degrees-of-freedom, range of motion, number of tools available and so on).

The requirements on system transport facilities depend on both the products and the stations. If stations are specialized, much transport and congestion can be expected, but station operation will be efficient. Many Flexible Manufacturing Systems (FMS) are designed this way. But FMS station times are long, so the analogy may break down. For assembly, we need to know if there are any generic transport layouts, whether layout detail can be decoupled from system performance, and what should be transported (partially completed assemblies, parts, fixtures, tools, complete parts kits for one unit, pallets full of identical parts, etc.).

iii. Operation Behavior — Our experience operating the assembly station in our laboratory showed that tool change time was a dominant feature of its behavior. There is little we can learn from it, however, concerning behavior of systems of such stations. Mathematical analysis and simulation are the tools available plus limited testing on a multi-station system described earlier.

Any such system presents scheduling (in what sequence should work be introduced) and queuing (do lines form at some stations) problems. Ideally scheduling problems can be eliminated if the work on the system provides equal load for all stations, even if

References pp. 320-321

each item on the system presents unequal station loads. The equal load criterion is so powerful that it is important to study its consequences. For example, if stations can do some of each others' work (due both to adequate station/tool design and prescient product design) then the load can be redistributed to achieve balance. But random fluctuations in station time, inevitable if sensory feedback is used, plus other fluctuations, will upset the neat balance. Do such upsets spread to other stations? Does steady state operation occur and how large must batches be for this to happen? Noting that such questions have been investigated for FMS, where the station times are long relative to transport and tool change times, do recent findings carry over to programmable assembly? Or are there critical relations between station time, transport time, amount of inprocess inventory, pallet and tool change time, and output rate?

These questions need to be looked at in several ways. Existing mathematical tools (such as Purdue's CAN-Q) [7] can be expanded as necessary to model the critical events in an assembly systems. Extremes of system type (lines, islands, loops) can be analyzed, using simple example layouts. Stochastic simulations of these same layouts are planned to determine other operating behavior to compare to CAN-Q.

More Speculative Research Questions

In addition to the ongoing research activities discussed above, there are several questions which have emerged but have not gelled into focused activities. These are discussed briefly:

1. Interface Motion Devices — The statistics on how the time is divided between gross, interface, and fine motion of the manipulator (Fig. 6) shows that gross motion time is only about 25 percent of the total time, whereas interface and fine motion represent over 40 percent of the total time. Since fine motion can be as short as 0.2 seconds per insertion then most of the 40 percent is due to interface motion.

Gross motion capability, at present, is the most expensive to provide. Fine motion, implemented by a passive system such as the RCC, is extremely cheap and fast, and can provide up to five additional degrees-of-freedom. The questions therefore are: Can interface motion devices be created that can dramatically reduce this time factor and still be economic? (i.e., increase the efficiency of the total system without a large cost increment). A cost increment below the price-time product relationship—would be the aim. Under price-time product, that means that even though you have doubled the system speed, you have not doubled its cost. A figure of merit therefore is how little you have to pay for significant increases in effective speed.

2. Other Applications for RCCs — Currently in the forming of materials by presses or forges the design philosophy is to provide massive-rigid structures to maintain exact alignment between dies. The RCC presents an opportunity for presses to be constructed where the form arising from material deformation would control the alignments of the dies.

Many products might benefit from the application of part mating analysis of the RCC principle in their design. The earlier example of an optimal chamfer is a case in point.

3. *The pedestal sensor offers a method of both measuring and monitoring exactly how people assemble things.* This is useful for documenting ease or difficulty of performing certain assemblies, for arriving at a definition of "force required to assemble", and for studying strategies by which people attack assembly problems.

4. *Adaptive Control and Learning by Assembly Robots via Force Sensing* — After a robot is taught a nominal assembly path, it begins executing this path on a succession of objects. The forces it encounters can be expected to vary as the parts, fixtures and the robot vary. Some of these variations will be random and some will be indicative of steady drift. It is reasonable to expect that the robot controller could take account of the drifts and either compensate for them by modifying the taught path or by flagging the drift as an error condition. Many questions arise such as separating and identifying drifts from multiple sources and setting thresholds and rates of correction.

Generic Research Problems

Assembly research offers many generic challenges which are beyond the scope of any one research team. We present here our summary and categorization of these problems, divided into two groups. In the first group we have the larger system issues such as product design for automation or for manual assembly; analytical tools and simulation for studying part mating and assembly systems; parts feeding and inspecting software and hardware organization and implementation studies; teaching methods; and economic studies and tradeoffs.

The second group is devoted to more elemental concerns. In this group are articulating devices; adaptive devices (both active and passive); sensing system tools and grippers; and part feeders and inspection devices.

1. System Issues — *a. Product Design* — Although the importance of this work has not really been highlighted it is quite obvious that maximum system efficiency will be realized only when system and product design are truly integrated. It doesn't matter what the system is. It can be manual or special machine or programmable system. The only thing that matters is that the limitations of the particular assembly techniques are compensated for in the product design. It has often been found in special machine design that once a product has been redesigned for that assembly technique it is often significantly cheaper to assemble manually.

The part mating studies at the present are our best technique for indicating system design requirements. But, this work is not complete so a complete programmable system product design specification cannot be written.

References pp. 320-321

b. Analytical Tools and Simulation — Three types of analytical tools are needed in this field:

- Part mating analysis techniques, especially to treat problems of non-rigid parts
- Assembly system design and operational tools, to specify system needs in the face of uncertain demands
- Assembly system financial analysis and justification methods which adequately take account of system flexibility.

c. Part Feeding — The parts to be assembled must be presented to the assembly system somehow. There are two crucial issues:

i. Should the parts be fed oriented in the proper way for the robot, as they are for other automated processes, or should they be fed (more cheaply) unoriented, with the robot having the responsibility for orienting them? The former makes for a simple robot system and pushes the responsibility off onto other parts of the factory. The latter requires a complex system but one that can more easily be introduced into present factories one worksite at a time.

ii. How to keep bad parts from entering the assembly system and how to carry out in process and final inspection? People are very good at both of these when their performance is up to par. Machines usually detect bad parts by jamming, taking time and requiring a person to free them. The required inspections can be quite subtle and often have not been quantified. ("Click means it's good, clunk means it's bad"). So far this problem has been addressed only by proposing pre-sorting of the parts by people or by supposing better and better computer-TV vision systems.

Table 1 lists some alternative feeding methods and comments on them.

d. Inspection Systems — Currently there are no systematic ways of looking at inspection—only singular solutions and devices. Some description of this current work is described below in the section on sensors under component issues.

e. Software and Hardware Control Systems — This section discusses design of computer controls for robot arms, while the next section discusses philosophies for teaching new assembly tasks to a robot.

A computer control system for a robot or robots must consist of one or more computers plus programs to run the arms, read sensors, input new assembly instructions, execute these instructions and interact with the operator. Data, consisting of points in space or forces, must be input and stored. The arms may be controlled in a variety of endpoint coordinate systems. Separate computers may be used for sensor processing, operator interaction, and so on.

Three approaches have been taken so far:

i. complex systems consisting of compiler-based languages, **IBM 360** or similar host computer, IBM System 3 or System 1 realtime controller, and perhaps additional

TABLE 1
Alternate Part Feeding Methods

Method	Comments	Examples
Conventional feeder tracks, vibratory bowls, etc.	The least programmable. Parts arrive oriented but often a person does the orienting. Can use simple robot arm (fewer than 6 axis could be sufficient).	Draper PAX U-6000 SIGMA Conventional assembly machines
Parts dumped randomly onto a vision station.	Parts arrive disoriented. Requires TV/computer. Requires 6 axis arm and time to reorient.	Bendix PAX U. of Rhode Island [8] SRI
Palletized kits of identical parts.	Parts arrive oriented. A person orients them (could be automated). Simple robot.	Westinghouse design [9].
Palletized kits of all parts needed for one product unit.	Parts arrive oriented. A person orients them (could be automated). Simple robot.	?
The same kit plus all the necessary tools.	Better for model mix assembly. Simple robot.	?

sensor processors. Examples include work at the Stanford University Artificial Intelligence Lab [10] and IBM [11].

ii. true minicomputer based systems (PDP 11-34 or Data General NOVA), stand-alone, usually with interpreter-based languages which are simpler than the above compilers. Examples include Bendix PAX and SIGMA [12].

iii. the simplest: hardwired logic set up to execute a sequence of taught spatial points one at a time. A computer may be used to input these points. Such systems have quite limited adaptability. The most they can do is stop if conditions are observed to change. The Unimate 6000 is an example.

Naturally, the simplest is the cheapest and most reliable. The major issue again is how much complexity is necessary. It is nice to contemplate an adaptable system but one must be able to show exactly what events the system is expected to be able to recover from. The hard fact is that many simple events (dropped part is a severe one) which a person could handle easily would require a dazzling robot system.

A second major issue is how to obtain the desired arm motions. In our work we have identified three motion regimes: gross motions (high speed, low accuracy transfers), fine motions (when parts are touching during assembly), and interface motions

References pp. 320-321

(between gross and fine, when the arm is lining up the parts). At present the robot performs all of these except if a Hi-Ti-Hand or Remote Center Compliance are in use. Computer control techniques for all three motion regimes are presently the same. Exceptions occur where binary force sensing (detecting contact) [12] or visual servoing [13] (using wrist TV to locate a hole) are used but even here the main control technique is for the computer to drive the arm to a position or through a sequence of positions. Forces, velocities, even frequencies, could be used as control references but so far have not. The required computer complexity for such methods has not been evaluated. Table 2 compares a number of current systems.

TABLE 2
Computer Control Systems of Various Robots

	Bendix PAX	Draper PAX (4)	U-6000	PUMA	Shuttle	SRI
Hardware Architecture	One mini plus disk.	One mini plus disk.	External servo hardwired control.	One mini plus diskette.	One computer	PDP-10 host plus 2 minis.
Control Complexity.	Several coord. systems. Stored points control of 2 arms. Program seq. with conditionals.	Two coord. systems for teaching. Stored points. Force feedback teaching aid.	Extra computer for teaching. Stored points. Two arms.	Several coord. systems. Stored points.	Several coord. systems. Stored points. Manual control.	Program sequence with conditionals Visual servo.
Multiple Arm Control	yes	no	yes	no	no	no
Trajectory Control	Nominal velocity between taught points.	Nominal velocity between points.	Position error.		Velocity ref.	Nominal velocity.
Vibration Suppression?	no	no	no	no	no	no
Use of Sensors.	Force, vision.	Force monitoring.	None	None yet.	None	Contact, force, vision.

NOTE: Draper PAX is an assembly station built around the prototype model of the Bendix PAX arm, but having a computer system and tooling developed by Draper Laboratory.

f. Teaching Methods — Teaching a robot a new assembly task involves determining the assembly actions required, their nominal sequence, and their detailed breakdown into robot motions, tests, loops, tool actions, contingency actions, and so on. This breakdown must then be communicated to the robot controller.

At present all robot programming is geometric. That is, points and orientations in space are the fundamental data used in programming. Force data is used mostly as a check or to conform to load requirements of the assembly itself ("Torque nut to 10 N-m"). Intellectually this means that all programming is physical, giving detailed "how to" guidance to the robot. There is as yet no functional programming ("This is what I want, you figure out how") although various approaches have been suggested in research labs [14].

The major current research issue in robot teaching involves what is termed explicit vs implicit programming. Explicit programming consists of expressing the desired robot actions in terms of explicit points in space and a sequence of logical actions. This technique is common among the simpler computer architectures. The implicit technique requires much more computer power and assumes that the robot and work comprise a surveyed environment which has been stored in the computer as a "world model". Locations and orientations can be referred to by name, and displacements caused by the robot's actions and the buildup of parts are kept track of automatically. If this is combined with trajectory planning then something closer to functional programming results, because the computer will calculate all the missing data.

Explicit programming can be compared to NC programming in APT, whereas implicit is more like projected CAD/CAM techniques in which the computer's internal model of the part is converted automatically into the data needed to support the NC programmer's description of the sequence of cuts. There is general agreement that implicit programming makes the programmer's job easier but only if he is a programmer by training. Explicit programming requires less of the programmer and the computer, but it will probably limit the degree of adaptability which can be incorporated into the program.

Table 3 compares several current systems.

TABLE 3
Comparison of Robot Control Languages

	Bendix PAX	Draper PAX	SIGMA	PUMA	Stanford A.I. Lab
Language Name	?	FSYS		VAL	AL
Explicit/ Implicit?*	Implicit	Explicit		Mix	Implicit
Compiler/ Interpreter?	Interp.	Interp.	Interp.	Interp.	Compiler
Features	Conditionals Multi-Arm Interlocks Link to vision station.	Macros link to force monitoring.	Conditionals Loops Interlocks Link to force monitoring.	Conditionals Interlocks Immediate execution of single commands.	World model coord. frames. Block structure Conditionals Two arm coordination.

*Judgment here grossly oversimplified.

    g. Economic Studies and Tradeoffs — By and large, assembly can be accomplished by people using their bare hands or hand held tools. This is not true of metal working and thus there has never been any doubt as to the need for metalworking machines. Assembly machines have in the past been justified economically only when the required production volume is so high that a special purpose machine can be built for

References pp. 320-321

much less than the payroll of the required assembly workers for one or two years. Technical justification (manual assembly is impossible or is of too low quality) is not usually included, although in retrospect it is often acknowledged that the machine does a better job. For these reasons, economic justification of assembly machines is often difficult.

There are technical reasons, too, for the fact that there is not much automation of assembly. Conventional assembly machines are dedicated to one product, are rigid in their operation and cannot alter their behavior spontaneously, and are almost always used to assemble very small items (less than about 10 cm cube).

Robot assembly systems must compete against the technically feasible alternatives. The advantages of robot assembly promise to be adaptability and programmability, but two caveats are in order. First, various system types proposed in the past make very different use of robot's programmability, which means that the economic value of that programmability may be quite different even for the same robot, depending on how it is used. Second, the economic values of adaptability and programmability are hard to capture in the abstract, and this makes economic modeling of assembly systems difficult.

Theoretical results in this area are still new. Lynch [2] modeled assembly systems in terms of the investment required to meet a given production volume and derived a relation between the cost and functional speed of a *robot* which shows if the proposed *system* will be economical compared to people or conventional machines. Boothroyd [9] extended this to include the life cycle of the machine as it is changed over from one product model or design to another, and showed that robot assembly is most economical when the product is made in many styles and undergoes many design changes during the machine's life.

2. Component Issues — a. Articulating Devices Such as Robot Arms — There are many concerns in this area. They range from kinematic design, number of degrees-of-freedom, the location of these degrees of freedom (in the arm or in the workplace), size, actuators, actuator location (shoulder or at respective joints) and accuracy. A paper by Whitney [15] goes into great detail in this with Table 4 summarizing the issues and Table 5 listing the various assembly robots presently available.

b. Adaptive Devices — We describe here a number of other devices used in assembly. With the exception of the conventional fixed mechanical workhead, all these are recent developments. They are mostly wrist or hand devices and their goal is to help a robot perform assembly under difficult conditions such as large part-to-part errors, very close clearances, and no chamfers or lead-ins.

i. Active (motorized) wrist with sensing — this device holds the workpiece, the robot carries it to the worksite and the device maneuvers the pieces together. The best known of these is the Hi-Ti-Hand [17]. It can assemble parts that have no chamfers, and do so

TABLE 4
Research Issues for Robot Arms

1. Kinematic design and number of axes. Included in this discussion is the question of should you have an elbow or slide axes like machine tools.

2. Size — Two issues here, namely how big a reach should an arm have and why are they so inefficient. Typically it takes a 500 to 1000 kg system consuming 40 kw to carry 2 kg loads a distance of 1 meter in 3-5 seconds.

3. Actuators — should they be hydraulic or electric.

4. Actuator location — should they be located at the joints or at the shoulder.

5. Accuracy — this issue is irrelevant if the arm is used only in a playback or incremental mode. In this case only repeatability and resolution are important. An arm when coupled with an RCC requires only enough accuracy to get to the chamfer (1 to 2 mm) which is much cruder than present systems and therefore possibly cheaper.

TABLE 5
Comparison of Assembly Robots

	Bendix PAX	Unimate 6000	PUMA	SIGMA	Shuttle Manipulator
Kinematics	cylindrical plus wrist	elbow	elbow	XYZ	elbow
No. of axes	6	6	5	3	6
Size	0.5 m × 0.5 m × 200°	~1.3 m reach	~1 m reach	~0.5 m^3	~18 m
Actuators	electric/gears, ball screws	hydraulic/gears ball screws	electric/gears	electric/ ball screws	electric, gears
Repeatability	~0.2 mm	0.2 mm	~0.1 mm	~0.1 mm	5 cm
Weight of moving parts	200 kg?	900 kg?	200 kg	100 kg	500 kg
Payload	2 kg	10 kg?	3 kg?	?	32 tonnes

NOTES: 1) PAX was a research effort by Bendix, now abandoned.
2) Unimate 6000 is a cooperative effort between Unimation, Inc., and Ford Motor Co.
3) PUMA is being built to General Motors Co. specifications.
4) SIGMA is being marketed by Olivetti, s.P.a.
5) The Shuttle Manipulator is being built by SPAR Ltd., Toronto, Canada.
6) All data in this table are estimates by the authors.

References pp. 320-321

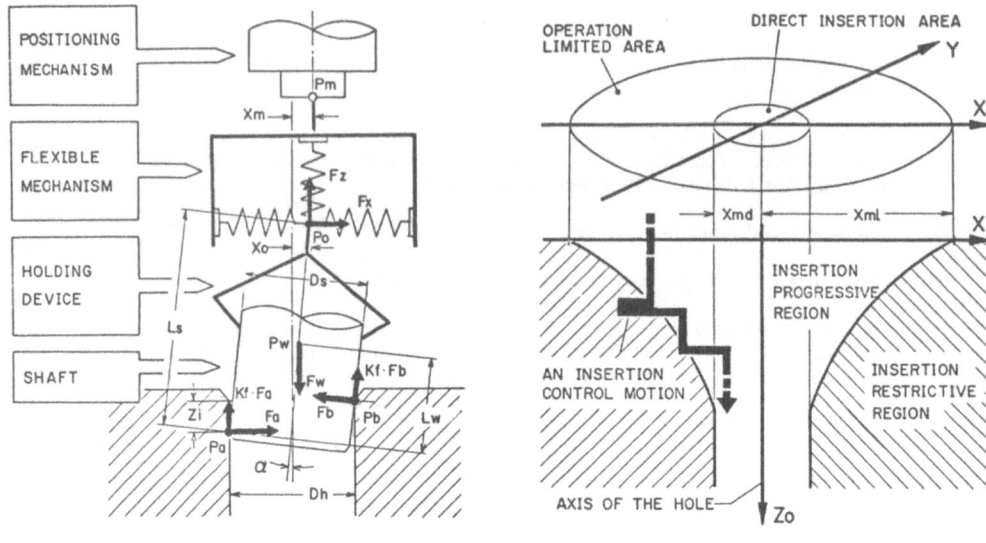

Two dimensional representation of inserting operation.

Insertion characteristics diagram.

Fig. 17 Hi-Ti-Hand.

with very light touch. Assembly usually takes 4 to 5 seconds, which is relatively long, and the device is rather fragile and in need of continual adjustment (Fig. 17).

ii. Active wrist with estimation theory — this as yet a theoretical technique with little lab verification. Modern control theory and mathematical models of the assembly operation are combined to yield a closed loop technique for assembling parts [5, 18].

In all of these approaches, the major issues involve how much complexity is enough to handle common assembly tasks and how can these tasks be modeled adequately so that effective devices can be designed. If one assumes the existence of chamfers and enough repeatability in robots, parts, grippers, jigs and feeders so that the chamfers will meet during assembly, then the simple Remote Center Compliance will assemble the parts. Interference fits as well as clearances in 0.01 mm range have been accomplished, starting from errors as large as 1.5 mm laterally and 2° angularly. For this reason robot repeatability better than 0.1 mm may not be needed.

When parts have 0.01 mm clearances and no chamfers, the Hi-Ti-Hand suffices, but often so does simple vibration [19], so the issue is still open. What does seem clear is that visual imaging is less useful in the 0.01-1.0 mm range than contact-based techniques.

Further description of this problem can be found under sensing in the next section.

c. Sensing Systems — It is generally agreed that sensing, especially monitoring, is essential to much of automation, and assembly is no exception. A conclusion of

Draper research is that quite a lot of assembly can be done without direct sensing of any kind except monitoring. In other areas of sensing, especially TV vision, there has not yet emerged an adequate balance between cost and effectiveness. An idea of the relative application areas of passive wrists, active wrists, robot arm inherent accuracy and vision is emerging, however, and this is expressed in Fig. 18.

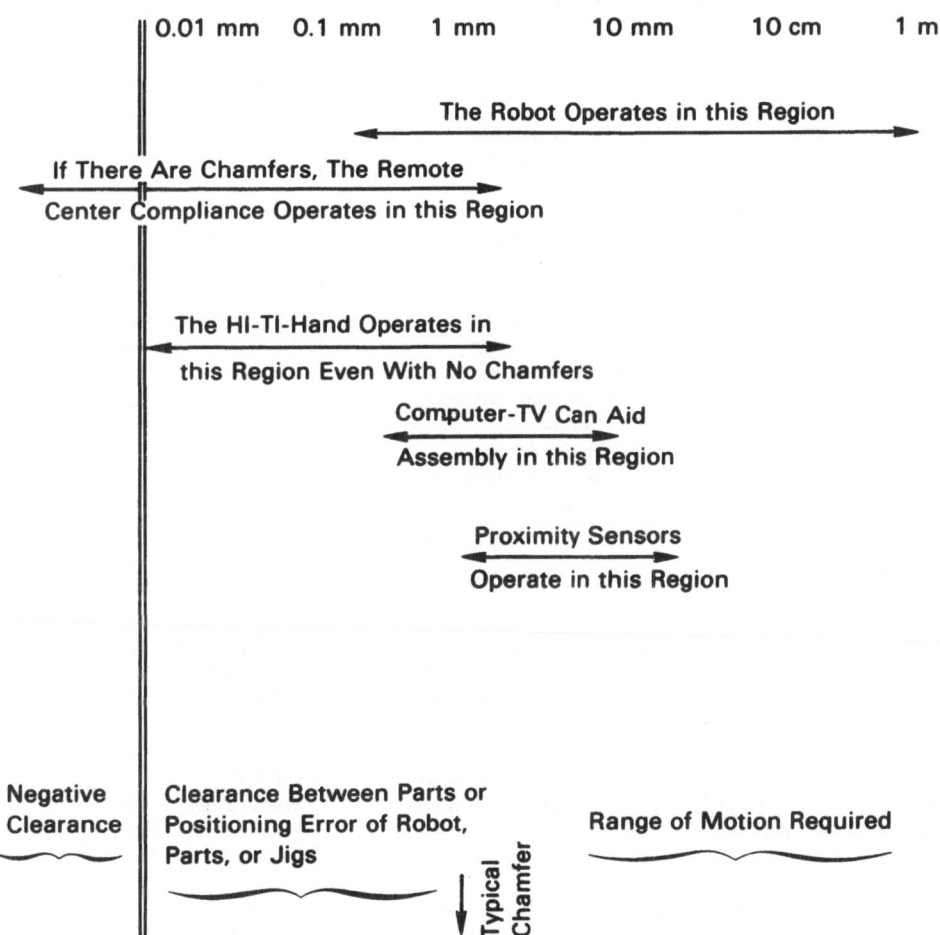

Fig. 18. Range of applicability of sensing modalities and assembly techniques.

A list of some current sensing techniques, with comments, appears as Table 6. For the time being, it would appear that force sensing is confined to a monitoring role while vision can be used for gross inspection and as an aid in part feeding if one does not ask too much. (A variety of parts, all jumbled together, is too much for people sometimes).

References pp. 320-321

TABLE 6
Sensing Options

Option	Comments	Examples
Active force sensing to guide assembly	Remote Center compliance, if applicable, will be much faster Can handle large errors	Hi-Ti-Hand Draper demo, 1972, 1975 Stanford A.I. Lab
Force sensing to monitor the progress of the assembly	Record history Detect drift Prevent damage Verify operations Abort failed operation	Draper bench unit SIGMA
Proximity sensor	Collision avoidance	Jet Propulsion Lab [16]
TV "vision"	Part orientation Gross inspection Product integrity Feeder/tool operation monitor Visual servo for endpoint guidance of robot	Bendix PAX U. of Rhode Island [8] SRI Stanford A.I. Lab

TABLE 7
Tool and Gripper Options

Options Currently in Use	Comments	Examples
Universal socket on robot plus tool changing and external tool storage	Take time to change tools Really universal within range of tools themselves	Draper PAX U-6000
Multiple arms or one arm at each workstation, each arm with one "permanent" tool	Limited range of tasks suited to intermittent chageover	U-6000 Bendix PAX Westinghouse Design SIGMA
Small tool turret on robot wrist	Limited range of tasks Could be heavy	U-6000
Other Possibilities:		
Parts fed with tool	Extends universality of wrist socket but takes time	
Truly universal tool	Possible?	
Product redesign to reduce number and variety of tools needed		

d. Tools and Grippers — An assembly robot will need at least one tool or gripper to do its job. How many more than one is the question. There are several options in use and others have been proposed. See Table 7. The issue is whether each part or operation should have its own tool, the problems being cost and time for tool changing. While both of these can be attacked the nagging philosophical issue is whether one has a reprogrammable system if there is a large fixed investment in specialized tools. On the other hand it is doubtful that a truly universal tool can be achieved. Even the human hand needs aid from other tools.

A similar problem arises in software system design. Should one expect future robots to come preprogrammed for any assembly operation or is it more reasonable to expect software facilities which can be used to compose programs for specific operations? In both tools and software it appears more reasonable to accept a certain degree of specialization. Hopefully this specialization will occur within a carefully designed general environment:

	General Apparatus	Interface	Specific Items
Software:	Compiler Editor Debugger	Real time operating system	Programs for specific assembly operations. Programs for specific products.
Tools:	The robot	Robot wrist with tool socket.	Specific tools

"Libraries" of useful programs and tool concepts can be expected to emerge. For example, four types of tools (6 tools in all) were needed by Draper to assemble the alternator [4].

- expanding gripper for small object I.D.
- contracting gripper for large object O.D.
- contracting gripper for small O.D.
- screwdriver

e. Part Feeders and Inspection Devices — There are many devices for part feeding and inspection. Some of these were described in previous sections and tables. Singular solutions and devices will be the order of the day until systematic ways of looking at the problems and defining the requirements (such as those being worked on by Birk [8] and Boothroyd [9] are fully developed.

CONCLUSIONS

We have tried to list here (but make no claim for completeness) the research issues remaining in this area. The intent was to show that this is truly a virgin research area.

References pp. 320-321

Our small successes, while significant, have mainly helped us get visibility into the larger issues and to help focus our research. We would be very interested in seeing if other researchers agree with our concerns.

REFERENCES

1. Drake, S.H. The use of compliance in a robot assembly system. IFAC Symp. on Information and Control Problems in Manufacturing Technology (1977).

   Drake, S. H. Using compliance in lieu of sensory feedback for automatic assembly. Ph.D. Th., MIT Mech. Eng. Dep. (1977).

   Watson, P. C. A multidimensional system analysis of the assembly process as performed by a manipulator. 1st North Am. Robot Conf. (1976).

2. Lynch, P. M. Economic-technological modeling and design criteria for programmable assembly machines. DL T-625, Ph.D. Th., MIT Mech. Eng. Dep. (1976).

3. Lynch, P. M. An economic guideline for the design of programmable assembly machines. ASME paper 77-WA/Aut-2.

4. Nevins, J., et al. Exploratory research in industrial moduar assembly. C. S. Draper Lab. Rep. R-1111, Cambridge, M. (1977).

   Nevins, J., and Whitney, D. Computer controlled assembly. *Sci. Am.* 238, 2 (1978), 62-74.

5. Simunovic, S. Ph.D. Th. in progress.

   Simunovic, S. Force information in assembly processes. Proc. 5th Int. Symp. on Industrial Robots (1975), 415-431.

6. Kondoleon, A. S. Application of technology-economic model of assembly techniques to programmable assembly machine configuration. S.M. Th., MIT Mech. Eng. Dep. (1976).

7. Barash, M.M. et. al. The optimal planning of computerized manufacturing systems. Sch. Ind. Eng., Purdue U. Rep. 3 (1976).

8. Kelley, R., Birk, J., and Wilson, L. Algorithms to visually acquire workpieces. Proc. 7th Int. Symp. in Industrial Robots, (Oct. 1977), 497-506.

9. Abraham, R.G., et. al. Programmable assembly research technology transfer to industry. Final Rep. Westinghouse Corp., Pittsburg, PA. 1977.

10. Finkel, R., et. al., AL, a programming system for automation. AIM-243, STAN-CS-74-456, A.I. Lab, Stanford U. 1974.

11. Lieberman, L. I., and Wesley, M.A. AUTOPASS, an automatic programming system for computer-controlled mechanical assembly. IBM Res. Dep. RC-5606, T.J. Watson Res. Cent., Yorktown Heights, N.Y. 1975.

12. d'Auria, A., and Salmon, M. SIGMA - an integrated general purpose system for automation manipulation. Proc. 5th Int. Symp. on Industrial Robots, (1975), 185-202.

13. Agin, G. Servoing with visual feedback. Proc. 7th Int. Symp. on Industrial Robots (1977), 551-560.

    Rosen, C., et al. Machine intelligence applied to industrial automation. Res. Rep. Series, SRI Int., Menlo Park, CA.

14. Whitney, D.E. State space models of remote manipulation tasks. *Trans. IEEE Autom. Contr.* AC-14, No. 6 (1969), 617-623.

15. Whitney, D. E. Discrete parts assembly automation - an overview. ASME 1978 Winter Annual Meeting.

16. Johnston, A. R. Optical proximity sensors for manipulators. JPL TM 33-612, Jet Propulsion Lab., Pasadena, CA.

17. Goto, T., Inoyama, T., and Takeyasu, K. Precise insert operation by tactile controlled robot "HI-T-HAND" Expert 2. Proc. 4th Int. Symp. on Industrial Robots (1974), 209-218.
18. Whitney, D. E. Force feedback control of manipulator fine motions. ASME J. Dyn. Syst. Meas. Control, (1977) 91-97.
19. Dunne, M. J. An assembly experiment using programmable robot arms. Proc. 7th Int. Symp. on Industrial Robots (1977), 387-397.

DISCUSSION

J. Albus *(National Burea of Standards)*

How may lines of code did you have in that program? *(In reference to a film showing the automatic assembly of automobile alternators.)*

Whitney

The first program had 140 lines. A typical line has 5 or 6 points in space in each one. That was early in the game. The program in the movie, I think, has about 60 or 65 lines. A lot of intermediate lines were eliminated.

FUTURE PROSPECTS FOR SENSOR-BASED ROBOTS

R. B. McGHEE

Ohio State University, Columbus, Ohio

ABSTRACT

First generation industrial robots have been largely characterized by being entirely preprogrammed with at most a capability to make continuation of motion conditional on a switch closure or other confirmation signal. In contrast, much research in artificial intelligence has been directed toward adaptive motion control in which information obtained from visual, tactile, and other types of sensors is used to modify the commands to a computer-controlled manipulator. A similar approach has been taken in studies of artificial-legged locomotion systems. With the continuing decline in the cost and physical size of small computers, sensor control is becoming increasingly attractive for industrial applications. This paper attempts to summarize the current status of this trend and offers some predictions regarding future prospects for industrial utilization of sensor-based robots.

INTRODUCTION

Although the notion of anthropomorphic machines is very old, the term *robot* is of quite recent origin having been introduced by Capek [1] around 1920, apparently derived from the Czech work *robotnik,* meaning *serf.* While science fiction has thrived on Capek's conception of a humanoid machine, and Asimov [2] has even proposed a moral code for robots, the actual realization of general-purpose machines capable of replacing human workers on a one-for-one basis is a still more recent phenomenon, belonging entirely to the latter half of the 20th Century. To be more specific, effective and reliable digital computers began to be generally available in the early 1950's and have subsequently replaced millions of humans in routine data analysis and information processing tasks. At about the same time, the first patents for the programmable material handling machines now known as industrial robots were issued, leading to the

References pp. 332-334

gradual introduction of such machines into factories beginning about a decade later [3]. Parallel developments in the field of powered artificial limbs and braces began with the pioneering work of Tomovic [4] in the early 1960's, approximately at the time the first Unimate robots were being delivered to industry. A little later, serious research began on the problem of developing walking vehicles as an alternative means of off-road transportation [5].

Digital computers are of course by now firmly entrenched as essential elements of industrial societies, performing work no longer regarded as suitable for human beings. On the other hand, the world-wide population of industrial robots still amounts to at most ten thousand or so machines, the number of powered artificial limbs in daily use is only slightly larger, and there are no walking trucks. This disparity in comparison to the success of digital computers suggests either a deficiency in technology or some inherently greater difficulties in building machines to interact with the physical world in comparison to their information processing counterparts. The author believes that in fact both of these factors have retarded the development of more adaptable and effective robotic systems. The balance of this paper is therefore devoted to a brief review of the current state of knowledge regarding coordination of joint motions in articulated mechanisms for locomotion and manipulation. Personal views on needed research and development are also included with particular emphasis on systems in which motions are not entirely pre-planned, but are rather to some extent driven by sensory inputs relating both to the relative motions of the segments of such linkages and to the interaction of a robot with its environment. In some instances, general concepts presented in the following discussion will be illustrated by their application to the problem of legged locomotion [6, 7], this being the particular area of robotics research with which the author is most familiar.

MOTION PLANNING

While it is conceivable that at some time in the future practical robots may be endowed with some capabilities to set their own goals, at present this seems to be neither desirable or feasible. Consequently, in all that follows it will be assumed that any robot will be to some extent under the control of a human operator who will specify the desired motions at some suitably high level. This arrangement is illustrated by Fig. 1 where it is to be understood that this is a *generic* decomposition of the robot control problem encompassing not only industrial robots, but also other types of manipulators, prostheses and orthoses, walking machines, and any other variety of programmable articulated mechanism. While this particular partitioning of the sensor-based control problem is admittedly somewhat arbitrary, it does have the advantage of having been found suitable for the successful development of a computer-control system for a hexapod walking machine [8].

In order to clarify the meaning of each of the blocks of Fig. 1, it is perhaps worthwhile to consider the evolution of industrial robots and to interpret the control techniques used at various times relative to this figure. While the block labeled "control computer" encompasses both motion planning and motion execution, the earliest

FUTURE ROBOTS

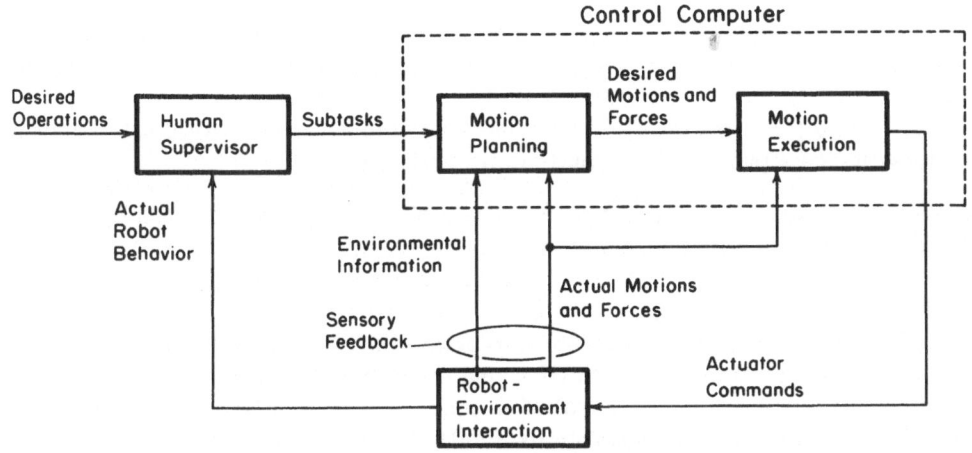

Fig. 1. Decomposition of general robot control problem.

remote manipulators, which in some sense were the forebearers of today's industrial robots, possessed neither a computer nor actuators. Rather, these manipulators were of the mechanical "master-slave" variety in which a human being not only planned all motions, but also executed them and provided all the necessary energy through a pantograph mechanism [9]. later, the pantograph was replaced by either electronic or hydraulic feedback systems and the slave was independently powered. However, even today, the control signals in such systems are ordinarily derived from transducers mounted on an exoskeleton or similar apparatus physically attached to the human operator [10]. The human therefore remains fully responsible for motion planning and must be continuously involved during manipulator operation. Thus, in this type of master-slave system, while motion execution has been turned over to the control computer, motion planning has not.

The next obvious advance in manipulator control is evidently to provide some means for storage of the motions executed by the human operator. When this is done, the human supervisor becomes a *teacher* rather than a *master* and he is relieved of the requirement of continuously participating in the control process once the *slave* has been *trained*. Since the introduction of the Unimate in the early 1960's, it has become customary to refer to a manipulator with this degree of computer-control as a "robot". This terminology certainly seems justified since such machines are indeed capable of replacing human workers on a cost-effective basis in certain routine materials handling tasks.

Relative to Fig. 1, one can say that the subtasks provided to the control computer for the original Unimate and similar industrial robots consists of the *teach-points* stored in the computer memory during training. In such *point-to-point* control schemes, the motion planning function is entirely finite-state in nature. That is, during motion between teach points, each manipulator axis is halted when it reaches the stored value, and progression of robot motion to the next teach-point occurs only after all axes have

References pp. 332-334

attained their desired values. Thus, this control concept is essentially that of an *asynchronous sequential machine* [11]. In addition to its use in industrial robots, this approach has also been used for control of robot locomotion [12] and for control of artificial limbs [13]. Its greatest appeal lies in its simplicity, which of course tends to reduce costs, ease programming, and improve reliability.

While the practicality of point-to-point control is amply demonstrated by the success of this type of robot in many industrial applications, there are other needs in industry which can be satisfied only by a more precisely controlled robot in which a continuous path for all joints is specified. This is the approach most often used, for example, in spray painting applications, as exemplified by the Trallfa robot [14]. An implication of continuous path control is that not only is the motion better controlled in space, but it is also better controlled in time. That is, continuous path control is *synchronous,* rather than *asynchronous.* Evidently, such control requires both more storage and more computation and tends to both increase the cost and complicate the programming of such controllers.

Until recently, the storage requirements of continuous path robot control systems could be economically met only by using analog or digital magnetic tape recordings. This hardware limitation carried with it an implication that robot motions must be entirely pre-programmed with the exception only that segments of motion could be triggered by external events such as limit switch closures. That is, while continuous path control utilizing magnetic tape storage allows for precise motion execution, the desired motions can be modified only in this trivial way. Thus, only *finite-state* environmental information can be used in such systems. Fortunately for the future of robotics, this limitation has at last been removed by the appearance of commercial machines such as the Cincinnati Milicron 6CH and the Unimate 500 (PUMA) in which a dedicated minicomputer accomplishes both the motion planning and motion execution functions [15, 16]. Perhaps the most dramatic effect of this hardware improvement has been the realization of a moving line capability in which a program prepared for a stationary work-piece can be *automatically* altered to account for the one-dimensional motion associated with a conveyor system of some sort. Referring to Fig. 1, this is evidently possible only because the control computer is provided with a certain amount of continuous information about the robot's environment and not merely finite-state information as in earlier robots.

While the author believes that the above discussion more or less summarizes the current motion planning capabilities of commercially available industrial robots, more adaptable robotic systems have been studied in a laboratory environment for more than a decade with perhaps the greatest attention being paid to the use of *visual* (television) information in motion planning. This is of course a major theme of this conference to which nothing can be added by this paper except for an expression of the opinion that the general visual control problem is still very far from solution. That is, realization of hand-eye coordination similar in quality to that of human workers must be regarded as a very distant goal despite impressive progress in this direction which has been made in situations involving rather structured environments [17, 18].

Certainly one of the most important consequences of the introduction of full computer control is that it allows elevation of the level of the dialogue between the human supervisor and the robot. An excellent example of this effect is provided by the short history of the class of robotic vehicles known as *adaptive walking machines* [7, 8]. Such vehicles achieve locomotion by means of articulated systems of levers rather than conventional wheels or tracks and are potentially capable of achieving improved off-road mobility with reduced ecological damage [7]. The joints of this type of machine are individually powered and are coordinated by some external agency to achieve the behavior desired by a human operator.

The first successful adaptive walking machine was the General Electric Quadruped Transporter [5, 19]. This machine possessed four hydraulically powered legs, each with three independently moveable joints (two at the hip, one at the knee) and was roughly the size of an elephant. It has no control computer; rather the hydraulic master-slave control concept was used with an operator wearing an exoskeleton mounted in the vehicle cab. While this machine first walked in 1967 and showed that the use of legs for support and propulsion did in fact lead to a great improvement in off-road mobility, it was also found that manual control of the twelve independent joints of this vehicle was an extremely demanding task which few could master and none could endure for more than a few minutes. Referring to Fig. 1, one can say that the control of this machine was inappropriately partitioned with the subtask specifications to the robot occurring at too low a level.

At about the same time as the GE machine was being tested, the first computer-controlled legged vehicle was developed at the University of Southern California. This machine was also a quadruped and, like its contemporary industrial robots, made use of point-to-point control. Its successful operation in 1968 verified that this technique could indeed coordinate joint motions for locomotion, but also made very apparent the need to use more environmental information to achieve adaptive behavior [12].

As with industrial robots, full computer control of walking machines has become possible only in the past several years with the introduction of low cost minicomputers and microcomputers. As a result of this development, at least two experimental computer-controlled legged vehicles habe been constructed and successfully tested to date [7, 8, 20]. Fig. 2 illustrates one of these machines, the OSU Hexapod Vehicle. Fig. 3 shows the organization of the control computer software for this vehicle. As can be seen, at this level of automation, the human operator (in this case remotely situated from the vehicle) is responsible for determining vehicle speed and direction more or less as in the control of conventional automotive vehicles. The computer then synthesizes a *support schedule* for the lifting and placing of the vehicle's legs and adjusts individual leg trajectories on every step to account for terrain irregularities [8, 21]. Since the vehicle structure is redundant (statically indeterminate) the distribution of forces among the supporting legs for a given motion is not unique, thus allowing for the imposition of constraints and/or optimization of some criterion function. For this system, the energy cost of locomotion is minimized while constraints on leg forces tending to equalize loads, prevent foot slippage, and avoid motor overloads are also enforced. An algorithm based on linear programming is used for this purpose [22].

References pp. 332-334

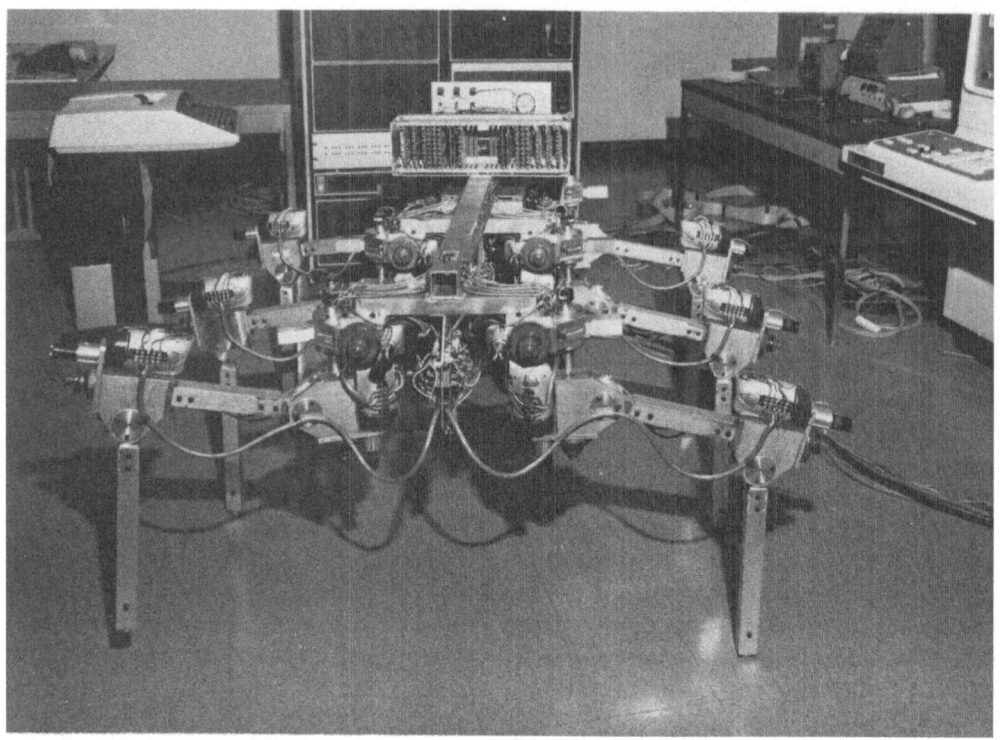

Fig. 2. Front view of OSU hexapod vehicle showing control computer in background. Vehicle length = 1.3 m., width = 1.4 m., total weight exclusive of cables = 103 kg.

Examination of Fig. 3 shows that both automation of terrain adaptability and minimization of the energy cost of locomotion require sensing of the forces involved in the interaction of each vehicle leg with its environment. This situation arises because of the presence of closed kinematic chains in any locomotion system in which two or more legs are simultaneously in contact with the ground. Other research has shown the value of such "force accomodation" in robot assembly tasks since such operations also produce closed chains [23, 24]. This is in contrast to the more typical material handling applications of contemporary industrial robots in which such chains are avoided except at the very beginning and end of the motion. The problem of motion planning in the presence of closed chains, including the use of redundant degrees of freedom, is explored in considerable depth in [25] and [26]. The question of choosing the most effective coordinate system for motion planning is addressed in [27] and [28].

To summarize this section of this paper, the work on visual sensing and scene understanding reported in other papers presented in this symposium certainly needs to be continued and expanded in order to extend the range of application of industrial robots and other computer-controlled manipulators. In addition, much more attention needs to be paid to the sensing and use of tactile information in adapting robot motion to changes in work-piece position and orientation as well as in assembly operations.

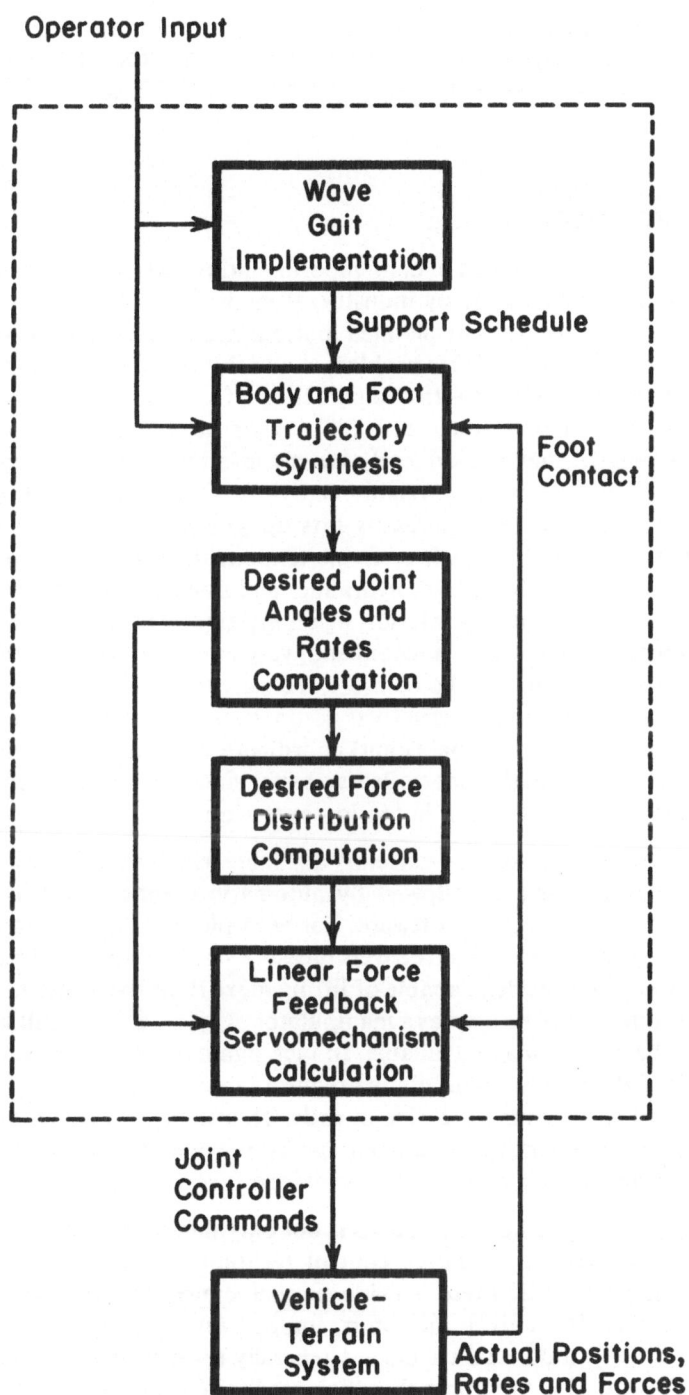

Fig. 3. Structure of interactive computer software system for control of OSU hexapod [7].

References pp. 332-334

The relative importance and utility of active and passive compliance in such systems is a question of prime importance to research in this area [29, 30]. This subject should in particular be investigated more thoroughly with regard to adaptive general purpose end-effectors [31].

MOTION EXECUTION

The excellent reliability record of contemporary industrial robots has of course been a key factor in their acceptance by industry. It might be inferred from this that the problem of effective execution of planned motions has been largely solved. Unfortunately, a closer examination of this problem shows this to be far from true. The most obvious way to see that this is so is to compare the performance of the typical industrial robot arm with that of the human arm [32]. While heavy-duty robots may possess greater brute strength, they are evidently grossly inferior to their biological counterparts in terms of their size and weight relative to their load handling capability. Specifically, while the human arm easily lifts up to ten times its own weight, most industrial robots are not able to lift even one-tenth of their weight, a hundred-to-one advantage for muscle and bone over hydraulics and steel. Furthermore, the dexterity of the robot arm is also painfully inferior, again in part due to its excessive weight, but also due to its reduced degrees of freedom and, very importantly, to its relative lack of internal and external feedback. To be more precise, while tactile and/or visual feedback adaptively guide the human arm to its destination, industrial robots rely solely on structural rigidity to relate internal (joint) coordinate to external coordinates. This is generally true even of visually controlled systems, since vision is typically employed only for motion planning and not in motion execution.

The author feels that sensors are currently available which would allow for measurement of manipulator flexure followed by automatic compensation in the motion execution portion of the control software. For example, in [33] it is shown that semiconductor strain gauges can very effectively measure small deflections in a light-weight electrically powered vehicle leg capable of lifting more than five times its own weight. This leg could equally well be used as a manipulator providing that a still greater degree of sensory feedback were made available. In fact, again observing nature, it may well be that flexible, light-weight manipulators require a micro-manipulator in the form of an adaptive hand as a terminal device in order to attain both precision and speed. Research on such kinematically redundant structures is badly needed if the size and weight of industrial arms is ever to become competitive with biological arms.

The relatively slow speed of robot arms is not only due to their excessive mass, but also to the comparatively primitive control techniques used in controlling joint motions. Specifically, while there is evidence that biological systems *anticipate* the joint torques needed for a given motion by implicit *inverse plant* (acceleration feed-forward) models [34], industrial robots are generally error driven, employing at most velocity feed-foward. This requires that high feedback gains be used to obtain even moderate response times. This situation is due in part to the modest computing power of current generation control computers, but also to a lack of theoretical understanding regarding computationally efficient techniques for acceleration feed-forward. Beginning with the work of Vukobratovic [35], there has been a gradual improvement in this situation and explicit real-time inverse plant computations based on the Newton-Euler formalism are now feasible for several degrees of freedom of manipulator motion using small minicomputers [36, 37, 38]. However, such computations have yet to find their way into control programs for commercially available industrial robots despite their potential for reducing cycle time. Other alternatives to increasing speed of motion for industrial robots, including nonlinear control and implicit modelling through computer learning, are also being studied [39], but relatively few results are as yet available.

In summary, the *mechanical* components of robot arms have generally been very heavily constructed in order to minimize structural flexure. An order of magnitude reduction in size and weight ought to be possible with a concomitant reduction in cycle time if the arm is treated computationally as a dynamic rather than a static component. Current trends in minicomputers and microcomputers should make this feasible in the near future providing that the necessary flexure and/or acceleration sensors are available and that redundant degrees of freedom are provided in the manipulator structure as needed.

CONCLUSIONS AND FUTURE PROSPECTS

Based on the above discussion of past history and current research relative to robotic systems, the author offers the following conclusions, predictions, and suggestions for further work:

1. The capabilities of inexpensive digital computers will continue to improve dramatically. This will allow the economical implementation of robot control schemes far more elaborate than those in use today, thereby greatly extending the range of application of robotic systems.

2. The weight and size of manipulator arms will be markedly reduced through on-line sensing of flexure and its computational correction.

3. The problem of motion planning will become better understood as more specific examples are studied. Special high-level languages for communication of sub-tasks to the control computer will come into general use [40].

4. The further understanding and use of both active and passive compliance in industrial robots will allow many assembly operations now performed by human workers to be automated.

5. Control programs will make more use of either explicit or implicit inverse plant models to reduce cycle time and thereby improve robot productivity. The use of nonlinear control schemes will also contribute to advances in this area.

6. The extensive use of visual and tactile feedback together with redundant degrees of freedom in manipulator joints will allow robots to function in much less structured environments than at present, especially with regard to manufacturing applications.

References pp. 332-334

7. Work on powered artificial limbs and braces will be helpful to the industrial robot field, especially in the area of man-machine communication. Likewise, industrial developments in the area of sensor-control will promote further advances in this field of medicine.

8. Mobile robots with many degrees of freedom will come into being. Some of these will use legged locomotion systems and will replace human beings to some extent in certain hazardous situations such as fire fighting, underground mining, explosive ordnance disposal, and nuclear reactor servicing. The development of new types of actuators capable of more efficient generation of commanded joint torques under computer control is crucial to the practical realization of such advanced systems.

In summary, providing that the necessary basic research is carried out, and that an appropriate dialogue is maintained between researchers, manufacturers, and users of robotic systems, the computer-controlled linkage systems we now call robots may eventually become as all pervasive as digital computers are today. Such machines have the potential for relieving human beings of physical drudgery and danger in much the same way that digital computers have improved the quality of the working environment in offices and in white collar service industries. There is, however, no room for complancency. The work to be done is difficult and the capabilities of contemporary robots in interacting with their physical environment is unimpressive in comparison both to that of their biological counterparts and to the information processing capabilities of digital computers.

ACKNOWLEDGEMENT

Basic research in computer control of robotic systems is supported at the Ohio State University by the National Science Foundation under Grant ENG74-21664.

REFERENCES

1. Capek, K. *R.U.R. (Rossum's Universal Robots)*. Samuel French, Ltd. London, 1923.
2. Asimov, I. *I, Robot*. Doubleday & Co., Inc., Garden City, N.Y., 1950.
3. Anon. Unimation, Inc.: expansion and advancement since the 1950's. *RIA Newsletter,* Robot Inst. of Am., April 1978.
4. Tomovic, R., and Boni, G. An adaptive artificial hand. *IRE Trans. on Automatic Control* AC-7 (April 1962), 3-10.
5. Liston, R. A. Walking machines. *J. Terramechanics* 1,3 (1964), 18-31.
6. McGhee, R. B. Control needs in prosthetics and orthotics. Proc. 18th Jt. Automatic Control Conf. (June 1977), 567-573.
7. McGhee, R. B. Control of legged locomotion systems. Proc. 18th Jt. Automatic Control Conf. (June 1977), 205-215.
8. McGhee, R. B., Chao, C. S., Jaswa, V. C., and Orin, D. E. Real-time computer control of a hexapod vehicle. Proc. 3rd CISM-IFToMM Int. Symp. on Theory and Practice of Robots and Manipulators (September 1978)

9. Johnsen, E. G., and Corliss, W. R. Teleoperators and human augmentation. NASA Rep. SP-5047, December 1967.
10. Corliss, W. R., and Johnsen, E. G. Teleoperator controls. NASA Rep. SP-5070, December 1968.
11. Tomovic, R., and McGhee, R. B. A finite state approach to the symthesis of bioengineering control systems. *IEEE Trans. on Human Factors in Electronics* 7,2 (June 1966), 65-69.
12. McGhee, R. B. Robot locomotion. In *Neural Control of Locomotion,* Herman, R. M., et al. (Eds.), Plenum Press, New York, 1976.
13. McGhee, R. B., Tomovic, R., Yang, P. Y., and MacLean, I. C. An experimental study of a sensor-controlled external knee locking system. *IEEE Trans. on Biomedical Engineering* 25, 2 (March 1978), 195-199.
14. Haugan, K. M., The De Vilbiss-Trallfa Spray Painting Robot, *Proc. 1st Conf. Industrial Robot Technology (1973)*
15. Anon., *The 6CH Arm Computer-Controlled Industrial Robot,* Cincinnati Milacron Inc., Cincinnati, Ohio, 1975.
16. Beecher, R. C., "PUMA: programmable universal machine for assembly. In *Computer Vision and Sensor-Based Robots,* G. G. Dodd and L. Rossol (Eds.), Plenum Press, 1979.
17. Perkins, W. A. A model-based vision system for industrial parts. *IEEE Trans. on Computers* 27, 2 (February 1978), 126-143.
18. Nitzan, D., and Rosen, C. A. Programmable industrial automation. *IEEE Trans. on Computers* 25, 12 (December 1976), 1259-1270.
19. Mosher, R. S. Test and evaluation of a versatile walking truck. Proc. Off-Road Mobility Research Symp. Int. Soc. for Terrain Vehicle Systems, Washington, D.C., 1968.
20. Gurfinkel, V. S., et al. Walking robot with supervisory control. Proc. 3rd CISM-IFToMM Int. Symp. on Theory and Practice of Robots and Manipulators (September 1978).
21. McGhee, R. B., and Iswandhi, J. Optimization of support state sequences for legged locomotion systems. Tech. Note 22, Comm. and Control Systems Lab., The Ohio State U., Columbus, OH, 1978.
22. McGhee, R. B., and Orin, D. E. A mathematical programming approach to control of joint positions and torques in legged locomotion systems. In *On Theory and Practice of Robots and Manipulators,* Polish Scientific Publishers, Warsaw, 1976, 221-230.
23. Whitney, D. E. Force feedback control of manipulator fine motions. Proc. 1976 Jt. Automatic Control Conf. (1976), 687-693.
24. Nevins, J. L., and Whitney, D. E. Research issues for automatic assembly. Proc. 1977 Jt. Automatic Control Conf. (1977), 139-147.
25. Liegeois, A. Automatic supervisory control of the configuration and behavior of multibody mechanisms. *IEEE Trans. on Systems, Man, and Cybernetics* SMC-7, 12 (December 1977), 868-871.
26. Orin, D. E., and Oh, S. Y. Automated motion planning for articulated mechanisms. Proc. Nat. Electron. Conf. (October 1978).
27. Paul, R. Cartesian coordinate control of robots in joint coordinates. Proc. 3rd CISM-IFToMM Int. Symp. on Theory and Practice of Robots and Manipulators (September 1978).
28. Chao, C.S. A software system for on-line control of a hexapod vehicle utilizing a multiprocessor computing structure. M.S. Th., The Ohio State U., August, 1977.
29. Jaswa, V.C. An experimental study of real-time computer control of a hexapod vehicle. Ph.D. Th., The Ohio State U., June 1978.
30. Houck, J.C., and Nichols, T.R. Improvement in linearity and regulation of stiffness that results from actions of stretch reflex. *J. Neurophysiology* 39, 1 (1976), 119-142.
31. Hanafusa, H., and Asada, H. Adaptive control of a robot hand with elastic fingers for mechanical assembly. Proc. 3rd CISM-IFToMM Int. Symp. on Theory and Practice of Robots and Manipulators (September 1978).

32. Paul, R., and Nof, S.Y. Human and robot task performance. In *Computer Vision and Sensor-Based Robots,* G.G. Dodd and L. Rossol (Eds.), Plenum Press. 1979.
33. Briggs, R.L. Real-time digital control of an electronically powered vehicle leg using vector force feedback. M.S. Th., The Ohio State U., December 1977.
34. Hemami, H., and Jaswa, V.C. On a three-link model of the dynamics of standing up and sitting down. *IEEE Trans. on Systems, Man, and Cybernetics* SMC-8, 2 (February 1978), 115-120.
35. Stepanenko, Y., and Vukobratovic, M. Dynamics of articulated open-chain active mechanisms. *Math. Biosci.* 28, 1,2 (1976).
36. Vukobratovic, M., Stokic, D., and Hristic, D. New control concept of anthropomorphic manipulators. *Mech. Mach. Th.* 12 (December 1977), 515-530.
37. Orin, D.E., McGhee, R.B., Vukobratovic, M., and Hartoch, G. Kinematic and kinetic analysis of open chain linkages utilizing Newton-Euler methods. *Math. Biosci* 30 (December 1978).
38. Paul, R., et al. Advanced industrial robot control systems. Rep. TR-EE 78-25, Purdue U., May 1978.
39. Klein, C.A., and Briggs, R.L. Minicomputer and microcomputer control of mechanical linkage systems. Proc. Nat. Electron. Conf. (October 1978).
40. Farnum, R.F., et al. A grammatical approach to automatic sensor control of artificial arms. Proc. 3rd CISM-IFToMM Int. Symp. on Theory and Practice of Robots and Manipulators (September 1978).

DISCUSSION

H. Freeman *(Rennselaer Polytechnic Institute)*

There was a 6 legged vehicle built in Italy.

McGhee

It has been converted to a desk. It never worked particularly well. It had only two degrees of freedom per leg. It was built by Professor Pedernella and he is proud that he found the first practical application of walking machines and now he uses it as a desk.

Symposium Speakers And Chairmen

SYMPOSIUM CHAIRMEN

G. G. DODD

L. ROSSOL

SESSION I: FUNDAMENTAL ISSUES IN VISION AND ROBOTICS

A. ROSENFELD

C. A. ROSEN

R. L. PAUL

R. L. GREGORY

B. K. P. HORN

SESSION II: VISION AND ROBOT SYSTEMS

P. M. WILL

S. W. HOLLAND

T. UNO

R. G. ABRAHAM

R. C. BEECHER

E. LAMONACA

SESSION III: FUTURE VISION SYSTEMS

H. G. BARROW

D. R. REDDY

Y. SHIRAI

A. D. GARA

J. M. TENENBAUM

SESSION IV: FUTURE ROBOT SYSTEMS

J. A. FELDMAN

J. F. ENGELBERGER

J. L. NEVINS

D. E. WHITNEY

R. B. McGHEE

SUMMARIZER & ADVISORS

P. H. WINSTON

M. L. MINSKY

A. ROSENFELD

C. A. ROSEN

SYMPOSIUM PARTICIPANTS

Abraham, R. G. *(Speaker)*
Westinghouse Electric Corporation
Pittsburgh, Pennsylvania

Agnew, W. G.
GM Research Laboratories
Warren, Michigan

Akeel, H.
GM Manufacturing Development
Warren, Michigan

Albers, W. A., Jr.
GM Research Laboratories
Warren, Michigan

Albus, J.
National Bureau of Standards
Washington, DC

Ambler, A. P.
University of Edinburgh
Edinburgh, United Kingdom

Baird, M. L.
GM Research Laboratories
Warren, Michigan

Bajcsy, R. R.
University of Pennsylvania
Philadelphia, Pennsylvania

Barrow, H. G. *(Session Chairman)*
SRI International
Menlo Park, California

Baumann, E. W.
McDonnell Douglas
St. Louis, Missouri

Beaman, R. T.
GM Research Laboratories
Warren, Michigan

Beck, G.
AMF Incorporated
Herndon, Virginia

Becker, R.
Chesebrough-Ponds, Inc.
Clinton, Connecticut

Beecher, R.C. *(Speaker)*
GM Manufacturing Development
Warren, Michigan

Bidwell, J. B.
GM Research Laboratories
Warren, Michigan

Binford, T. O.
Stanford University
Stanford, California

Birk, J. R.
University of Rhode Island
Kingston, Rhode Island

Bolles, R.
SRI International
Menlo Park, California

Branaman, L.
General Electric Co.
Syracuse, New York

Brown, C.
University of Rochester
Rochester, New York

Buchholtz, R. F.
Central Foundry Division, GMC
Saginaw, Michigan

Buzan, L. R.
GM Research Laboratories
Warren, Michigan

Caplan, J. D.
GM Research Laboratories
Warren, Michigan

Chenea, P. F.
GM Research Laboratories
Warren, Michigan

Chern, B.
National Science Foundation
Washington, DC

Chien, R. T.
University of Illinois
Urbana, Illinois

Corbato, F. J.
Massachusetts Institute of Technology
Cambridge, Massachusetts

Curtis, K. K.
National Science Foundation
Washington, DC

Cwycyshyn, W.
GM Manufacturing Development
Warren, Michigan

Davis, R.
Chesebrough-Ponds, Inc.
Clinton, Connecticut

Dewar, R.
GM Manufacturing Development
Warren, Michigan

Dixon, J. K.
Naval Research Laboratories
Washington, DC

Dodd, G. G. *(Symposium Co-chairman)*
GM Research Laboratories
Warren, Michigan

Dolik, Z.
Ford Motor Company
Dearborn, Michigan

Druffel, L.
Advanced Research Projects Agency
Arlington, Virginia

DuVair, J.
GM Assembly Division
Warren, Michigan

Eichler, R. H.
Chevrolet Bay City Plant, GMC
Bay City, Michigan

Ejiri, M.
Hitachi, Ltd.
Mountain View, California

Engelberger, J. F. *(Speaker)*
Unimation, Inc.
Danbury, Connecticut

Feldman, J. A. *(Session Chairman)*
University of Rochester
Rochester, New York

Fennema, C. L.
Control Data Corporation
Minneapolis, Minnesota

Flint, M.
Chevrolet Manufacturing R&D
Warren, Michigan

Ford, B. R.
ASEA
White Plains, New York

Fredkin, E.
Massachusetts Institute of Technology
Cambridge, Massachusetts

Freeman, H.
Rennselaer Polytechnic Institute
Troy, New York

Fu, K. S.
Purdue University
West Lafayette, Indiana

Gardels, K. D.
GM Research Laboratories
Warren, Michigan

Gara, A. D. *(Speaker)*
GM Research Laboratories
Warren, Michigan

SYMPOSIUM PARTICIPANTS

Gaschnig, G.
Raytheon Company
Bedford, Massachusetts

Geschke, C. C.
University of Illinois
Urbana, Illinois

Glorioso, R.
Digital Equipment Corporation
Maynard, Massachusetts

Gregory, R. L. *(Speaker)*
University of Bristol
Bristol, United Kingdom

Hanify, D. W.
IIT Research Institute
Chicago, Illinois

Hart, D. E.
GM Research Laboratories
Warren, Michigan

Haskell, R.
Oakland University
Rochester, Michigan

Hazard, J. E.
Scott Paper Co.
Philadelphia, Pennsylvania

Hohn, R. E.
Cincinnati Milacron Inc.
Cincinnati, Ohio

Holland, S. W. *(Speaker)*
GM Research Laboratories
Warren, Michigan

Hollyer, R. N.
GM Research Laboratories
Warren, Michigan

Holzwarth, J. C.
GM Research Laboratories
Warren, Michigan

Hoover, A.
GM Manufacturing Development
Warren, Michigan

Horn, B. K. P. *(Speaker)*
Massachusetts Institute of Technology
Cambridge, Massachusetts

Houchens, A.
GM Manufacturing Development
Warren, Michigan

Iadipaolo, R. M.
GM Manufacturing Development
Warren, Michigan

Jamerson, F. E.
GM Research Laboratories
Warren, Michigan

Jarvis, J. F.
Bell Telephone Laboratories
Holmdel, New Jersey

Jessop, D. E.
Detroit Diesel Allison Division, GMC
Detroit, Michigan

Joyce, J. D.
GM Research Laboratories
Warren, Michigan

Kanal, L.
University of Maryland
Silver Spring, Maryland

Kaplit, M.
GM Research Laboratories
Warren, Michigan

Kavetsky, E. M.
GM Manufacturing Development
Warren, Michigan

Kelley, R. B.
University of Rhode Island
Kingston, Rhode Island

Kendrick, J. C.
Delco Electronics Division, GMC
Kokomo, Indiana

King, S.
Ford Motor Company
Dearborn, Michigan

Klein, C. A.
Ohio State University
Columbus, Ohio

Knappmann, R.
Fraunhofer Gesellschaft
Stuttgart, Germany

Krull, F. N.
GM Research Laboratories
Warren, Michigan

Kuehn, R., Jr.
Boeing Commercial Airplane Co.
Seattle, Washington

Lamonaco, E. *(Speaker)*
Olivetti
Torino, Italy

Larson, D. F.
GM Manufacturing Development
Warren, Michigan

Levine, M. D.
McGill University
Montreal, Canada

Lewis, R.
GM Research Laboratories
Warren, Michigan

Luh, J. Y. S.
Purdue University
West Lafayette, Indiana

Macri, G. C.
Ford Motor Company
Dearborn, Michigan

Martin, A. L.
GM Research Laboratories
Warren, Michigan

McCarthy, J.
Stanford University
Stanford, California

Messinger, R. C.
Cincinnati Milacron, Inc.
Cincinnati, Ohio

Minsky, M. L. *(Advisor)*
Massachusetts Institute of Technology
Cambridge, Massachusetts

Miller, R. C.
GM Manufacturing Development
Warren, Michigan

Muench, N. L.
GM Research Laboratories
Warren, Michigan

Mundy, J. L.
General Electric R & D Center
Schenectady, New York

Nevatia, R.
University of Southern California
Los Angeles, California

Nevins, J. L. *(Speaker)*
Charles Stark Draper Laboratory
Cambridge, Massachusetts

Newcomb, T.
Central Foundry Division, GMC
Saginaw, Michigan

Nitzan, D.
SRI International
Menlo Park, California

Nof, S.
Purdue University
West Lafayette, Indiana

SYMPOSIUM PARTICIPANTS

McCormick, B. H.
University of Illinois - Chicago Circle
Chicago, Illinois

McDonald, R. J.
GM Research Laboratories
Warren, Michigan

McGhee, R. B. *(Speaker)*
Ohio State University
Columbus, Ohio

Olsztyn, J. T.
GM Research Laboratories
Warren, Michigan

Orin, D. E.
Case Western Reserve University
Cleveland, Ohio

Ossenberg, K.
Fraunhofer Gesellschaft
Karlsruhe, West Germany

Pastorius, W. J.
Diffracto
Windsor, Canada

Page, C.
Michigan State University
East Lansing, Michigan

Paul, R. L. *(Speaker)*
Purdue University
West Lafayette, Indiana

Perlett, R.
Rochester Products Division, GMC
Rochester, New York

Perkins, W. A.
GM Research Laboratories
Warren, Michigan

Pohl, Frederik *(After-dinner Speaker)*
Red Bank, New Jersey

Polgreen, R. W.
3M Company
Saint Paul, Minnesota

Popplestone, R. J.
University of Edinburgh
Edinburgh, United Kingdom

Pratt, W. K.
University of Southern California
Los Angeles, California

Rabischong, P.
Institute National de la Sante
et de la Recherch Medicale
Montpellier, France

Raibert, M. H.
California Institute of Technology
Pasadena, California

Ray, J.
Cincinnati Milacron Inc.
Cincinnati, Ohio

Reddy, D. R. *(Speaker)*
Carnegie Mellon University
Pittsburgh, Pennsylvania

Rosen, C. A. *(Speaker, Advisor)*
SRI International
Menlo Park, California

Rosenfeld, A. *(Session Chairman, Advisor)*
University of Maryland
College Park, Maryland

Rossol, L. *(Symposium Co-chairman)*
GM Research Laboratories
Warren, Michigan

Roth, S. D.
GM Research Laboratories
Warren, Michigan

Rutledge, G.
Delco Electronics Division, GMC
Kokomo, Indiana

Ryckman, G. F.
GM Research Laboratories
Warren, Michigan

Sallot, B. M.
Robot Institute of America
Dearborn, Michigan

Saraga, P.
Philips Research Laboratories
Redhill, United Kingdom

Scheinman, V.
Unimation Inc.
Mountain View, California

Schwing, R. C.
GM Research Laboratories
Warren, Michigan

Seres, D. A.
E.I. DuPont DeNemours & Company
Wilmington, Delaware

Shirai, Y. *(Speaker)*
Electrotechnical Laboratory
Tokyo, Japan

Shulhof, W. P.
Central Foundry Division, GMC
Saginaw, Michigan

Stepchuk, Z.
3M Company
Saint Paul, Minnesota

Stoddard, K.
GM Research Laboratories
Warren, Michigan

Stone, R.
GM Assembly Division, GMC
Warren, Michigan

Tenenbaum, J. M. *(Speaker)*
SRI International
Menlo Park, California

Tracy, J. C.
GM Research Laboratories
Warren, Michigan

Tuesday, C. S.
GM Research Laboratories
Warren, Michigan

Uno, T. *(Speaker)*
Hitachi, Ltd.
Tokyo, Japan

Vamos, T.
Hungarian Academy of Sciences
Budapest, Hungary

Ward, M. R.
GM Research Laboratories
Warren, Michigan

Weisel, W.
PRAB Conveyors, Inc.
Kalamazoo, Michigan

Weyland, R. G.
GM Manufacturing Development
Warren, Michigan

West, J. K.
GM Manufacturing Development
Warren, Michigan

West, P.
GM Manufacturing Development
Warren, Michigan

Whitney, D. E. *(Speaker)*
Charles Stark Draper Laboratory
Cambridge, Massachusetts

Will, P. M. *(Session Chairman)*
Thomas J. Watson Research Center
Yorktown Heights, New York

Williams, D. S.
Jet Propulsion Laboratory
Pasadena, California

SYMPOSIUM PARTICIPANTS

VanderBrug, G. J.
 National Bureau of Standards
 Washington, DC

Voorhis, G.
 Delco Electronics Division, GMC
 Kokomo, Indiana

Wagner, G.
 Delco Electronics Division, GMC
 Kokomo, Indiana

Waltz, D. L.
 University of Illinois
 Urbana, Illinois

Willis, W.
 Central Foundry Division, GMC
 Saginaw, Michigan

Winston, P. H. *(Summarizer)*
 Massachusetts Institute of Technology
 Cambridge, Massachusetts

Wipson, J. W.
 Reticon Corporation
 Sunnyvale, California

SUBJECT INDEX

A

Accuracy, 26-27.
Actuators, 24, 26, 325, 332.
Analysis of algorithms, 172.
APAS (Adaptable Programmable Assembly System), 117-136.
Architecture
 - algorithm driven, 173.
 - computer, 169-185.
 - for vision, 169-185.
 - memory driven, 173-176.
 - parallel computer, 179-185.
 - software, 172.
Area measurement, 89, 110.
Arm, see Robot, 153-163, 330-331.
ARPA network, 182.
Array processors, 179-181.
Artificial intelligence, 3, 4, 51-53, 323.
Assembly, 11-13, 117, 141, 144, 153, 156, 160, 275-276, 289, 293.
Assumed invarience, 54.
Autocorrelation, 211.

B

Batch assembly, 11, 118.
Binary image, 105, 107, 117, 134-135, 243-244.
Binary video signal, 107.
Bin-of-parts problem, 7-9, 73, 240-241.
Biological vision, 73, 170, 243.
Blocks world, 72.
Brightness constancy, 56.

C

C.mmp, 181.
Camera
 - array, 13, 15, 105, 133, 134.
 - CCD (Charge Coupled Device), 101, 176.
 - line, 14-15, 218, 229-230, 232.
 - vidicon, 13.
Cellular logic image processor, 180.
Channel distortion, 59.
Cincinnati-Milacron robot, 326.
CLIP (Cellular Logic Image Processor), 180.
Closed-loop sequential recognition, 101, 113.
Cm*, 182.
Coherent light, 211-212, 224-227, 229.
Compliance, 5, 27-28, 330-331.
Computational complexity, 171.
Computer architecture, see Architecture, 169-185.
 - parallel, 179-183.
Computer vision systems, 81-96, 101-114.
 - three dimensional, 187-205.
 - with geometric models, 200-203.
Conditional probabilities, guidance by, 54.
Cones, generalized, 201.
Continuous path, 326.
Control, adaptive, 323.
Control computer, 324-328.
Conveyors, 6, 240.
Convolution, 176, 190, 195, 207, 222.
Coordinate transformation, 189, 190.
Correlation, 176, 193.
Correspondence problem, 193.
Curved objects, representation of, 198.
Cycle time, 159, 331.

D

Data mismatching, 67.
Degrees of freedom, 153, 155, 165, 328, 330-332.
Depth measurement, 71.
Depth perception, 61.
Directed graph, 198.
Disparity, 61-63.

E

Economics, 118, 269, 289.
Edges, 191-192, 197, 250-252, 255.
- detection, 175-176, 251.
- enhancement, 208, 210, 216, 222-225.
Electric Quadruped Transporter, 327.
Electron density maps, 183.
Encoder, 26, 91.
End effectors and tools, 26, 128-130.

F

Features
- extraction, 89, 243.
- points, 193-195.
Feature-locking distortions, 64.
Feedback, 325, 328, 330-331.
- force, 5.
Feeders, 9.
Feed-forward, 330-331.
Filtering, 189, 191.
- Fourier transform, 207-210, 216, 227.
- gradient, 223-224.
- holographic, 222-223, 226.
- matched, 207, 228.
Force feedback, see Feedback, 5.
Force, 155, 159, 328.
Fourier transform, 207-210, 216, 218, 226-229.

G

Generalized cylinders, 254.
Geometric modeling, 201-203.
Gradient, 191.

Gradient filter, 223-224.
Gray-level image 7, 244, 248, 251, 255.
Grippers, 153-163.

H

Hand-eye coordination, 326.
Harmonic gear reducers, 26.
Hexapod Vehicle, 327-329.
Histogram, 190, 191.
Holographic filter, 222-223, 226.
Homogenous transformations, 25, 27.

I

IC (Integrated Circuit)
- chip manufacturing, 190.
- lead bonding, 72, 188.
Illumination, 104.
Illusions
- ambiguity, paradox, created, hallucinatory, 58.
- cafe wall, 65, 67.
- cognitive, 65.
- creating and destroying size-depth distortion, 59-60.
- destroying perspective, 60-61.
- distortion, 65.
- error, 55.
- physiological and cognitive errors, 58.
Illusion-destruction, apparatus for, 61.
Image
- converter, see Image transducer.
- processing, 101-113.
- processing, definition, 71.
- processing, hardware, 188-190.
- processing, radar, 183.
- rotator, 226, 228-229.
- stripe, 86, 196.
- transducer, 211-216, 225, 227-228, 231, 233.
- understanding, science of, 69-77.
Image dissector camera, 196.
Incoherent-to-coherent image converter, see image transducer.
Industrial eye, 101-113.
Industrial vision, 239-259.

SUBJECT INDEX

Interferences, 53.
Insertion, 27.
Inspection, 13, 17.
  - qualitative, 15-16.
  - semi-quantitative, 15-16.
Instruction sets, custom, 177-179.
Interlace, 105, 110.
Intrinsic characteristics, 246, 248-255.
Intrinsic image, 246-255.
Inverse plant, 330-331.

J

Jigging, 12.
Joint, 25-26.
Junction dictionary, 197-199.

K

Knowledge
  - informal, 170.
  - sources, 172.

L

Label reading, 16.
Labels, Huffman, 198.
Lambert's Law, 249.
Landsat imagery, 171.
Laser, 224-225.
  - pulsed, 197.
  - range finder, 246, 248, 253.
  - spot controller, 196.
  - tracker, 196.
Lateral inhibition, 60.
Lens, Fourier transform, 209, 216, 218, 225
Limited-sequence robot, see Robot, 15.
Linear diode array, see Camera, 218, 229-230, 232.
Line drawing, labelled, 199.
Liquid crystal transducer, see Image transducer.
LSI processor, 251.

M

Machine vision, 3-22, 70-71.
  - history, 71-72.
  - techniques, 70.
Macroscopic processing, 105-111.
Manipulator, see Robot.
  - arms, 127-128.
  - computer-controlled, 323-324, 328.
  - remote, 325.
Masking, 176.
Massively Parallel Processor, 180.
Material handling, 325, 328.
Mean, 195.
Methods Time Measurement (MTM), 30-32.
Microcode, 177-179.
Microscopic processing, 105-111.
Model
  - acquisition, 170.
  - geometric, 187-188.
  - wire, 60-63.
  - 2-dimensional, 240, 242, 246.
  - 3-dimensional, 242, 255.
Motion, 27.
  - parallax, 52.
  - planning, 324-328, 330.
Muller-Lyer and Ponzo figures, 59, 62, 63.
Multiprocessors, 181-182.

N

Natural intelligence, 51-52.
Networks, 182-183.
Newton-Euler formalism, 331.

O

Object recognition, 101-102.
Optical
  - birefringence, 212-213.
  - computing, 207-233.
  - correlation, 208, 210-211, 226-233.
Optimization, 171-172.
Orientation, 24, 26-27, 46.

P

Paint-spraying, 4, 10.
PAL, 29-42.
Pantograph, 325.
Parallel processing, 251-253.
Parts feeding equipment, 118.
Path control, 127.
Pattern classification, definition, 71.
Pattern matching, 101, 108, 111.
Perception
- as hypotheses, 53.
- biological and machine visual, 51.
- experimental study of, 52.
- mechanisms of, 51-68.
- scientific hypotheses, compared with, 54-59.
- three-dimensional, 64.
Perspective convergence, 67.
Photometric stereo, 73.
PIPS project, 188, 203.
Point mapping, 189, 191.
Point-to-point, 325-327.
Polarized light, 213.
Polaroid glasses, 62.
Position, 24, 26, 39.
Position-controlled, 27.
Potentiometers, 26.
Powered artificial limbs, 324, 332.
Processing elements, custom, 172-177.
Processor-memory-switch, 179-183.
Processor
- parallel, 193.
- parallel pattern, 189.
Production control computer, 104.
Productivity, 4, 24, 158-159, 162.
Productivity improvement, 117.
Product quality improvement, 118.
Programmable
- assembly, 117-136.
- automation, 10.
- belt feeder, 130.
- fixture, 117, 130-132.
- parts presenters, 117, 130-132.
- systems, 153-163.
- x-y table, 10, 15.
Projection, monocular or stereoscopic, 61-63.
PUMA, 128, 326.

Q

Quadruped, 327.
Quality control, 3.

R

Range sensing
- by phase shift, 197.
- data acquisition, 193, 197.
- data, by using local constraints, 197-200.
- data, processing, 197-200.
- data, segmentation of, 197.
- finder, 7, 8, 197.
- time-of-flight methods, 196.
RCC (Remote Center Compliance), 118, 285-287.
Region, 243, 248, 255.
- connected, 191.
- growing, 72.
- labelled, 197.
- segmentation by, 197.
Relaxation, parallelism, 74.
Repeatability, 26-27.
Representation of objects, 74.
Retinal
- perspective, 59.
- signals, 64.
Robot, 3, 4, 9, 11, 23-50, 101, 153-166, 323-332.
- Cincinnati-Milacron, 326.
- limited sequence, 15.
- PUMA, 128, 326.
- Servo-controlled, 117, 124.
- Stanford arm, 24-50.
- Time and motion (RTM), 40-43.
- Unimate, 231.
- vision, 210.

S

Safety, 3-4, 17.
Scene analysis, 9.
- definition, 71-72.
- desk, 191-193.
Scene description, 200.

SUBJECT INDEX

Segmentation, Ohlander's program, 171.
Sensing, 24.
- auditory, 3.
- force, 3, 23.
- non-contact, 70.
- olfactory, 3.
- tactile, 3, 5, 46.
- torque, 3.
Sensors, 159, 323, 328, 330-331.
Sensory feedback, 5, 37.
Sensory systems, 132-135.
Servo-controlled robot, 117, 124.
Sheet of light, 196.
SIGLA, 155.
SIGMA, 153-162.
Size
- constancy, 58-59, 67.
- distortion, 60.
Software architecture, 172.
Sorting, 8.
Spatial
- frequency spectrum, 209, 211, 216.
- move, 33.
- reach, 31.
Sphere, 25.
Stanford arm, 24-50.
Stereo
- depth, 61.
- images, 60.
- photometric, 73.
- vision, 61, 193-195, 203.
Stereopsis, 195, 252, 253.
Stimulus patterns, 52.
Structured light, 7, 86.
Subjective contour, 243, 245.
Surface orientation, 244, 246-253, 255.
Surface reflectance, 246-253, 255.
System organization, 172.

T

Template, 101, 190.
- selection of, 107.
Texture gradients, 52.
Thresholding, 243-244, 248.
Threshold logic unit, 134.
Time Measurement Unit (TMU), 30-38.
Tomography, computer aided, 175, 183.

Tool wear, 15.
Transducers, 54.
Triangulation
- accuracy of, 196.
- passive and active, 193.
- range data obtained by, 198.

V

Variance, 193, 195.
Verification, 112.
Vision, see Computer vision
- non-purposive, 74.
- systems, 117, 124.
Visual
- device, 101.
- feedback, 3, 5.
- servoing, 124, 241-242.

W

Walking machines, 324, 327.
Welding
- arc, 5, 12.
- spot, 4.
Wire-bonding system, see IC lead bonding
Working volume, 154-155.
Workpieces, 5-9, 12.
Workplace, 24, 45.
Writable control store, 177-179.

X

X-ray photographs, analysis of, 190.